Springer Series in
**Surface Sciences**

13

Editor: Robert Gomer

# Springer Series in **Surface Sciences**

Editors: Gerhard Ertl and Robert Gomer


Volume 1: **Physisorption Kinetics**
By H.J. Kreuzer, Z.W. Gortel

Volume 2: **The Structure of Surfaces**
Editors: M.A. Van Hove, S.Y. Tong

Volume 3: **Dynamical Phenomena at Surfaces, Interfaces and Superlattices**
Editors: F. Nizzoli, K.-H. Rieder, R.F. Willis

Volume 4: **Desorption Induced by Electronic Transitions, DIET II**
Editors: W. Brenig, D. Menzel

Volume 5: **Chemistry and Physics of Solid Surfaces VI**
Editors: R. Vanselow, R. Howe

Volume 6: **Low-Energy Electron Diffraction**
Experiment, Theory and Surface Structure Determination
By M.A. Van Hove, W.H. Weinberg, C.-M. Chan

Volume 7: **Electronic Phenomena in Adsorption and Catalysis**
By V.F. Kiselev, O.V. Krylov

Volume 8: **Kinetics of Interface Reactions**
Editors: M. Grunze, H.J. Kreuzer

Volume 9: **Adsorption and Catalysis in Transition Metals and Their Oxides III**
By O.V. Krylov, V.F. Kiselev

Volume 10: **Chemistry and Physics of Solid Surfaces VII**
Editors: R. Vanselow, R. Howe

Volume 11: **The Structure of Surfaces II**
Editors: J.F. van der Veen, M.A. Van Hove

Volume 12: **Diffusion at Interfaces: Microscopic Concepts**
Editors: M. Grunze, H.J. Kreuzer, J.J. Weimer

Volume 13: **Desorption Induced by Electronic Transitions, DIET III**
Editors: R.H. Stulen, M.L. Knotek

R. H. Stulen M. L. Knotek (Eds.)

# Desorption Induced by Electronic Transitions DIET III

Proceedings of the Third International Workshop
Shelter Island, New York, May 20–22, 1987

With 181 Figures

Springer-Verlag Berlin Heidelberg New York
London Paris Tokyo

Dr. Richard H. Stulen

Surface Science and Chemical Physics, Division 8343,
Sandia National Laboratories, Livermore, CA 94550, USA

Dr. Michael L. Knotek

National Synchrotron Light Source, Brookhaven National Laboratory,
Building 725B, Upton NY 11973, USA

*Series Editors*

Professor Dr. Gerhard Ertl

Fritz-Haber-Institut der Max-Planck-Gesellschaft, Faradayweg 4–6
D-1000 Berlin 33

Professor Robert Gomer

The James Franck Institute, The University of Chicago, 5640 Ellis Avenue,
Chicago, IL 60637, USA

ISBN-13:978-3-642-73730-5 e-ISBN-13:978-3-642-73728-2
DOI: 10.1007/978-3-642-73728-2


Softcover reprint of the hardcover 1st edition 1988



2154/3150-543210

# Preface

These proceedings are the result of the third international workshop on Desorption Induced by Electronic Transitions, DIET III, which took place on Shelter Island, NY, May. 20–22, 1987. The work contained in this volume is an excellent summary of the current status of the field and should be a valuable reference text for both "seasoned" researchers and newcomers in the field of DIET. Based on the success of the meeting it seems clear that interest and enthusiasm in the field is strong. It is also apparent, from the many lively discussions during the meeting, that many unanswered questions (and controversies) remain to be solved. It was particularly pleasing to see many new participants from new and rapidly advancing fields, ranging from gas phase dynamics to semiconductor processing. The resulting cross-fertilization from these separate but related fields is playing an important role in helping us understand desorption processes at solid surfaces.

In general, the topics covered during the course of the workshop overlapped those of both DIET I and DIET II. However, clear advances have been made and in general there is a much more sophisticated understanding of the physics and chemistry of stimulated desorption. Of particular note in this regard is the gas phase research highlighted in this workshop by the work of Nenner et al., where new results indicate that in the gas phase, photodissociation can precede or compete strongly with autoionization and other electronic relaxation pathways. Although we do not yet have any direct experimental evidence that this occurs for adsorbates on surfaces, it is certainly possible that such effects may be important, for example, in the desorption of hydrogen isotopes from larger molecular species that are only weakly coupled to an underlying surface. Other gas phase results presented at this workshop underscore the tremendous advances in electron-ion and ion-ion coincidence experiments that are now being used to elucidate the important photodissociation pathways and related electronic excitations in molecules. The understanding that these experiments provide will give us a valuable conceptual framework for understanding DIET at surfaces. Condensed phase studies (of rare gases and other physisorbed molecules) provide a link between the gas phase experiments and DIET from more strongly bound chemisorbed molecules.

There is now a new generation of complex experimentation utilizing lasers, synchrotrons, digital ESDIAD, and neutral detection. Applications of these tools will no doubt give us a much clearer picture of the desorption dynamics and relevant excited states. The work of Burns et al. shows that we are now able to look at internal state distributions following desorption, and as a result theory is now being developed to understand the dynamics of DIET including electronic and nuclear effects. With these and similar results in hand, current theoretical studies continue to push ahead in understanding relaxation, screening, and deexcitation effects following initial excitation at the surface. In the area of ESDIAD there have been significant advances in instrumentation, providing new capabilities and resulting in dramatic new observations. Measurements by Yates et al. have elucidated the nature and energetics of complex hindered rotor adsorbate-adsorbate interactions, which opens the way for significant new areas of exploration in the dynamics of molecules on surfaces. The insights drawn from these measurements would be exceedingly difficult to derive with any other experimental probe.

As the modelling of the mechanisms of DIET has matured, the importance of relaxation and screening mechanisms which govern the lifetime of the excitations which produce desorption has become more apparent. Charge transfer (CT) processes are the most effective of relaxation mechanisms and are also the most difficult to understand. Measurements and modelling by Cowin and co-workers have given us new insight into the dynamics of CT for generalized adsorbate-surface systems, and the effect it can have on electronically induced phenomena on surfaces. The work of Rabalais et al. is extending the ideas in DIET to chemical sputtering phenomena by relating chemical effects in sputtering to electronic interactions important to bonding.

Finally, we feel a few comments and conjectures about future directions in DIET are appropriate. Broady speaking, the work presented in this and past DIET workshops represents a mixture of basic and applied. As the field matures there will be greater emphasis on directing the techniques described in these proceedings to new areas in technology. Applications of DIET spectroscopy are now found in areas ranging from biological systems to semiconductor processing, and they are having an important impact in the areas concerned with radiation effects on surfaces, such as tokamaks, storage rings, and glasses for high-power laser applications. As new materials technology moves toward the use of radiation-enhanced chemistry in semiconductor processing, plasma materials processing, and diamond coatings, this body of knowledge from DIET will become critical in understanding and optimizing processing variables.

Most importantly, we would like to express our gratitude to Brookhaven National Laboratory, the Office of Naval Research through Dr. L.R. Cooper, and the Air Force Office of Scientific Research through Dr. L.W.

Burggraf for providing the financial support which made this workshop possible. We also thank Sandia National Laboratories and the National Bureau of Standards for providing administrative services during the preparation of the workshop and proceedings.

Shelter Island, NY
January 1988

*M.L. Knotek*
*R.H. Stulen*

# Contents


Part I **Introductory Lecture**

Some Recent Developments in DIET
By D. Menzel (With 5 Figures) . . . . . . . . . . . . . . . . . . . . . . . . 2

Part II **Gas Phase**

Photodissociation and Photoionization Processes of Core Excited Molecules: The $CH_3Br$ Case
By I. Nenner, P. Morin, M. Simon, P. Lablanquie, and G.G.B. de Souza (With 14 Figures) . . . . . . . . . . . . . . . . . . . 10

Auger Electron-Ion Coincidence Studies on Free Molecules
By W. Eberhardt and R. Murphy (With 3 Figures) . . . . . . . . . . 32

Electron-Stimulated Desorption of $D_n^+$ Clusters from Condensed Deuterium Layers on Pd(111) Between 5 and 20 K
By R.H. Stulen (With 5 Figures) . . . . . . . . . . . . . . . . . . . . . . . 39

Desorption Processes in Planetary Science
By R.E. Johnson . . . . . . . . . . . . . . . . . . . . . . . . . . . . . . . . 47

Atom Selective Photochemistry with Monochromatic Soft X-Rays
By D.M. Hanson . . . . . . . . . . . . . . . . . . . . . . . . . . . . . . . . 50

Part III **Neutrals and Negatives**

Photon Stimulated Desorption of Neutrals by Synchrotron Radiation from Chemisorbed, Physisorbed, and Condensed Layers
By P. Feulner, S. Auer, T. Müller, A. Puschmann, and D. Menzel (With 7 Figures) . . . . . . . . . . . . . . . . . . . . . . . . . . . . . . . . 58

Rovibrational Laser Spectroscopy of ESD Neutrals from Chemisorbed Species
By A.R. Burns, E.B. Stechel, and D.R. Jennison (With 5 Figures) 67

Electron Stimulated Desorption of Excited Hydrogen Atoms from an Alkali Promoted Surface
By P.D. Johnson, X. Pan, J. Tranquada, S.L. Hulbert, and E. Johnson (With 2 Figures) . . . . . . . . . . . . . . . . . . . . . . 73

Dissociative Attachment in ESD from Condensed Molecules
By L. Sanche (With 8 Figures) . . . . . . . . . . . . . . . . . . . . . . . . 78

Image Charge, Coherent Scattering, and Symmetry Effects in Electron Stimulated Desorption: $O^-$ from $O_2$ Condensed on Metals
By D.E. Ramaker and H. Sambe (With 6 Figures) . . . . . . . . . . . 85

Techniques for the Detection of Photodesorbed Negative Ions
By C.E. Young, E.L. Schweitzer, M.J. Pellin, D.M. Gruen, Z. Hurych, P. Soukiassian, M.H. Bakshi, and A.S. Bommannavar (With 3 Figures) . . . . . . . . . . . . . . . . . . . . . . . . . . . . . . . . 94

Part IV **Electron Stimulated Desorption Ion Angular Distribution (ESDIAD)**

Observation of Molecular Rotors on Surfaces by ESDIAD: Studies of $PF_3$ and $NH_3$ Chemisorption on Ni Surfaces
By J.T. Yates, Jr., M.D. Alvey, M.J. Dresser, A.-M. Lanzillotto, and K.J. Uram (With 8 Figures) . . . . . . . . . . . . . . . . . . . . . . 100

The Influence of Surface Structure on Mechanisms of Stimulated Desorption
By R.L. Kurtz, R. Stockbauer, and T.E. Madey (With 5 Figures) 109

DIET in the Second Layer: An ESDIAD Study of $NH_3$ on a CO Layer on Ni(111) and Ni(110)
By M.J. Dresser, A.-M. Lanzillotto, M.D. Alvey, and J.T. Yates, Jr. (With 3 Figures) . . . . . . . . . . . . . . . . . . . . 115

ESDIAD of Small Molecules on Surfaces: A Few Caveats
By T.E. Madey, M. Polak, A.L. Johnson, and M.M. Walczak (With 5 Figures) . . . . . . . . . . . . . . . . . . . . . . . . . . . . . . . . 120

An Ellipsoidal Mirror Analyzer for the Study of Photon Stimulated Desorption
By R. Stockbauer, R.L. Kurtz, and T.E. Madey (With 1 Figure) . 126

Digital Video Data Acquisition/Analysis for Existing ESDIAD Apparatus
By A.L. Johnson, R. Stockbauer, D. Barak, and T.E. Madey (With 3 Figures) . . . . . . . . . . . . . . . . . . . . . . . . . . . . . . . . 130

## Part V Theory


Dynamics in Neutral DIET from Chemisorbed Molecules
By E.B. Stechel, D.R. Jennison, and A.R. Burns (With 3 Figures) 136

The Nature of Repulsive States and the Role of Nuclear Dynamics in Desorption Induced by Electronic Transitions
By Ph. Avouris, R. Walkup, R. Kawai, D.M. Newns, and N.D. Lang (With 7 Figures) . . . . . . . . . . . . . . . . . . . . . . . . 144

The Dynamics of $O^+$ Desorption from $TiO_2$
By R.E. Walkup and R.L. Kurtz (With 4 Figures) . . . . . . . . . . . 160

DIET and the Electronic Structure of Chemisorbed Molecules and Physisorbed Rare Gases
By D.R. Jennison, E.B. Stechel, and A.R. Burns . . . . . . . . . . . . . 167

Theory of Desorption of Neutrals by Temporary Ionization of Physisorbed Species at Metal Surfaces
By Z.W. Gortel, H.J. Kreuzer, P. Feulner, and D. Menzel (With 6 Figures) . . . . . . . . . . . . . . . . . . . . . . . . . . . . . . . . . . . . 173

The Magnitude of Secondary Electron Contributions in Photon Stimulated Desorption
By D.E. Ramaker, T.E. Madey, R.L. Kurtz, and H. Sambe . . . . . 182

Interatomic Auger Rates for the Sodium Fluoride Crystal
By T.A. Green and D.R. Jennison . . . . . . . . . . . . . . . . . . . . . . . . 185


## Part VI Spectroscopy


Ultraviolet Surface Photochemistry and Charge Transfer
By E.P. Marsh, F.L. Tabares, M.R. Schneider, and J.P. Cowin (With 3 Figures) . . . . . . . . . . . . . . . . . . . . . . . . . . . . . . . . . . . . 190

Chemically Induced Desorption of Nickel Oxide and Carbide by 5-200 eV $C^+$ and $O^+$ Ion Bombardment
By H. Kang, S.R. Kasi, O. Grizzi, and J.W. Rabalais (With 6 Figures) . . . . . . . . . . . . . . . . . . . . . . . . . . . . . . . . . . . . 196

PSID SEXAFS Studies of Silicon Surface Chemistry
By G. Thornton, D.R. Warburton, R. McGrath, I.T. McGovern, and D. Norman (With 3 Figures) . . . . . . . . . . . . . . . . . . . . . . 205

Photon Stimulated Desorption from Oxidized Al(110)
By E.D. Johnson, R.F. Garrett, M.L. Knotek, and F. Sette (With 9 Figures) . . . . . . . . . . . . . . . . . . . . . . . . . . . . . . . . . . . . 210

Stimulated Desorption from CO Chemisorbed on Cr(110): Sensitivity to Bonding Changes
By N.D. Shinn and T.E. Madey (With 2 Figures) ........... 217

Temperature-Dependent PSD of $H^+$ from $H_2O/Si(111)$ (7×7)
By R.A. Rosenberg, C.-R. Wen, and D.C. Mancini
(With 4 Figures) ........................... 220

Electronic Sputtering and Its Application to SIMS Analysis of Biomolecular Monolayers
By B.U.R. Sundqvist, A. Hedin, P. Håkansson, G. Jonsson, and G. Säve (With 3 Figures) .................... 225

Limits of Applicability of the Popcorn Model for Desorption of Large Biomolecules by Fast Ion Impact
By P. Williams and B.U.R. Sundqvist .................... 230

Photon-Stimulated Desorption in High Electric Fields
By S. Jaenicke, A. Ciszewski, J. Dösselmann, W. Drachsel, J.H. Block, and D. Menzel (With 4 Figures) .............. 236

Observations of Photon-Induced Gas Desorption from Technological Materials Using Synchrotron Light
By E.M. Williams (With 3 Figures) .................. 242

## Part VII Insulators

The Origin of Positive Ions and Excited Neutrals in Electron Stimulated Desorption from Alkali Halides
By R.E. Walkup, Ph. Avouris, and A.P. Ghosh (With 5 Figures) 248

Ion Desorption Induced by Core Exciton States in MgO
By R.L. Kurtz, R. Stockbauer, F. Senf, R. Nyholm, and S. A. Flodström (With 3 Figures) ................ 258

Studies of Sapphire Photoablation by Laser-Induced Fluorescence and Photothermal Deformation Measurement
By R.W. Dreyfus, F.A. McDonald, and R.J. von Gutfeld
(With 5 Figures) ........................... 262

Multiphonon-Induced Desorption from $BaF_2(111)$
By H.B. Nielsen, J. Reif, E. Matthias, E. Westin, and A. Rosén
(With 9 Figures) ........................... 266

PSD of Excited Hydrogen From KCl
By L.T. Hudson, A.V. Barnes, N.J. Halas, R.F. Haglund, M.H. Mendenhall, P. Nordlander, N.H. Tolk, Y. Wang, and R.A. Rosenberg (With 4 Figures) ................ 274

Ground-State and Excited Atom Production by Electron and Ion Bombardment of NaCl and $CaF_2$
By G. Betz, E. Wolfrum, P. Wurz, K. Mader, B. Strehl, W. Husinsky, R.F. Haglund, Jr., and N.H. Tolk (With 4 Figures) 278

Optical Radiation from Electron, Photon and Heavy Particle Bombardment of Lithium Fluoride and Lithium-Dosed Surfaces
By N.H. Tolk, R.G. Albridge, A.V. Barnes, R.F. Haglund, Jr., L.T. Hudson, M.H. Mendenhall, D.P. Russell, J. Sarnthein, P.M. Savundararaj, and P.W. Wang (With 5 Figures) . . . . . . . . . 284

Long Lifetime of Bulk Luminescence Observed in Spectrosil Glasses Under Electron Bombardment
By P.W. Wang, R.F. Haglund, Jr., L.T. Hudson, D.L. Kinser, M.H. Mendenhall, N.H. Tolk, and R.A. Weeks (With 6 Figures) . 289

**Index of Contributors** . . . . . . . . . . . . . . . . . . . . . . . . . . 295

# Part I

## Introductory Lecture

# Some Recent Developments in DIET

*D. Menzel*

Physik-Department E20; Technische Universität München,
D-8046 Garching, Fed. Rep. of Germany

## 1. The General Situation

The field of DIET, the investigation of electronically stimulated desorption processes, which has been active and alive now for many years, appears to still gain in interest. There are many reasons, of which I want to emphasize three. One is, of course, the fact that our understanding of the underlying physics is continuously improving due to progress both in experimentation and in theory. The increasing application of synchrotron radiation as the means of primary excitation - which makes it possible to prepare very well-defined initial excitations, both in terms of energy and symmetry - figures strongly in the first instance. The introduction of laser methods into our field for the investigation of internal state distributions of desorbing particles, but also for the primary excitations, increases the detail of available information. But also in less obvious respects, the sophistication of experimentation is progressing, for instance in the detection of neutral or negative particles. All this poses a challenge to theory which is increasingly being met. A further very helpful circumstance is the mature state of electron spectroscopies which have also profited considerably from synchrotron techniques and which facilitate the interpretation of desorption data through the improved knowledge accumulating about the formation, evolution and decay of electronic excitations. So we can say that on its own ground DIET is healthily progressing.

The two other factors I want to single out have to do with the connections to other fields. A very fruitful connection exists to the very impressively developing area of molecular photodissociation, of which we will have proof during this workshop. Looking at the possibilities that exist in the investigation of isolated molecules, with all the sophisticated multiple coincidence techniques, a surface scientist can only be envious. But even if we cannot, because of fundamental limitations, apply similar techniques to our systems, the general knowledge about processes in isolated systems, as well as the system-specific information that is gained here, will be of great value to us who try to understand what happens if these molecules are coupled to the surface of an extended bulk, or if they join to form a bulk system themselves. Let me just mention one aspect that has impressed me personally. The finding that there are cases where photodissociation of a core-excited molecule can be faster than the electronic de-excitation, so that the potential surface which determines the motion of nuclei is that of the core-excited system and not of its multiply valence-excited Auger decay product /1/ has come as a surprise to me and certainly to many of us. While I do not yet know of a similar case in DIET, we will have to watch out for such a possibility.

Finally, the other important connections to other fields - which have, of course, always been there, but appear to become more obvious recently - have to do with more practical parallels. There have always been direct applications of DIT, from pressure measurement to plasma machines, and new ones appear. I have been told that the cost of the Superconducting Super Collider may well be strongly influenced by the extent of photodesorption in it, via the tube size which may be determined by the required pumping conductance. The rapidly developing field of atomic resolution electron microscopy, both of solid materials /2/ and of biological specimens /3/, appears to be strongly influenced by radiation damage in the necessary high electron fluxes which is largely connected with DIET and similar processes. The formation and liberation of molecules in space /4/ may, at least in part, be governed by processes similar to the ones of interest to us. And the many techniques in semiconductor technology connected with the use of energetic particles, from lithography to beam-induced etching and deposition, contain - even though they are much more complex than our desorption processes - many similar steps and mechanisms /5/.

It appears, therefore, that intrinsic progress in experiment and theory, and the cross-linking to fundamental as well as applied other fields makes for a situation which stimulates further development and contains promise for the future.

## 2. A Selective View of Some Progress

### 2.1 Two Extremes of DIET

Having been given license to select for a closer look what may be close to my heart, I may be forgiven for having picked out some of our own work and the conclusions we would draw from it. While the latter will certainly not be uncontroversial, they may stimulate further discussion.

Trying to systematize what we know about DIET /6/, I would conclude that the important aspect that sets off DIET from molecular photodissociation, the coupling to the substrate, can in principle have two effects. In one extreme case (Fig. 1), the system to be dissociated would have to be excited directly by the primary agent (electron or photon), while the coupled substrate would act as an excitation sponge which can take up energy and/or charge such that it is rapidly delocalized and will never come back to the primary spot. In such a case the localization of excitation becomes of utmost importance, and compared to a corresponding isolated system, the desorption probability or efficiency (the ratio of desorption and excitation cross sections) will be much smaller than 1 and depend drastically on various parameters (see Fig.1

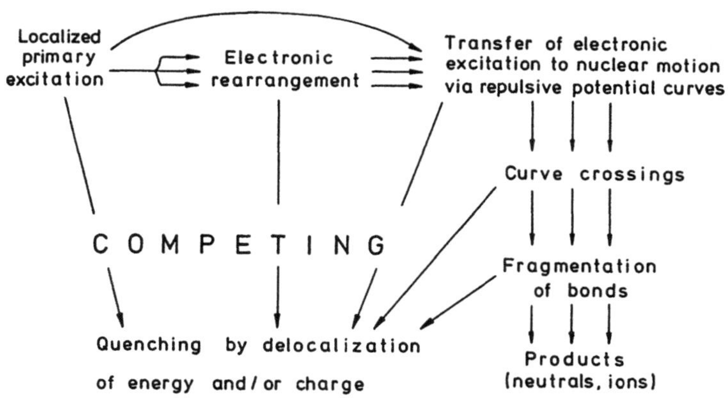


Fig. 1: Sketch of the sequential and competing processes important for DIET in a system with strong tendency to degradation of excitation, e.g. molecular adsorbates on a metal surface

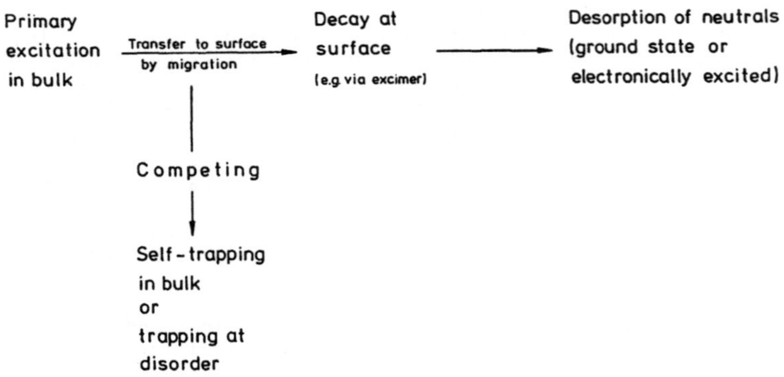


Fig. 2: Sketch of the processes leading to DIET in a system with easy migration of excitation without degradation, e.g. a bulk insulator

and below). In the other extreme (Fig. 2), a comparatively thick layer of the substrate can gather up primary excitation, channel it selectively to the surface and lead to desorption effects which are much larger than the corresponding dissociation of isolated systems and which respond positively to the transfer rates in the bulk. In principle, both aspects should always be present; which of them dominates depends on the dispersion of excitation. While there are no totally pure examples of these limiting cases, the closest representations are adsorbates on metals or semiconductors on the one hand, and bulk insulators on the other. Because of the high density of low-lying electronic excitations in the former, and their strong coupling, the delocalization of an adsorbate excitation leads to its rapid dispersion in space and energy, so that indeed it will never come back. On the other hand excitations in wide-gap materials can travel intact over rather long distances without degradation, so that - if mechanisms exist which let them decay dissociatively at the surface - the bulk can act as collector as indicated. No case is pure, of course: metals can in some sense act as collectors too, as secondary electrons created in the bulk can become dissociatively effective at the surface as well; and as dispersion mechanisms will act even in wide gap materials. But the leading influence will be much stronger and determine the general behavior. Some examples for the results on which this view is based will be given in the following.

## 2.2 Photodesorption from Molecules on Metal Surfaces

The prototype here is, of course, CO on transition metals, but others like NO, $N_2O$ and $H_2O$ have been investigated. Synchrotron radiation has made it possible to study the effectiveness of various well-defined primary excitations for desorption, and for CO on Ru(001) we have now completed a survey of the whole range from the lowest valence excitations to multiple core excitations, part of which has been reported before /7,8/. It turns out that here - apart from a possible very small neutral desorption effect via low-lying particle-hole excitations - two-hole excitations at least are necessary for desorption of neutral molecules, molecular ions and fragment ions. Furthermore, the more complex the primary excitations are, the higher their desorption efficiency. This can be understood in terms of the strong screening of the holes by the metal which leads to a decrease of two-hole Coulomb localization as compared to the isolated CO and which in turn can be counteracted by increased correlation in a more complex excitation. The consequence is the relative magnification of increasingly higher excitations. A careful analysis of core excitation lineshapes with resonant and non-resonant Auger spectra /9/ and comparison with desorption efficiencies has shown that (1) the primary core excitation need not lead to fully relaxed states; (2) the majority of excitations which show up in Auger spectra have fully relaxed before Auger decay takes place; (3) the very complex excitations which do show up in fragment desorption must correspond to the tail of the lifetime distribution not contained in (2) (Fig. 3). An alternative would be that the latter come about by dissociation before Auger decay (see above); there are arguments against this, and comparison to data for isolated molecules would be interesting. While in general this favors the predominance of high excitations, we have recently seen a case of strong preferential desorption via a state very close to (core) threshold for the case of $H_2O$, both in chemisorbed and condensed layers (Fig. 4 /10/); in this case the core-excited state may indeed contribute directly to dissociation.

In any case, the drastic influence of the primary excitation on desorption efficiency, above that known from molecular branching ratios, shows the high importance of localization for desorption. Coupling and screening will decrease it, correlation increase it. Other influences on the desorption efficiency fit this picture. There is a clear coverage dependence in the CO/Ru system with constant desorption cross section up to $\sqrt{3}$ coverages and decline above /11/ which is correlated with the constant intermolecular distance in $\sqrt{3}$

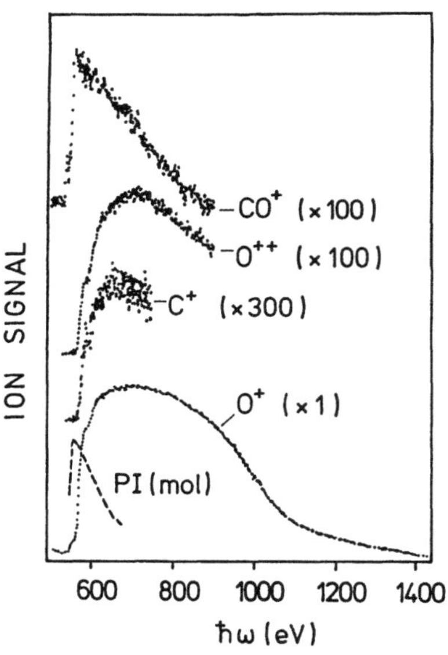


Fig. 3: Wide range photon energy dependence of ionic desorption products from CO/Ru(001) above O 1s, compared to the molecular photoionization cross section (PI (mol)). For fragment desorption, the range of high energy multiple excitations is strongly enhanced. (After /9/)

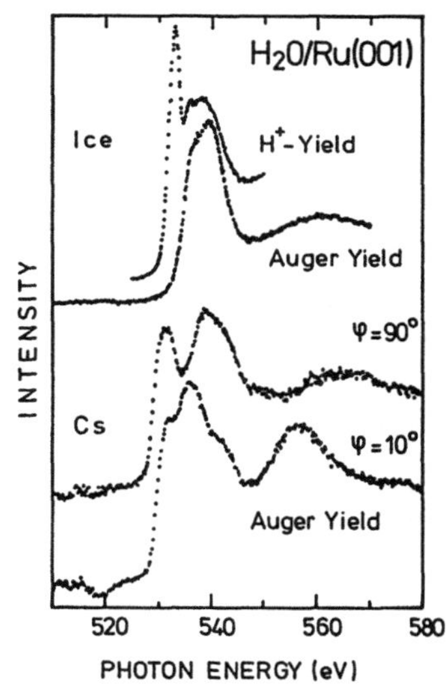


Fig. 4: Comparison of $H^+$ and O KLL Auger yields above O 1s for thin ice layers (two top curves), and Auger yield for chemisorbed $H_2O$ on Ru(001), at two polarizations. The $H^+$ yield from the latter is similar to that from ice (besides strong general attenuation) and in particular also contains the high narrow peak at about 534 eV. None of the Auger yields show a corresponding structure which proves that the narrow $H^+$ peak corresponds to an excitation with particularly enhanced dissociation probability /10/

islands and the stronger lateral overlap (also obvious from the dispersion of adsorbate states /12/) and consequent additional lateral delocalization at high coverages. On the other hand, the existence of "minority states" with increased cross sections correlated to impurities or surface disorder /13/, which do not depend as strongly on localization by the primary excitation corroborates the general picture, as do the well-known isotope effects.

These considerations have not differentiated between the various products which a detailed discussion would have to do. In particular, one has to distinguish between energy delocalization and charge transfer. In terms of the MGR mechanism, there is a certain part of the excitations for which energy and charge is delocalized on the time scale of the primary excitation, so that these could simply be considered as coupled excitations, i.e. the excitation itself is partly delocalized. Another part may delocalize after electronic rearrangement (e.g. Auger decay) and another after some nuclear motion has occurred. For the latter, charge can be transferred at a later stage which can be viewed as curve-crossing events. Such processes will lead to desorbing neutrals, as already outlined in the original MG paper /14/. A parameter which can be used to influence these branching ratios is a high electric field /15/. A detailed discussion would have to take all these differences into account.

## 2.3 Desorption from Insulator Surfaces: Rare Gas Solids

The desorption from insulators has been investigated over the years using the example of alkali halides, and while some controversy in detail exists /16/, there appears to be consensus that in this case the gathering-up effect of the solid works mainly with primary excitations of the halogen sublattice which can get to the surface by a replacement cascade or as a mobile excited state ($V_k$ + e state which can decay into H + F centres) or be trapped. At the surface the decay leads to halogen desorption. The alkali metal either accumulates or desorbs thermally, if the temperature is high enough /17/.

Another example where the active primary excitations have recently been investigated in detail are the rare gas solids. At the last DIET workshop a lively discussion about the prevalent desorption mechanism for these systems took place /18/, and agreement existed about the exciton-excimer path. Recent synchrotron work by Zimmerer et al. /19/ and by us /8,20/ showed clearly that bulk and surface excitons as well as ionic primary excitations, and probably also two-step processes (photoelectron-induced excitons) are active in desorption (Fig. 5). The detection of fluorescence light from desorbing atoms by

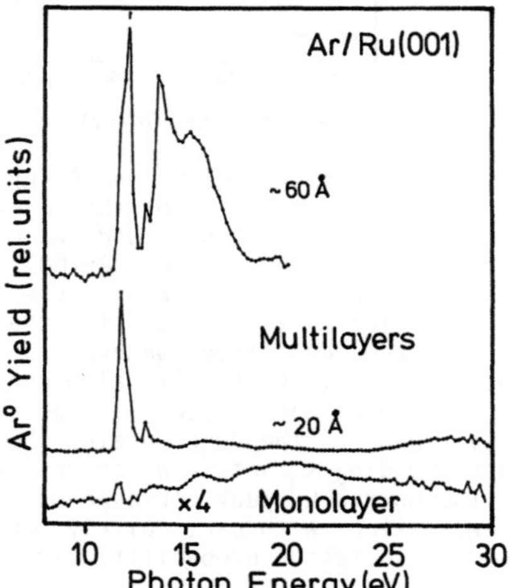


Fig. 5: Yield of neutral Ar atoms desorbed by photons in the excitonic and ionic range for Ar monolayers, and multilayers of two different thicknesses. A surface exciton in the former case, bulk excitons in the latter, and ionic excitations in all cases, with changing weights, are obvious (after /20/)

Zimmerer et al. indicates that electronically excited atoms desorb also, and led them to propose an additional desorption mechanism. Our evidence for a thickness-dependence of the desorption yields via the various excitations and for radiation damage effects emphasize the importance of transport processes in the bulk. The obvious next step for more accurate information on the contributing states will be measurements of kinetic energy distributions of the desorbing neutrals.

An interesting case which brings us back to sect. 2.2 is that of rare gas monolayers on metals /8,20/. While the yields appear to be more strongly attenuated than would correspond to the thickness decrease compared to multilayers (indicating the take-over of delocalization into the bulk metal as a quenching path), and bulk excitons are of course missing, the surface exciton can still be observed for an Ar monolayer, besides ionic excitations (Fig. 5). The latter have so far been discussed in terms of an Antoniewicz-type mechanism, which will have to be modified in order to include neutral excitations. The interpretation is under active discussion /20-22/ and would again be helped by kinetic energy measurements.

## 3. Concluding Remarks

This is a very personal and selective snap-shot of the situation in DIET. Its shortcomings are obvious, and quite a few colleagues may not agree with some of the attempted generalizations. Nevertheless, it hopefully shows that the field of DIET, while now coming of years, still or renewedly is dynamic and shows potential yet unexploited. There is more DIET to come!

## References


1. P. Morin, I. Nenner: Physica Scripta T 17, 171 (1987);
   I. Nenner et al., this volume
2. See for instance: Study of Surfaces and Interfaces be Electron Optical Techniques, ed. by A. Howie and U. Valdré; Plenum Publ. Co., New York (in press)
3. See for instance: Cryomicroscopy and Radiation Damage, ed. by E. Zeitler, vol. I and II, North-Holland 1982/84
4. R.E. Johnson: this volume
5. J.W. Coburn: this volume, and references therein.
6. Desorption Induced by Electronic Transitions - DIET-I; ed. by M.M. Traum, N.H. Tolk, J.C. Tully, T.E. Madey, Springer, Berlin 1983;
   Dto., DIET-II; ed. by W. Brenig, D. Menzel, Springer, Berlin 1985.
7. R. Treichler, W. Riedl, W. Wurth, P. Feulner, D. Menzel: Phys. Rev. Letters 54, 462 (1985);
   dto.: DIET-I (ref. 6), p. 68;
   S. Auer, P. Feulner, D. Menzel: Phys. Rev. B35, 7752 (1987)
8. P. Feulner et. al.: this volume
9. W. Wurth, C. Schneider, R. Treichler, E. Umbach, D. Menzel: Phys. Rev. B35, 7741 (1987);
   D. Menzel, P. Feulner, R. Treichler, E. Umbach, W. Wurth: Physica Scripta T17, 166 (1987)
10. D. Coulman, A. Puschmann, H.P. Steinrück, P. Feulner, D. Menzel, presented at ECOSS-9 (April 1987), and to be published.
11. P. Feulner, H.A. Engelhardt, D. Menzel: Appl. Phys. 15, 355 (1978);
    T.E. Madey: Surface Science 79, 575 (1979)
12. P. Hofmann, J. Gossler, A. Zartner, M. Glanz, D. Menzel: Surface Sci. 161, 303 (1985)

13. P. Feulner, S. Auer, W. Riedl, A. Cassuto, D. Menzel: Surface Sci. 182, L 221 (1987);
D. Menzel: Nucl. Instrum. Methods Phys. Res. B13, 514 (1986)
14. D. Menzel, R. Gomer: J. Chem. Phys. 41, 3311 (1964)
15. S. Jaenicke et al.: this volume, and references therein
16. R.E. Walkup, P. Avouris, A.P. Gosh: Phys. Rev. Letters 57, 2227 (1986) and work cited therein
17. See the articles by N. Itoh, DIET-I, p. 229; N.H. Tolk et al., DIET-II, p. 152; M. Szymonski et al., DIET-II, p. 160; (ref. 6), and work cited therein
18. See the articles by H. Haberland (p. 177), R. Pedrys et al. (p. 180), M.L. Brown et al. (p. 199) in DIET-II (ref. 6)
19. F. Coletti, J.M. Debever, G. Zimmerer: J. Physique Lett. 45, L 467 (1984), and to be published
20. P. Feulner, T. Müller, A. Puschmann, D. Menzel: Phys. Rev. Letters (1987, in press)
21. Z.W. Gortel, H.J. Kreuzer, P. Feulner, D. Menzel: Phys. Rev. B35, 8951 (1987)
22. D.R. Jennison: this conference

# Part II

# Gas Phase

# Photodissociation and Photoionization Processes of Core Excited Molecules: The $CH_3Br$ Case

*I. Nenner[2], P. Morin[2], M. Simon[2], P. Labianquie[1], and G.G.B. de Souza[3]*

[1]LURE, Laboratoire mixte CNRS, CEA and MEN,
Université de Paris-sud, F-91405 Orsay Cedex, France
[2]CEA- IRDI/DESICP, Département de Physico-Chimie,
Centre d'Etudes Nucléaires de Saclay,
F-91191 Gif sur Yvette, France
[3]Instituto de Quimica,
Universidade Federal do Rio de Janeiro,
RJ, 21190, Brazil

## 1. INTRODUCTION

Innershell excitation of isolated molecules is commonly studied by photoabsorption or electron energy loss spectroscopy (EELS) [1]. Unlike valence excitation, significant absorption is found concentrated in well defined energy regions, each of them, being characteristic of one given component of the molecule. Details of the resonance pattern usually observed below and above the core ionization limit, is however very much dependant upon the valence electronic properties of the considered molecule [2,3]. This is illustrated, in Fig. 1 by the EELS of the hydrogen bromide molecule, HBr [4] and the methyl bromide molecule, $CH_3Br$ [5] measured near the bromine 3d edge, and by the photoabsorption spectrum of the silane molecule, $SiH_4$ [6] measured near the silicon 2p edge. In these three examples, the most intense feature seen in the discrete spectrum, corresponds to a transition from the core level to the first unoccupied antibonding valence ($\sigma^*$) orbital often localised along a specific bond of the molecule. In contrast, the narrower features, lying closer to threshold correspond to transitions from the core to Rydberg orbitals, i.e. the excited electron lies at larger distances from the nuclei, than for valence orbitals. Another broad resonance (~ 20 eV wide) is found in the continuum of the core ionization spectrum in the three molecules under study. This is illustrated in Fig.1 only for $CH_3Br$, which is of specific interest in this paper. These are atomic-like shape resonances. For HBr and $CH_3Br$, this resonance is of the bromine 3d $\rightarrow \epsilon f$ type and for $SiH_4$, it is of the silicon 2p $\rightarrow \epsilon d$ type. Such resonances are of different nature than molecular shape resonances, which reflect the anisotropy of the molecular field [3]. Here, the bromine or silicon atoms are not very much perturbed by their ligands as far as their potential is concerned.

In the present paper, we show that the $\sigma^*$ resonances are described by strongly repulsive states which account for a large Franck Condon envelope in the photoabsorption spectra and also explain that the core hole can relax at large internuclear distances. In other words, we show that photodissociation may precede (or compete) with autoionization. We briefly review the most important results of our recent photoelectron studies on HBr [7] and Silane [8] performed in the photon energy region indicated in Fig. 1 and we report new photoelectron and photoion spectrometry measurements in the $CH_3$ Br molecule photoexcited in the region of the bromine 3d ionization edge, i.e. from 68 to 120 eV photon energy. We will

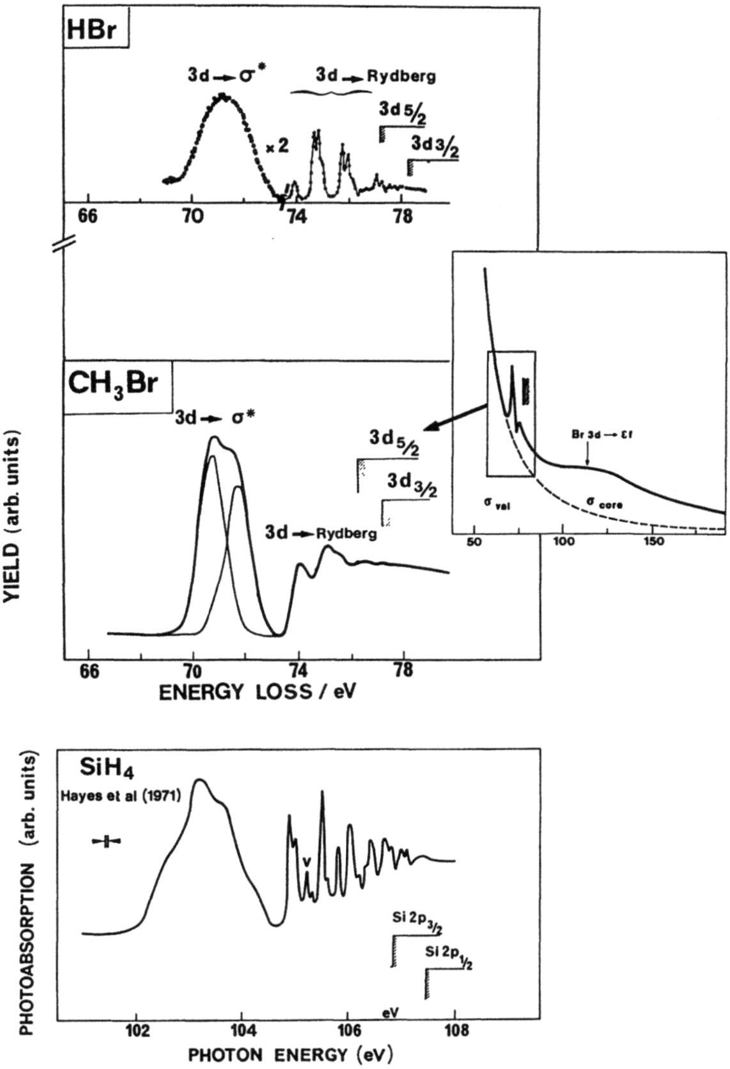


Figure 1. Electron energy loss spectra of HBr (top) ; $CH_3Br$ (middle) ; $SiH_4$ (bottom), reproduced from references 4, 5 and 6 respectively

consider first, the problem of fast dissociation and the core hole relaxation ; then, considering $CH_3Br$ as a model polyatomic molecule, we will see how this process competes with molecular autoionization and, lastly, we will analyse the effect of the site selectivity of the photoexcitation into $\sigma^*$ or the continuum of the bromine atom, on the final ionic fragmentation pattern.

2. EXPERIMENTAL

Two different experiments have been used with the same photon source, one is an angular resolved photoelectron spectrometer which has been described in details elsewhere [9], another is a time of flight (TOF) mass spectrometer also described earlier [10]. Here we briefly present the overall arrangement and procedures. We present in Fig. 2, a schematical diagram of the toroidal monochromator which disperses the VUV synchrotron radiation emitted from the ACO storage ring at LURE, Orsay. The 60 to 120 eV photon energy range, which is of interest in this work is covered with two different toroidal gratings. It is refocused at the center of the interaction chamber where it crosses at right angle an effusive jet of gas. Then we detect either electron or ions using two different analysers which are mounted separately in the experimental chamber.

When the electron analyser is used, it is operated at a constant pass energy of 10 eV (0.7 eV resolution) and positioned at the pseudo-magic-angle which eliminates angular effects. In addition to conventional photoelectron spectra (PES) we have measured partial photoionization cross sections, continuously with photon energy, using the constant ionic state (CIS) method [9]. All spectra are corrected for the incident light intensity (monitored by a gold mesh photoemission) and for the electron transmission of the analyser.

When the TOF spectrometer is used, it is oriented in the horizontal position, i.e. the main axis coincides with the main polarization vector of the light. This spectrometer is used either to obtain conventional mass spectra or coincidences between two or three ionic fragments. In the first

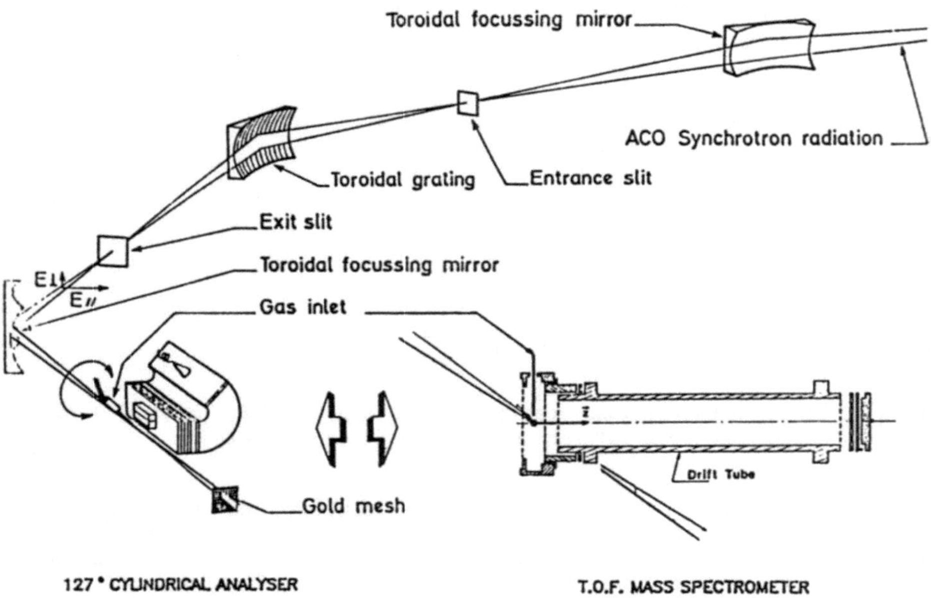


Figure 2. Schematical diagram of the toroidal monochromator, electrostatic analyser and time of flight mass spectrometer

operation mode, the ions are extracted from the ionization region by a pulsed voltage (520V at a 50kHz repetition rate). Secondly, in the coincidence mode, a DC field of 520 V extracts the ions continuously in the drift region on which a 1090 V potential is applied. Ionic fragments which are issued from the same ionization event such as dissociative double ionization, $m^{2+} \to m_1^+ + m_2^+$ arrive at the detector with a constant TOF difference. A time to amplitude converter and a pulse height analyser provides a photo-ion photo-ion coincidence (PIPICO) spectrum. In the case of triple dissociative ionization events, $m^{3+} \to m_1^+ + m_2^+ + m_3^+$, the same DC field is applied in the ionization chamber. Assuming that $m_1^+$ , $m_2^+$ and $m_3^+$ correspond to ion masses in increasing order, the first ion $m_1^+$ generates a gate pulse which allows subsequent ions to start a time to amplitude converter, after an interval corresponding to the difference in flight time between $m_1^+$ and $m_2^+$ . The gate width is chosen to match exactly the whole $m_2^+$ arrival time spread. The triple coincidence spectrum (3PICO) is then obtained by recording $m_3^+$ after the $m_1^+/m_2^+$ pair. Quantitative information can be deduced from these coincidence experiments in the following way. If we define $N_1$ as being the percentage of events leading to a single ionic fragments, $N_2$ the percentage of events leading to an ion pair and $N_3$ the percentage of events leading to an ion triplet, then their relative proportion can be deduced from the number of coincidences per unit time in a PIPICO ($C_1$) and 3PICO ($C_2$) experiment with the following formulae

$$N_2 = (R_1 f - R_2(3-2f))/\left(f^2 - R_1 f + R_2(1-2f)\right) \qquad \{1\}$$

$$N_3 = R_2/(f^2 - R_1 f + R_2(1-2f)) \qquad \{2\}$$

$$N_1 + N_2 + N_3 = 1, \qquad \{3\}$$

where $R_1 = C_1/N_i$, $R_2 = C_2/N_i$ with $N_i$ is the total number of ions detected per second and f the detection efficiency. The f value is very sensitive to the experimental conditions and has to be evaluated the very day when the experiment is performed.

## 3. RESULTS AND DISCUSSION

### 3.1 Atomic autoionization following very fast dissociation in HBr, $SiH_4$ and $CH_3Br$

The first direct evidence of purely repulsive core excited states being responsible for a very fast dissociation compared to the electronic relaxation of the core vacancy, has been offered by Morin and Nenner [7] through a photoelectron study of HBr excited near the bromine 3d edge. The relaxation of the Br3d → $\sigma^*$ transition in HBr can be described in two steps, as shown in Fig. 3. i) The core excited HBr molecule dissociates into a ground state H atom and an excited bromine atom. ii) The latter autoionizes into the $Br^+ + e^-$ and $Br^{2+} + 2e^-$ continua. This process gives rise to atomic bromine lines in the molecular photoelectron spectrum recorded at the $\sigma^*$ resonance. Such spectrum is shown in Fig. 4 together with the spectrum recorded off resonance. The energy of atomic lines have been calculated knowing the 3d → 4p excitation energy and ionization energies of ground state bromine from independent experimental sources. These values reported in Fig.4b on an electron kinetic energy scale are in

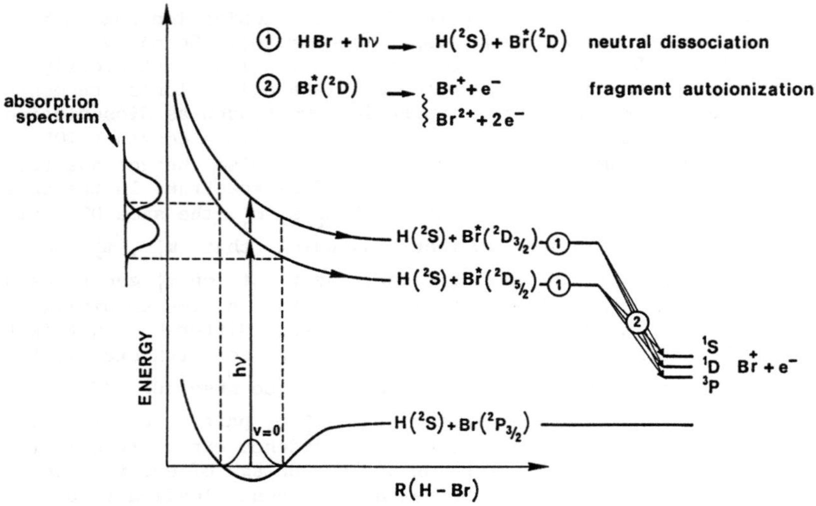


Figure 3. Schematics of relevant potential curves of HBr describing the two-step relaxation process

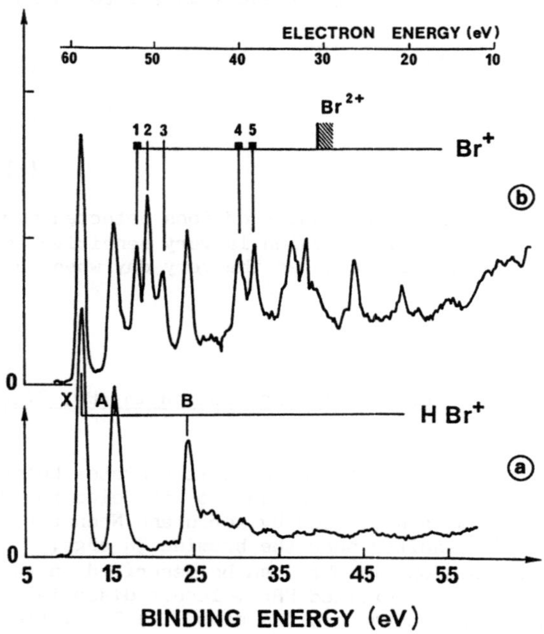


Figure 4. PES recorded off (a) and on (b) resonance respectively at 68.2 and 70.6 eV photon energy.

good agreement with the position of extra lines. The latter are clearly distinguished from molecular bands, especially the three sharp lines observed around 18 eV binding energy, in which $Br^+$ is produced with a 3p outer vacancy i.e. in the $^3P_{2,1,0}$ $^1D_2$ and $^1S_0$ states. Another fingerprint of this two-step process is that by scanning the photon energy along the resonance profile, the bromine bands are found to appear at a constant electron kinetic energy on the photoelectron spectrum (not shown here, see ref 7), contrary to normal molecular bands. This shows that such electrons are indeed secondary, (see step 2 in Fig. 3). very much like an Auger process.

On a real time scale, this description is realistic for the simple following reasons : one can assume that the kinetic energy released in the dissociation, which equals the resonance energy minus the energy of the dissociation limit, i.e. $E_R = 2.47$ eV is entirely transferred to the H atom because of the high H/Br mass ratio of 80. Knowing that the Br 3d hole lifetime, $\tau$, is larger than $7.10^{-15}$ seconds, we can estimate the internuclear distance d at which the Br inner hole relaxes, after this time $\tau$. We obtain

$$d = R_e + (2E_R / m_H)^{1/2} \tau, \quad \{4\}$$

where $R_e = 1.41$ Å, is the equilibrium internuclear distance. The 3 Å value thus estimated is large enough to justify a-posteriori the description of the relaxation process with two independent steps. Consequently we predict that, at the $\sigma^*$ resonance, one should observe in a mass spectrometry study, $Br^+$ and $Br^{2+}$ as dominating fragments compared to $HBr^+$ or $HBr^{2+}$ species. Such studies are presently underway.

The same process has also been shown, by de Souza et al (8), in the case of a polyatomic hydride, the silane molecule. From the photoabsorption spectrum of Fig. 1, obtained near the silicon 2p edge, we observe that the discrete spectrum is dominated by a wide and intense resonance around 103.2 eV. Actually from the theoretical analysis of Schwarz [13], it results from a transition from the Si 2p core orbital into the $\sigma^*$(Si-H) unoccupied orbitals. In the tetrahedral symmetry group, such $\sigma^*$ orbitals correspond to two components, $a_1$ and $t_2$ each of them being split by the spin-orbit interaction of the 2p core hole. De Souza et al [8] have shown that the decay of this resonance takes place in two steps, as schematically represented in Fig. 5. The first one is the rapid ejection of a ground state H atom and the formation of a core excited $SiH_3^*$ radical. Using the Z+1 core equivalent model (13), one can replace the core excited silicon atom by phosphorus. The core excited silane molecule is then equivalent to $PH_4$ . This species is known to be highly unstable and dissociates without energy barrier into $PH_3$ + H with an excess energy of 1.4 eV. The second step is the autoionization of $SiH_3^*$ into $SiH_3^+ + e^-$ and $SiH_3^{2+} + 2e^-$. In Fig.6, we have reproduced two PES of silane obtained on and off resonance, by de Souza et al (8). One observes in Fig 6a a broad feature around 15 eV, underlying the two valence bands ($2t_2$ and $3a_1$). De Souza et al.(8) have interpreted it as the result of autoionization of $SiH_3^*$ into the ground state ionization continuum, $SiH_3^+ + e^-$. Using the experimental first ionization energy of this radical and an estimation of its core excitation from the energetics of Fig. 5, the calculated electron energy is expected to be equal to 89.4 eV, i.e. 13.8 eV on the binding energy scale of Fig. 6a. The agreement with the position of the broad band seen in the inset of Fig.6a is fair enough to conclude the existence of a two-step process. However, the broadness of the band prevents any detailed analysis such as the one made in the HBr case. During the autoionization process, one expects large changes in vibrational excitation because the core excited radical and the final $SiH_3^+$ ion have certainly quite different equilibrium geometries.

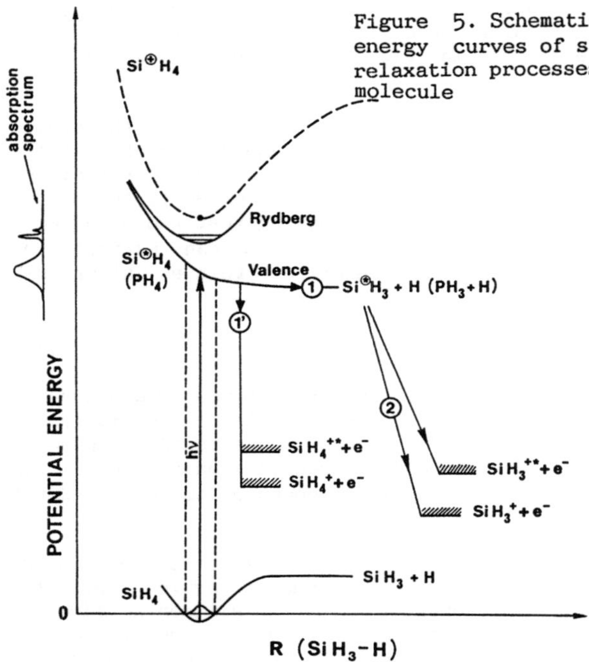


Figure 5. Schematics of relevant potential energy curves of silane for describing the relaxation processes of the core excited molecule

On a real time scale we may estimate, a posteriori, what would be the internuclear separation $d(SiH_3^*\text{-}H)$ at a time $\tau$ following the Si2p $\sigma^*$ transition, $\tau$ being the core hole lifetime. An upper value of $\tau$ is obtained from the width of Rydberg lines observed in the photoabsorption spectrum (see Fig.1). Their potential surface likely resembles those of their ionic core. Because the Si2p electron is non-bonding, the core ionized silane ion has a negligible Franck-Condon envelope. Consequently, the width of pure Rydberg lines are only ruled out by lifetime effects. Here, the 0.06 eV width corresponds to a lower limit of $\tau$ of $10^{-14}$s. As a result, knowing the equilibrium distance $H_3Si\text{-}H$ as being 1.48 Å [14], we obtain from the simple equation {4} , that d equals 3.15 Å. This value exceeds $R_e$ by a factor of 2 and is fully consistent with a sequential two-step relaxation of the $\sigma^*$ resonance. Notice, that the resonant part of the spectrum shown in Fig. 6a(the $3a_1$ band and high lying satellites) must be due to the relaxation of $SiH_3$ excited species rather than $SiH_4^*$. This point has not been checked yet, because the spectrum of $SiH_3^+$ excited states is not known.

The methyl bromide molecule, which is now analysed in detail , is of particular interest because it can be compared to the HBr molecule if we consider it as a diatomic - like species in which the bromine atom is now linked to a radical 15 times heavier.

The $CH_3Br$ molecule belongs to the $C_{3v}$ symmetry group with the following ground state configuration

| | |
|---|---|
| $Br[1s^2 2s^2 2p^6]$ $C[1s^2]$ $Br[3s^2 3p^6 3d^{10}]$ | $1a_1^2 2a_1^2 1e^4 3a_1^2 2e^4$ |
| Core orbitals | valence MO's |

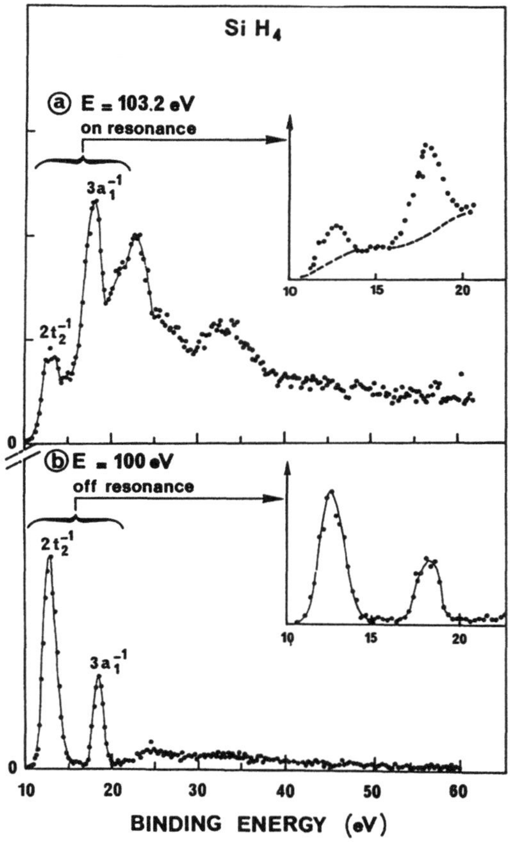


Figure 6. PES of silane recorded on (a) and off (b) resonance

We show in Fig. 7a a photoelectron spectrum of methyl bromide recorded at 68 eV i.e. below the 3d→σ* resonance. The peaks labelled X, A, B, C and D correspond to the ionization of the 2e, $3a_1$, 1e, $2a_1$ and $1a_1$ molecular orbitals respectively, following previous PES studies with He I and He II [15-17] and theoretical analysis [17,18]. Briefly, These MO's are described as follows : the outer valence 2e corresponds to the Bromine 4p lone pair, the $3a_1$ and 1e correspond respectively to the σ(C-Br) and σ(C-H) bonding orbitals, essentially built with the C(2p) orbital. The two inner valence $1a_1$ and $2a_1$ previously detected in He II PES [17] are here found with significant intensity and correspond to σ(C-Br) bonding MO's. The $2a_1$ MO likely has a significant C(2s) character and $1a_1$ is mostly Br(4s) in close analogy with the Br(4s) band of HBr [7], following Mulliken [18] and previous analysis of methyl halides [15]. Notice that the inner valence ionization lines are also distributed over a large set of unresolved satellite lines which reflects the breakdown of the one electron picture [19]. The $3d_{5/2}$ ionization limit is taken at 76.2 eV and the $3d_{3/2}$ one at 77.2 eV, following high resolution XPS measurements [20].
In Fig. 7b, we show a PES recorded at 71 eV, i.e. on the σ* resonance. the most striking observations are i) several sharp and broad lines appear in addition to normal valence bands, especially at very low binding energy

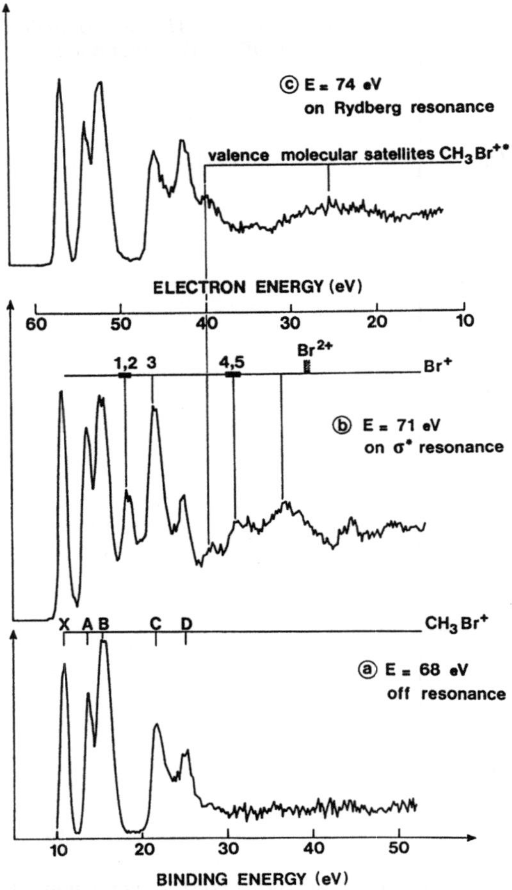


Figure 7. PES of $CH_3Br$ recorded off (a) and on resonances

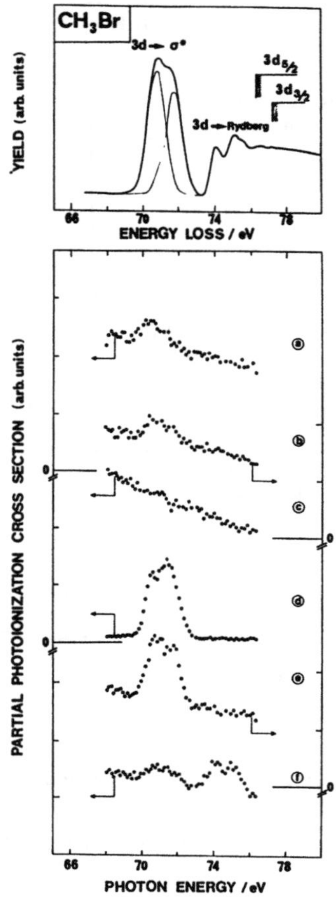


Figure 8. Top: EELS from ref 5
Bottom: CIS curves for the X(a), A(b) B(c), 18.7 eV(d), C(e), D(f) bands.

(18.7 eV) ii) the C band is greatly enhanced. Figure 7c shows a PES recorded at 74 eV, i.e. on one of the Rydberg state. In contrast to Fig. 7b, there is no low lying sharp lines around 18.7 eV, only an enhancement of the D band and much broader lines in the high binding energy region. It is then clear that the decay of the σ* resonance is very different from the decay of the Rydberg resonance, despite the fact that one common feature between the two spectra is found at 40 eV binding energy. In order to analyse the detailed energy dependence, we have measured, using the CIS method, the relative partial cross section for the main features below 28 eV binding energy, as seen in Figs. 8a (X state), 8b (A state), 8c (B state), 8d (supplementary band), 8e (C state) and 8f (D state). It is very striking to observe that the supplementary and C bands resonate strongly at the energy of the resonance, whereas the X, A and D bands are only slightly enhanced and the B one decreases monotonically. Notice also the specific behaviour of the D band which is found to be largely enhanced at the energy of the Rydberg resonance compared to all other bands. These observations remind one greatly of those of HBr and can be interpreted in a similar way.

We are offering in Fig. 9, a schematic of the relevant potential energy curves in $CH_3Br$, along the $CH_3$-Br internuclear distance. The $\sigma^*$ resonance state is assumed to be purely repulsive, leading to a core excited bromine atom and a $CH_3$ radical in its ground state, $^2A_2''$. Knowing the 3d→4p excitation energy E ($Br^*$) of the bromine atom from photoabsorption measurements [21], and the valence ionization energies E(val) of Bromine from experiment [22] and calculations [23], we evaluate the energy spectrum of secondary electrons E(second) produced from the relaxation of the 3d hole in $Br^*$, after the $CH_3$ radical has been ejected. We obtain : E(second.) = E($Br^*$) - E(val). Those values have been reported on top of Fig. 7b and are found in excellent agreement with the supplementary bands. Lines 1-3 correspond to Br4p ionization i.e. $Br^+$, $^3P_{2,1,0}$, $^1D_2$, $^1S_0$ . Notice that line 3 overlaps exactly the C band of the molecule. Lines 4 and 5 result from multiplets issued from simple 4s ionization and various excited $Br^+$ configurations (see Ref 7 for more detailed assignments.

We now analyse the phenomenon on a real time scale. The lifetime $\tau$ of the 3d hole can be estimated from the width of the 3d→4p photoabsorption line of atomic bromine [21]. the 90 meV value corresponds to $\tau > 7.10^{-15}$s. We assume that this value is not far from that of the molecule. The

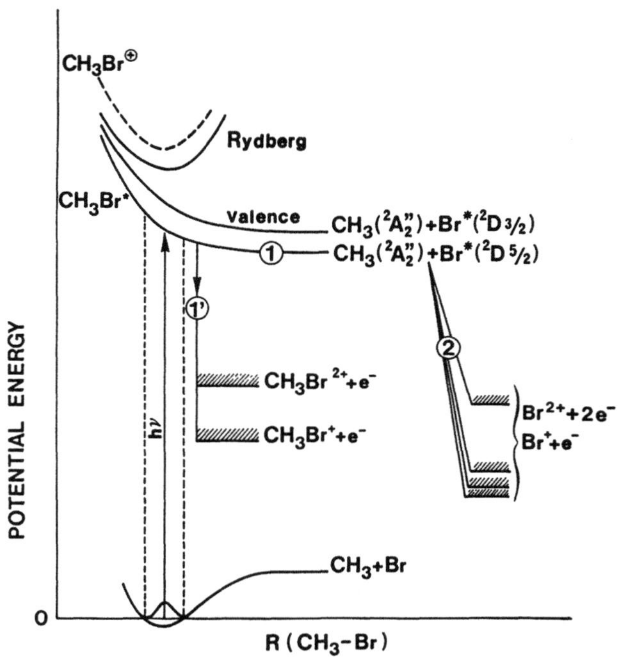


Figure 9. Schematics of the relevant potential curves of $CH_3Br$ for describing the relaxation processes of the core excited molecule

kinetic energy $E_R$ released in the dissociation is given by the difference between the photon energy (70.6 eV) and the dissociation limit, which is

$$D\left(CH_3\left(^2A_2''\right)+Br\left(^2D_{5/2}\right)\right) = D\left(CH_3\left(^2A_2''\right)+Br\left(^2P_{3/2}\right)\right) + E(Br^*),$$

where the dissociation limit of the ground state molecule $D\left(CH_3\left(^2A_2''\right)+Br\left(^2P_{3/2}\right)\right)$ equals 2.9 eV from Hertzberg [24] and $E(Br^*)$ equals 64.38 eV [21] . We obtain $E_R$ = 3.32 eV. This energy is shared between the $CH_3$ and $Br^*$ fragments. As for HBr and $SiH_4$ , we can estimate the internuclear distance d at which the bromine inner hole relaxes after the time $\tau$. Extension of formula {4}, gives

$$d = R_e + \left(v_{CH_3} + v_{Br^*}\right)\tau,$$

where $v_{CH_3}$ and $v_{Br^*}$ are respectively the velocity of each fragment. The equilibrium distance equals 1.939 Å according to Hertzberg [24]. We obtain d = 2.39 Å. This distance is 0.45 Å larger than the equilibrium distance. We readily conclude that the Br3d core hole relaxation occurs outside the Franck-Condon zone defined along the C-Br coordinate i.e.

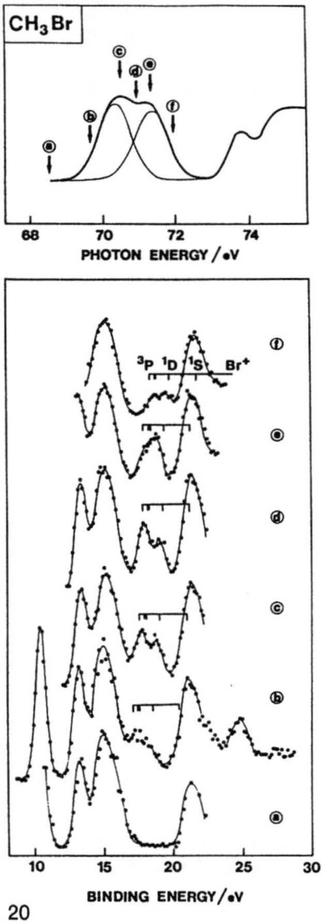


Figure 10. Top : threshold electron spectrum of $CH_3Br$ ; bottom : PES recorded at selected photon energies as indicated on the TES spectrum

$1.920 < R < 1.958$ Å. In other words, this means that a possible decay of the $\sigma^*$ resonance is a fast dissociation (labelled 1 in Fig.9). We now give further experimental evidence of the existence of bromine autoionization as a secondary step. Figure 10 shows a series of PES recorded in the 10-25 eV binding energy region, at selected photon energies along the resonance profile. The normal molecular bands are found at a constant binding energy, whereas the $Br^+$ bands are shifted when the photon energy changes. We have reported in Fig.10 the calculated position of those bands, assuming the two-step model of Fig.9. The good agreement shows that those electrons have a fixed kinetic energy and are issued from the secondary event, labelled 2 in Fig. 9. However, in contrast to the HBr case [7], the maximum distance d, compared to $R_e$ is much smaller than for HBr (d=3.15 Å compared to $R_e$=1.48 Å). This implies now that the dissociation of the core excited molecule and the autoionization of the core excited bromine atom are no longer separable processes. In other words, the Br3d hole may relax in the presence of the molecular field, following normal molecular electronic relaxation pathways (labelled 1' in Fig. 9). This point is now analyzed in more detail . This could explain why the $^3P$ and $^1D$ atomic bands are barely resolved in Fig.10.

## 3.2 Core hole relaxation pathways during the $CH_3$ departure in $CH_3Br$

The specific problem to solve at this point is to search for the relaxation pathways which compete with the fast dissociation. Generally, one expects that a core hole (whithin an atom or a molecule) decays via autoionization rather than an X-ray emission, as long as the atom under consideration is of low Z value. Autoionization processes are usually separated into spectator and non - spectator transitions as schematically summarized :

$$E_{h\nu} + (c^2\ v^n\ u^0) \longrightarrow (c^1\ v^n\ u^1),$$

where c, v and u represent respectively core, valence and unoccupied orbitals

$$(c^1\ v^n\ u^1) \longrightarrow (c^2\ v^{n-1}\ u^0) + e^- \qquad (\alpha_1)$$

$$\longrightarrow (c^2\ v^{n-2}\ u^1) + e^- \qquad (\beta_1)$$

$$\longrightarrow (c^2\ v^{n-2}\ u^0) + 2e^- \qquad (\alpha_2)$$

$$\longrightarrow (c^2\ v^{n-3}\ u^1) + 2e^- \qquad (\beta_2)$$

The $\alpha_1$ and $\alpha_2$ processes correspond to participator transitions with the ejection of one or two electrons ($\alpha_1$ is also called autoionization in the restricted sense). The $\beta_1$ and $\beta_2$ processes are spectator transitions, often referred to as simple (or double) resonant Auger. Notice that $\beta_1$ and $\beta_2$ produces respectively singly and doubly charged ions with excited configurations. In cases where those are imbedded in the double or triple ionization continua, one or two electrons may be ejected further. We deal then with cascade autoionization. For example, the $\beta_1$ process may be followed by

$$(c^2\ v^{n-2}\ u^1) \longrightarrow (c^2\ v^{n-2}\ u^0) + e^- \qquad (\alpha\beta)$$

In contrast, $\alpha_1$ produces a singly charged ion undistinguishable from the same ion produced by direct valence ionization. The same is true for $\alpha_2$. Most processes associated with the ejection of a well defined kinetic energy electron ( $\alpha_1$ , $\beta_1$ and $\alpha\beta$) are, in principle, detectable in a conventional photoelectron spectrum. In contrast, the processes associated with the ejection of two electrons ($\alpha_2$ and $\beta_2$) only appear as a continuum in a photoelectron spectrum. This is due to the quasi uniform distribution in kinetic energy of the two electrons, from zero and the maximum excess energy. Consequently, threshold photoelectron spectroscopy [see ref 27] or coincidences between the two outgoing electrons or between one electron and ions [25] are more appropriate tools to show the existence of such multiionization processes. Moreover, their complexity is such that it is extremely difficult to separate the contribution of the decay of a core excited atom (or radical) from the decay of a core excited molecule.

More information on multiionization processes can be obtained from mass spectrometry measurements. It is possible to measure the number of positive charges carried by the residual ion. More importantly, in the case of a competition between direct dissociation of a core excited molecule and core hole autoionization, the mass of the ionic fragments brings complementary information. Following the diagram of Fig. 9, if the core excited $CH_3Br$ molecule dissociates rapidly through the 2nd channel, we should only detect $Br^+$ or $Br^{2+}$ as residual ions ii) in the normal molecular relaxation decay (1'), one should find $CH_3Br^+$ and $CH_3Br^{2+}$ and also $CH_3^+$, $Br^+$ and $Br^{2+}$ as a result of the dissociation of these primary ions. In the present work, we have measured a threshold electron spectrum (TES) and performed a series of mass spectrometry experiments, to clarify this problem.

At the top of Fig.10, we show a threshold electron spectrum measured with our electrostatic analyser in the 68-76 eV photon energy region, with an energy resolution of 0.5 eV. The spectrum is found to mimic faithfully the shape of the photoabsorption spectrum (Fig.1), except that the relative intensity of the $\sigma^*$ resonance, compared to Rydberg lines, is roughly smaller by a factor of 2. As seen above, those low energy electrons come from specific channels such as $\alpha_2$ or $\beta_2$. Other relaxation processes like the two step autoionization process ($\alpha\beta$), the two ejected electrons have a finite kinetic energy generally with non zero values [27], and thus are not counted in the threshold electron spectrum. As a result, the TES of Fig.10 shows that the $\alpha_2$ and $\beta_2$ decay channels are relatively more effective on Rydberg resonances than on $\sigma^*$.

We show in Fig.11a, three typical mass spectra of $CH_3Br$ recorded at 68.2 eV (off resonance), 70.5 eV on the $\sigma^*$ resonance and in the continuum at 117 eV. As seen in Fig. 11a and 11b, which show the relative variation of each fragment as a function of photon energy, we observe that $Br^+$ and $Br^{2+}$ fragments resonate strongly along the $\sigma^*$ resonance. In contrast, $CH_i^+$ (and $H^+$) resonate weakly and $CH_iBr^+$ stays essentially constant. In the present experimental conditions, we discriminate ions with high kinetic energy, because of the loss of some energetic fragments produced between extraction pulses [28] . This affects significantly the intensity measurements of $H^+$ and to a less extend $CH_i^+$ but not the heavier fragments. If we assume the discrimination to be constant with photon energy, we deduce the previous conclusion to be quite valid. From these measurements,

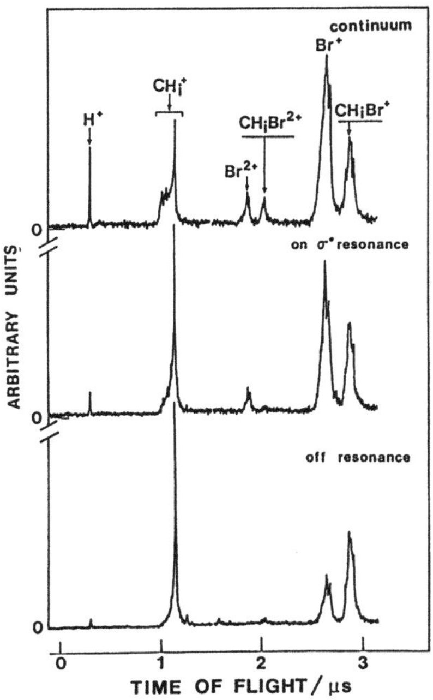


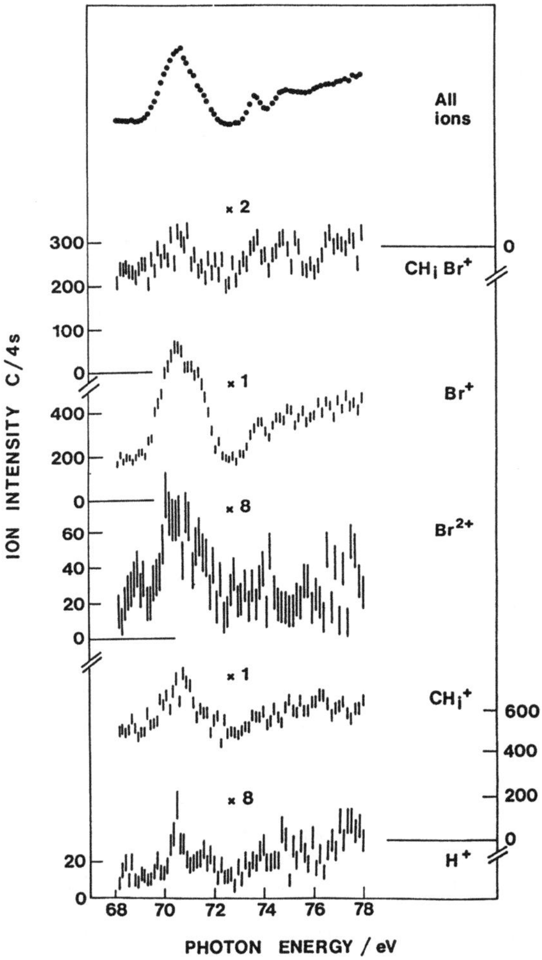


Figure 11.a) TOF mass spectra of $CH_3Br$ recorded off (68.2 eV) and on (70.5 eV) resonance, and in the continuum (117eV)

Figure 11.b) fragment ion intensity as a function of photon energy

we observe also that the molecule fragments very efficiently through core excitation, in contrast to valence ionization which is responsible for the whole of the parent ion production and only a part of the fragment intensity. The weakness of $CH_3Br^{2+}$ at the energy of the σ* resonance does not mean that double ionization is negligible. On the contrary, this rather means that the multiply charged ions produced are not stable on a microsecond scale and dissociate rapidly. The difficulty to identify the number of positive charges carried by these ions is that all of them produce the same singly charged fragments. We illustrate this point in Table I, which shows a compilation of fragmentation processes of the singly, doubly or triply charged parent molecular ion.

Table I - Dissociative ionization processes in $CH_3Br$

| | | | | |
|---|---|---|---|---|
| $h\nu$ + $CH_3Br$ | $\longrightarrow$ | $CH_3Br^+$ | | a) |
| | $\longrightarrow$ | $CH_i^+$ + Br + H | | b) |
| | $\longrightarrow$ | $CH_i$ + $Br^+$ + H | | c) |
| | $\longrightarrow$ | $CH_i$ + Br + $H^+$ | | d) |
| | $\longrightarrow$ | $CH_3$ + $Br^*$ | | |
| | | $\hookrightarrow$ | $Br^+$, $Br^{2+}$ | e) |
| | $\longrightarrow$ | $CH_3Br^{2+}$ | | f) |
| | $\longrightarrow$ | $CH_i^{2+}$ + Br + H | | g) |
| | $\longrightarrow$ | $CH_i$ + $Br^{2+}$ + H | | h) |
| | $\longrightarrow$ | $CH_i^+$ + $Br^+$ + H | | i) |
| | $\longrightarrow$ | $CH_i^+$ + Br + $H^+$ | | j) |
| | $\longrightarrow$ | $CH_i$ + $Br^+$ + $H^+$ | | k) |
| | $\longrightarrow$ | $CH_3Br^{3+}$ | | l) |
| | $\longrightarrow$ | $CH_i^+$ + $Br^{2+}$ + H | | m) |
| | $\longrightarrow$ | $CH_i$ + $Br^{2+}$ + $H^+$ | | n) |
| | $\longrightarrow$ | $CH_i^+$ + $Br^+$ + $H^+$ | | o) |

Notice that sequential events are possible ; for instance, reactions d), j) and k) can be considered as the second step of reactions b),g) and i) respectively.

In order to identify these reactions, we have performed coincidence experiments. We show in Fig.12a, three typical PIPICO spectra recorded at the same photon energies as for Fig. 11a. Here, there are no ion discrimination effects. The relative intensity of pairs is drastically different from those measured at 40.8eV (HeII discharge lamp) by Curtis and Eland [29] . In particular, the $H^+/CH_i^+$ and $H^+/Br^+$ pairs are more intense (compared to $CH_i^+/Br^+$) at high photon energy than at lower one. This confirms that core excitation favors the production of small singly charged fragments. This statement is even more evident in showing Fig. 12b where we present the photon energy dependence of each ion pair. The most striking observation is that all pairs resonate at the energy of discrete resonances, except the $H^+$ + $CH_iBr^+$ pair which stays essentially flat. From these two sets of results, we can conclude that reactions e), f), h), i), j), k) are likely to occur. The absence of reaction g) is not surprising since $CH_3^{2+}$ is certainly not stable. The identification of reactions i), j), and k) at the energies of discrete resonances, shows that the $CH_i$ $Br^{2+}$ ion is produced in the Br3d hole relaxation. In other words, there is a competing channel to the primary fast dissociation of the $\sigma^*$ resonance. In addition, one observes that ion pairs are very intense in the continuum. This is easily explained in terms of Auger processes in which the residual ion is expected to have a double positive charge.

The production of three positively charged fragments, such as in reaction o) can contribute to three different PIPICO peaks. Then a triple

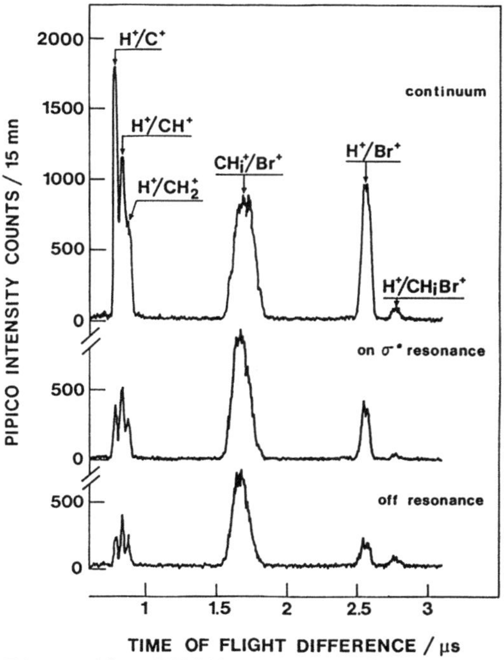


Figure 12a. PIPICO spectra of $CH_3Br$ off (68.2 eV) and on (70.5 eV) resonance, and in the continuum (117eV)

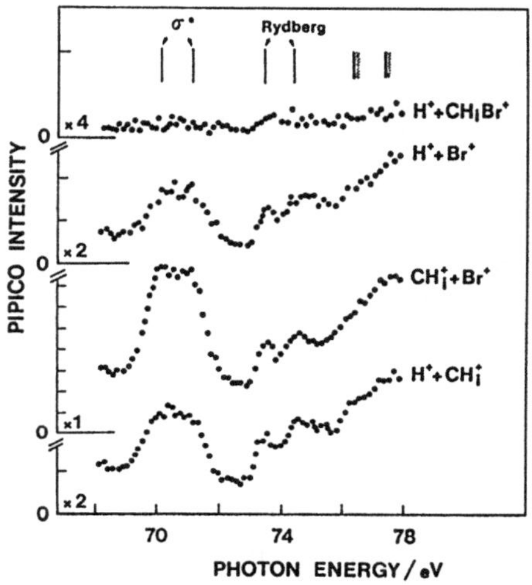


Figure 12b. Excitation spectra of ion pairs in the PIPICO mode

coincidence experiment is necessary to show its existence. We show in Fig. 13 three 3PICO spectra of $CH_3Br$ recorded at the same photon energies as for the spectra of Fig.11a and Fig.12a, and in the same extracting conditions as for the PIPICO experiments. We readily see that reaction m) exists at the $\sigma^*$ resonance and in the continuum, but not at a photon energy corresponding to the direct valence ionization. This shows that reaction m) is not negligible at the $\sigma^*$ resonance and more pronounced in the continuum. In this latter case, a triple positive charge of the ion originates from a cascade Auger process.

The major task is now to evaluate the branching ratios of ionization events defined in section 2, such as $N_1$ (reactions a) to h) and l)) $N_2$ (reactions i), j),k), m) and n)) and $N_3$ (reaction o). These suppose the determination of the detection efficiency f. We have then adopted the following procedure :

In a first approximation, we neglect reactions f, g, h, and l on the basis of our mass spectra (Fig. 11a). We also neglect reactions m) and n) for the same reasons. Consequently, $N_1$ represents the proportion of absorption events that lead to single ionization. Generally at photon energies in the core ionization continuum, $N_1$ can only come from the underlying valence ionization continuum. The idea is that the ejection of a core electron is necessarily followed by an Auger-like decay, thus by multiionization. Consequently, knowing the relative magnitude of the core excitation one, in the photon energy of interest, $E_{h\nu}$, the $N_1$ value at $E_{h\nu}$ is deduced from

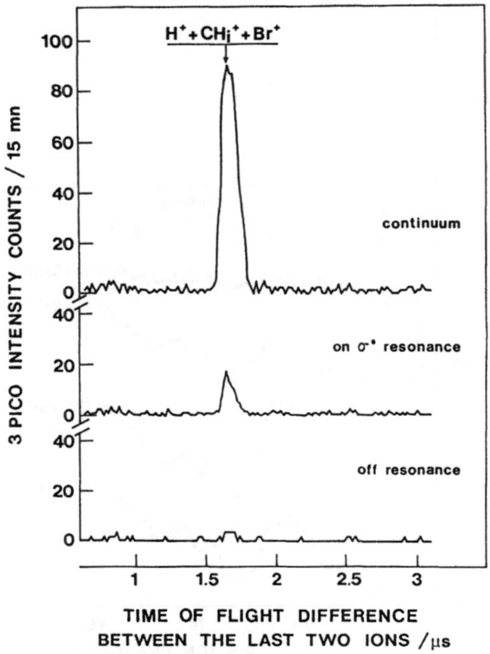


Figure 13. Triple ion coincidence spectrum (3PICO) of $CH_3Br$ recorded off (68.2 eV) and on (70.5 eV) resonance, and in the continuum (117 eV)

the $N_1$ value at $E'_{h\nu}$= 68 eV, for which the photoabsorption cross section is equal to the valence ionization one. We obtain the following formula:

$$N_1(E_{h\nu}) = N_1(E'_{h\nu})\frac{\sigma(\text{val},E_{h\nu})}{\sigma(\text{val},E_{h\nu})+\sigma(\text{core},E_{h\nu})} \quad . \qquad \{5\}$$

The partitioning of the valence and core cross sections has been made by extrapolating the EELS curve of Hitchcock and Brion [5], as indicated in Fig.1, from the shape of the spectrum in the low energy region. Using formula {5} for several selected photon energies, we obtain f = 5 ± 0.5 % for the detection efficiency. This large uncertainty originates from the determination of the valence cross section in the EELS of Fig. 1. The $N_1$, $N_2$ and $N_3$ values are now estimated using formulas {1}, {2} and {3} and are reported in table II for selected photon energies. We estimate that these results are affected by a 30 % uncertainty due to the inaccuracy of the f value.

We now analyse in more detail the $N_1$, $N_2$ and $N_3$ values associated only to the decay of the $\sigma^*$ and of the Br3d continuum. They are obtained in Table II after substracting the contribution of σ(val), and shown in Table III. We estimate the uncertainty of the values of Table III to about 50 %. As seen above, in this evaluation, reactions f, g, l, m and n are neglected and $N_i$ is reduced to the $i^{th}$ ionization probability. Table III shows that :

i) Below the 3d excitation threshold (68 eV), $N_1$ and $N_2$ represent the percentage of single and double valence ionization events. At this particular photon energy, which is 55 eV higher than the first ionization

Table II - Relative coincidence counting rates in PIPICO (C2) and 3PICO (C3) experiments. Branching ratios for $N_1$, $N_2$ and $N_3$ events, in %.

| Photon energy/eV | 68 | 70 | 90 | 117 |
|---|---|---|---|---|
| $R_2 \cdot 10^2$ | 1.05 | 1.57 | 2.25 | 3.23 |
| $R_3 \cdot 10^5$ | 0 | 3.25 | 19.1 | 36 |
| $N_1$ | 79 | 58 | 51 | 23 |
| $N_2$ | 21 | 40 | 37 | 47 |
| $N_3$ | 0 | 2 | 12 | 30 |

Table III. $N_1$, $N_2$ and $N_3$ values of the valence excitation at 68 eV and of the $\sigma^*$ (70 eV) and Br3d continuum (90 and 117 eV) for which the $\sigma$(val) contribution has been subtracted

| Photon energy/eV | 68 | 70 | 90 | 117 |
|---|---|---|---|---|
| $N_1$ | 79 | 36 | 9* | 0 |
| $N_2$ | 21 | 59 | 56 | 56 |
| $N_3$ | 0 | 5 | 34 | 44 |

** should be zero according to our hypothesis (see text). This is due to errors in the valence contribution to the total photoabsorption.

onset, the ejection of 2 electrons because of electron correlation is very efficient. Such observations have already been found in other molecules [10,28]. In contrast the direct ejection of three electrons is found negligible.

ii) In the Br3d continuum (90 and 117 eV), the importance of $N_2$ is simply explained by normal Auger which leads to double ionization in a two step-process.

$$(c^2\ v^n\ u^0) + E_{h\nu} \longrightarrow (c^1\ v^n\ u^0) + e^-(\text{core})$$

$$(c^1\ v^n\ u^0) \longrightarrow (c^2\ v^{n-2}\ u^0) + e^-(\text{Auger})$$

Triple ionization is likely due to cascade Auger. The significant increase of $N_3$ events from 90 to 117 eV is probably explained by the opening of shake up excitation i.e. the simultaneous excitation of a core and a valence electron [28].

iii) For the $\sigma^*$ resonance, one expects that the fast dissociation (reaction e of Table I ) favors the production of $Br^+$ and $Br^{2+}$, therefore the $N_1$ value. The competing molecular core hole relaxation pathways which produce mostly dissociative $CH_3Br^+$ and $CH_3Br^{2+}$ are equally expected in $N_1$, $N_2$ and $N_3$ events. The figures of Table III show that this competing channel is extremely important since it represents at least 64 % of the total, and the direct dissociation of the $\sigma^*$ resonance represents at most 36 %. Lastly, it is interesting to notice that the $\sigma^*$resonance produces a large amount double ionization events ($N_2$) compared to single ionization ones ($N_1$). This observation agrees with previous studies in atoms [30] but contrasts with studies on diatomic molecules [28] and can be explained as follows : The most abundant relaxation channels leading to doubly charged ions are generally found among spectator transitions such as double resonant Auger ($\beta_2$) and cascade autoionization ($\alpha\beta$). The only difference we find between earlier systems and the present $CH_3Br$ case is that the $\alpha\beta$ process is here enhanced because the first double ionization onset is especially low in energy (30 eV) and most of the singly charged ions with excited configurations are imbedded in the double ionization continua.

### 3.3 Site selectivity in the final ionic fragmentation pattern

The $CH_3Br$ molecule is a model polyatomic system because it has only two kinds of bonds arranged in sequence. In other words, the bromine atom is only linked to the carbon atom and sees the hydrogens as second neighbours. The excitation of the bromine atom via the 3d shell offers here the possibility of looking at the present results as follows. Is the initial localisation of the exitation conserved throughout the relaxation of the core hole, and is there any selectivity in the final fragmentation products ? The answer is contained in the mass spectrometry results of Fig. 11, 12 and 13. Indeed the C-Br bond breaking is especially enhanced through core excitation, at the $\sigma^*$ and Rydberg resonances. However, the breaking of the C-H bonds are by far not negligible as seen in the PIPICO spectra of Fig.12b ($H^+/Br^+$ and $H^+/CH_i^+$ ). This shows that the efficient multiionization obtained by core excitation produces highly unstable parent ions with multiholes in the valence orbitals. Those ions are very often found to dissociate in breaking more than one bond [10,11,29] either instantaneously or in sequence. In contrast, the non resonant behavior of the $H^+/CH_iBr^+$ pair in the PIPICO results of Fig.12b is of particular interest because it reflects the fact that the core excitation does not produce $CH_3Br^{2+}$ with internal energy associated to the breaking of a single C-H bond alone, whereas direct double ionization populates them significantly. As a result, the excitation of a core electron of bromine enhances the population of certain doubly charged ions, but breaks not only the C-Br bond but also the C-H one. These observations complement the earlier observations of selective fragmentation of polyatomic molecules, using conventional mass spectrometry measurements [31-34]. In the region of discrete resonances, the strong variation of fragment intensity points out the quite different ionization processes between direct valence ionization and core excitation. The effect of site- selectivity on the fragmentation, associated strictly to the core excitation exists but it is rather small and found more pronounced in dissociative double ionization events, as already shown in tetramethylsilane [35].

In contrast, the ionization of the 3d electron destroys most of the selectivity found in the discrete spectrum. This is illustrated in Fig. 14 by a 3PICO spectrum obtained at 117 eV. The gate pulse which selects the

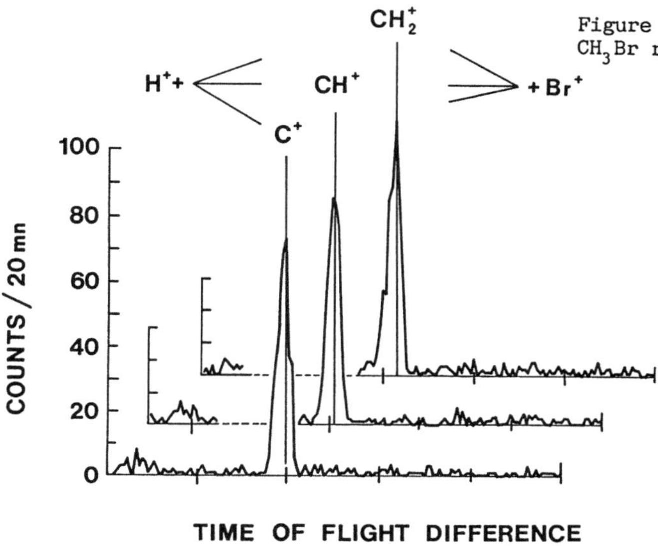


Figure 14. 3PICO spectra of $CH_3Br$ recorded at 117 eV

$m_2^+$ ion (here $CH_i^+$) out of the ion triplet is narrower than in the measurements of Fig. 13. Consequently, the number of hydrogen atoms in the $CH_i^+$ fragments is determined. We see that triple ionization, which is found more intense in the continuum (Table II), favors the Coulomb explosion [36] of the molecule into $H^+ + CH_i^+ + Br^+$ (i=0 to 2) with similar probabilities.

## IV - CONCLUDING REMARKS

The excitation of a core electron into a low lying valence unoccupied orbital in the two hydride molecules HBr and $SiH_4$ and in $CH_3Br$, has been found to produce very specific decay channels i.e. an unexpectedly fast dissociation sometimes faster than autoionization. The generalisation of this phenomenon can be done by searching molecules for which the core hole lifetime is rather long, typically superior to 5 femtoseconds, and for which the relative mass ratio of its atomic components is big. This constraint favors primarily systems with low Z elements and rules out deep electrons of heavy elements.

The complexity of electronic relaxation processes which has been analysed in details for $CH_3Br$, shows that photoelectron spectroscopy and mass spectrometry in the coincidence mode are required. Moreover, multicoincidence experiments between photoelectrons and ions, such as the ones developed recently [37-39], are highly desirable in the future.

The site selectivity of the ionic fragmentation pattern following core exitation of a molecule is clearly evidenced in $CH_3Br$, although it is a small effect. We show that the selectivity is found in double ionization pathways. The doubly charged ions which are not populated, decrease somewhat the probability of breaking the C-H bond, compared to the direct double valence ionization processes. Further investigations on larger systems will elucidate how far from the absorbing atom the breaking goes.

ACKNOWLEDGEMENTS

We are grateful to A. P. Hitchcock for helpful discussions. We thank the LURE staff for operating the ACO storage ring and for general facilities.

REFERENCES

[1] - A bibliography of inner-shell spectra of molecules is given by A.P. HITCHCOCK. J. electron Spectrosc. rel. Phen. 25, 245 (1982).

[2] - I. NENNER and J.A. BESWICK in Handbook on Synchrotron Radiation, Vol. II2 G.V. Marr ed. (North-holland, Amsterdam) Chap. 6 (1987).

[3] - J.L.DEHMER, A.C. PARR and S.H. SOUTHWORTH, Resonances in molecular photoionization, Handbook on synchrotron Radiation, Vol. II, G.V. Marr ed. (North-holland, Amsterdam) (1987).

[4] - D.A. SHAW. D. CVEJANOVIC, G. C. KING and F.H. READ, J. Phys. B 17, 1173 (1984)

[5] - A.P. HITCHCOCK and C.E. BRION, J. Electron. Spectrosc. Relat. Phenom. 13, 193 (1978)

[6] - W. HAYES and F.C. BROWN, Phys. Rev. A 6, 21 (1972)

[7] - P. MORIN and I. NENNER, Phys. Rev. Lett. 56, 1913 (1986)

[8] - G.G.B. DE SOUZA, P. MORIN and I. NENNER, Phys. Rev. A 34, 4770 (1986)

[9] - P. MORIN, M.Y. ADAM, I. NENNER, J. DELWICHE, M.J. HUBIN-FRANSKIN and P. LABLANQUIE, Nucl. Instrum. Methods, Phys. Rev. 208, 761 (1983)

[10]- P. LABLANQUIE, I. NENNER, P. MILLIE, P. MORIN, J.H.D. ELAND, M.J. HUBIN-FRANSKIN and J. DELWICHE, J. Chem. Phys. 82, 2951 (1985)

[11]- J.H.D. ELAND, F.S. WORT, P. LABLANQUIE and I. NENNER, Z. für Physik D, Atom, Mol. and Clusters, 4, 31 (1986)

[12]- A. HITCHCOCK, P. LABLANQUIE, P. MORIN, E. LIZON A LUGRIN, M. SIMON, P. THIRY and I. NENNER. Phys. Rev. A (Submitted)

[13]- W.H.E. SCHWARZ, Chem. Phys. 11, 217 (1975)

[14]- D.R.J. BOYD, J. Chem. Phys. 23, 922 (1955)

[15]- L. KARLSSON, R. JADRNY, L. MATTSSON, F. TIM CHAU and K. SIEGBAHN, Physica Scripta, 16, 225 (1977) and refer therein.

- U. HEINZMANN, B. OSTERHELD, F. SHAPERS and G. SCHOHENSE, J. Phys. B : At. Mol. Phys. 14, L 79 (1981)

- C. UTSUNOMIYA, T. KOBAYASKI and S. NAGAKURA, Bull. Chem. Soc. Jpn 53, 1216 (1980)

[16]- A.W. POTTS, H.J. LEMPKA, D.G. STREETS and W.C. PRICE, F.R.S. Phil. Trans. Roy. Soc. Lond. A. 268, 59 (1970)

[17]- W. VON NIESSEN, L. ÅSBRINK and G. BIERI, J. Elec. Spect. Related. Phenom. 26, 173 (1982)

[18]- R.S. MULLIKEN, Phys. Rev. 4, 413 (1935)

[19]- L.S. CEDERBAUM, W. DOMCKE, J. SCHIRMER ans W. VON NIESSEN, Adv. in Chem. Phys. 65, 115 (1986).

[20]- W.B. PERRY and W.L. JOLLY, Inorg. Chem. 13, 1211 (1974)

[21]- M. MAZZONI and M.PETTINI, Phys. Lett. 85 A, 331 (1981)

[22]- B. RUCCI, J.P. GREENE and J. BERKOWITZ, J. Phys. B. 17, 1503 (1984)

[23]- C.E. MOORE, Atomic Energy Levels as derived from Analysis of Optical Spectra, National Bureau of Standards, circular NO 467 (U.S. GPO, Washington P.C. 1949) p. 159.

[24]- G. HERTZBERG, Electronic Spectra of polyatomic molecules, (D. Van Nostrand company, N.Y. 1966).

[25]- P. LABLANQUIE, J.H.D. ELAND, I. NENNER, P. MORIN, J. DELWICHE and M.J. HUBIN-FRANSKIN, Phys. Rev. Lett. 58, 992 (1987)

[26]- P. MORIN and I. NENNER, Phys. Scrip., T17, 171 (1987)

[27]- P. MORIN in Giant resonances in Atoms, Molecules and solids. Proceedings of the NATO Advanced Study Institute. Les Houches, 1987 (Plenum, N.Y.) p 291

[28]- A.P. HITCHCOCK, P. LABLANQUIE, P. MORIN, E. LIZON A LUGRIN, M. SIMON, P. THIRY and I. NENNER, submitted to Phys. Rev. A (1987)

[29]- M.A. CURTIS and J.H.D. ELAND, private communication

[30]- P. MORIN, LURE Activity Report (1987)

[31]- W. EBERHARDT, T.K. SHAM, R. CARR, S. KRUMMACHER, M. STRONGIN, S.L. WENG and D. WESNER, Phys. Rev. Lett. 50, 1038 (1983)

[32]- W. EBERHARDT and T.K. SHAM, Proc. of S.P.I.E. 447, 143 (1984)

[33]- K. MULLER-DETHLEFS, M. SANDER, L.A. CHEWTER and E.W. SCHLAG, J. Phys. Chem. 88, 6098 (1984)

[34] - M.C. NELSON, J. MURAKAMI, S.L. ANDERSON and D.M. HANSON, J. Chem. Phys. 86, 4442 (1987)

[35] - P. MORIN, G.G.B. DE SOUZA, I. NENNER and P. LABLANQUIE, Phys. Rev. Lett. 56, 131 (1986)

[36]- T.A. CARLSON and R.M. WHITE, J. Chem. Phys. 44, 4510 (1966)

[37]- W. EBERHARDT, E.W. PLUMMER, I.W. LYAO, R. REININGER, R. CARR, W.K. FORD and D. SONDERICKER, Aust. J. Phys. 39, 633 (1986).

[38]- J.H.D. ELAND, Mol. Phys. 61, 725 (1987)

[39]- L. FRANSINSKY, M. STANKIEWICZ, K. RANDALL, P. HATHERLY and K. CODLING, J. Phys. B 19, L819 (1986)

# Auger Electron-Ion Coincidence Studies on Free Molecules

*W. Eberhardt*[1] *1 and R. Murphy*[2]

[1]Exxon Research and Engineering Co.,
Route 22 East, Annandale,
NJ 08801, USA
[2]Physics Dept., Univ. of Pennsylvania
Philadelphia, PA 19103, USA

Coincidence studies between Auger electrons and ions produced in the decay of a core hole excited state in free molecules reveal the individual pathways of soft x-ray induced fragmentation processes. We are able to identify the electronic configuration leading to stable or unstable ionic states. Measuring the kinetic energy of the ions, we also get data about the potential energy curves of the electronic states populated in the Auger decay of the core hole.

## INTRODUCTION

The absorption of a soft x-ray photon by a free molecule with an energy sufficiently high to excite a core electron predominantly results in the production of ionic and neutral fragments of the parent molecule [1-3]. The same processes are also involved in the desorption of ions and neutrals from surfaces, which is the main topic of this conference. The first step in the chain of events leading to the production of fragment particles is the creation of a core hole in the system. This can be done by ionizing the core electron or alternately by exciting it into an unfilled molecular orbital. This core hole excited state is usually a stable configuration of the molecule [4]. Subsequently the core hole excited state decays into various valence hole configurations by fluorescence or, in the light elements we are dealing with, preferentially by an Auger-type transition. This is the unstable molecular configuration, which leads to the production of ionic fragments. Thus measuring the ions in coincidence with energy-selected specific Auger final states, we are able to identify the pathways of the fragmentation [3,5]. This is analogous to the Knotek-Feibelman mechanism in desorption from ionic solids. A core hole is created by absorption of the photon and the following Auger transition depletes the bonding structure of electrons such that the complex becomes unstable and falls apart.

As example for this type of coincidence spectroscopy we are going to discuss the fragmentation within the manifold of the singly charged states of $N_2^+$ populated in the electronic decay of the N 1s → $1\pi_g$ core to bound state resonance. This is an extension of the previous work on $N_2$ [5], where we studied the ionic decay products in coincidence with the doubly charged electronic configuration populated in the normal Auger decay of $N_2$.

In order to clarify the language, we will use the term Auger decay only in connection with the processes following the complete ionization of a core electron in the initial absorption of a soft x-ray photon. The electronic decay of a neutral core hole excited state will be described as deexcitation. Technically this is an autoionization process, which is described by an Auger type matrix element. The difference to the normal Auger decay is that the initial state is a neutral configuration and the final state a single ionized configuration of the molecule.

The electronic configurations populated in these decays are substantially lower in energy than the core hole excited states. Most of the energy of the core hole excited state is carried away by the emitted electron. Therefore these states also can be excited by lower energy photons or in electron scattering experiments. However, in general, the singly ionized outer valence hole configurations of the diatomic and triatomic molecules are stable configurations. Only two hole-one electron and higher excited electronic states are dissociative. The cross section for direct excitation of these states by photons or electron scattering is rather small and is superimposed on the background of the continuum of the lower energy single electron ionization events. Therefore it is advantageous to study these fragmentation channels by populating and characterizing the electronic configuration via Auger decay or deexcitation of a core hole excited state. In the discussion of our results we will obviously include the previous findings connected with a direct excitation into these electronic configurations [6-8].

## EXPERIMENT SETUP

The general experimental arrangement, including the differential pumping stages, has been described previously [9]. The electrons are detected by a commercial cylindrical mirror analyzer (CMA). Opposite the CMA and in line with its axis a TOF spectrometer detects the produced ions. This detector and the coincidence electronics setup are shown schematically in Fig. 1. The photon beam, the gas beam, and the axis of the CMA-TOF spectrometer combination are all at right angles with respect to each other. The intersection of the photon and gas beams is located in the center of the extraction region of the TOF mass spectrometer. The ions produced have large kinetic energies (up to more than 20 eV per particle) such that only ions travelling along trajectories of type d will be detected. All others will be discriminated against by the various apertures. Thus for each ionic species we expect to observe two peaks centered around the arrival time of an ion of that specific mass but with no, or only thermal, initial kinetic energy. The separation, $\Delta T$, of the two peaks is proportional to the kinetic energy released in the fragmentation process.

The coincidence experiments are conducted by starting a Time to Digital Converter (TDC) with the electron spectrometer set to detect electrons of a specific core hole decay channel. The time difference between the (fast) electron pulse and the ion pulse yields, as explained, both the mass to charge ratio and kinetic energy of the ions, which are produced in this specific event. The total count rate has to be small enough such that the detectors are not swamped by accidental coincidences. This is described in more detail in Ref. 10.

The experiments were conducted using the quasimonochromatic radiation of the mini-undulator installed at X17T at the X-ray ring at the NSLS-Brookhaven as excitation source. Even though we had to reduce the undulator intensity to lower the background signal due to accidental coincidences, we have found that an undulator is very helpful if not essential to perform this type of experiment in the soft x-ray region. In our experience, counting rates in the electron channel on bending magnet lines are too low to tune the detection system for optimum performance in the coincidence mode. Typically we ran at a counting rate of several hundred c/sec in the electron channel and about $3{*}10^4$ c/sec in the ion channel. Under these conditions the typical coincidence rate was between $10^{-2}$ and 0.5 (c/sec).

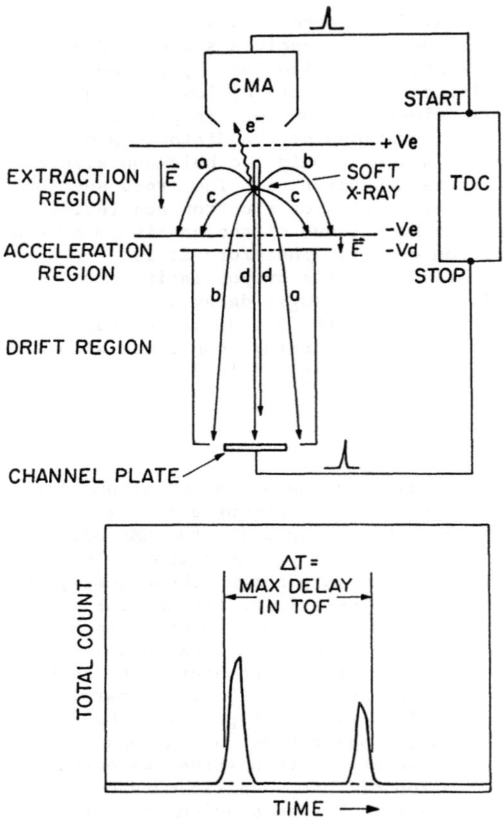


**Fig. 1** Schematic view of the coincidence setup. The interaction region is given by the intersection of the photon beam and the gas beam. The electrons are detected on one side of this interaction region by a commercial CMA, whereas the ion mass-to-charge ratio is determined by the time-of-flight spectrometer.

RESULTS AND DISCUSSION

$N_2$ Core Hole Deexcitation

Tuning the fundamental of the undulator to coincide with the excitation energy of the N 1s→1$\pi_g$ ($\pi$*) transition at 401 eV, we create a large number of neutral core electron excited $N_2$ molecules. The width of the undulator fundamental is very large since it has only 8 periods, such that the core ionization is still the predominant event. However the electron emission spectrum generated in the decay of these core hole excited states as shown in Fig. 2, exhibits several lines (shaded) which originate from the deexcitation of the $\pi$* transition. The individual decay spectra for both the $\pi$* deexcitation and the Auger decay following N 1s ionization have been published previously [11].

The ion spectra taken in coincidence with the electron emission at a kinetic energy around 384 eV (A), 376 eV (B), and 363 eV (C) are shown in

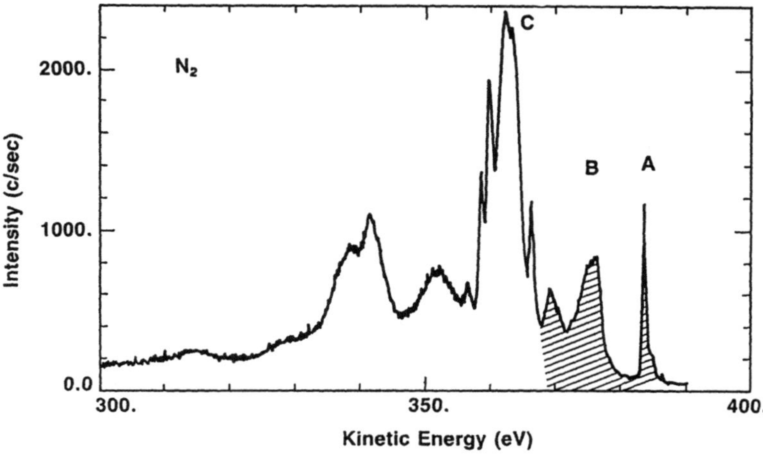


**Fig. 2** Electron kinetic energy distribution for $N_2$ produced in the decay of a core ionized or core excited state (shaded region).

Fig. 3. Curve A only shows the presence of $N_2^+$, whereas in curve B $N^+$ and $N_2^+$ are observed. Curve C is shown for comparison only. The electronic configurations curve C is taken in coincidence with are a result of the Auger decay. The products of these electronic states are $N_2^{++}$ with thermal energy and $N^+$ with approximately 4.5 eV kinetic energy per ion [5].

We have to add here that the coincidence measurements are taken with a static ion extraction field applied across the ionization region. Due to the vertical width of the photon beam, which was restricted by apertures in the coincidence mode, this will cause an energy smearing of the energy distribution curve of the detected electrons. This uncertainty is estimated to be on the order of ±3 eV under the present conditions.

The electrons emitted in region A of the spectrum (Fig. 2) correspond to the $X^2\Sigma_g^+$ and $A^2\Pi_u$ states of $N_2^+$. In a single particle notation these states are described as a $3\sigma_g^{-1}$ or $1\pi_u^{-1}$ hole state respectively. These are stable ionic configurations and consequently we only observe $N_2^+$ in the coincidence spectrum.

More interesting is region B of the electron emission spectrum. Now we are observing both $N^+$ and $N_2^+$ in the ion spectrum. The energy remaining in the molecule after emission of an electron with 376 eV kinetic energy is 25 eV. This is well below the double ionization potential of $N_2$ (43.2 eV). Therefore, even though there is a technical ambiguity in the degeneracy of $N_2^{++}$ and $N^+$ in any TOF measurement, the peak at a flight time of 5 μsec in spectrum B has to be unambiguously assigned as $N^+$. We also want to add here that the $N_2^+$ peak observed in the same spectrum is

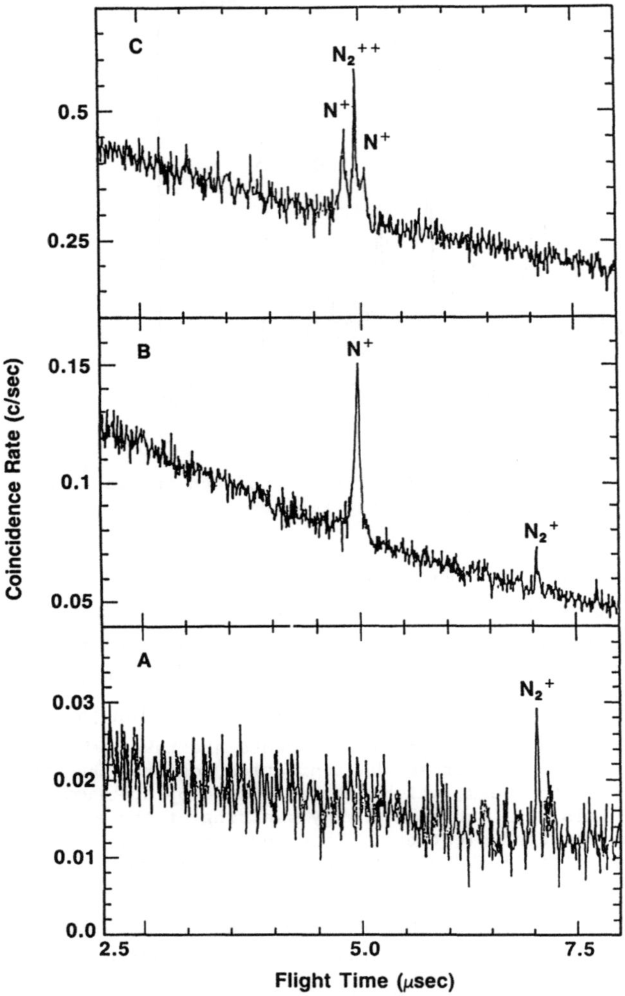


**Fig. 3** TOF ion spectra measured in coincidence with the observation of an electron having a kinetic energy of 384 eV (A), 376 eV (B), and 363 eV (C).

not caused by the energy smearing in our electron detection channel due to the ion extraction field. Otherwise we would have to observe a peak around 5 μsec flight time in spectrum A.

From the width of the $N^+$ peak in Fig. 3B we conclude that the kinetic energy of the $N^+$ fragments is about or less than 1 eV. For comparison, the $N^+$ fragments observed under the conditions of Fig. 3C have a kinetic energy of about 4.5 eV [5].

The potential energy curves for states of $N_2^+$ in this energy region all yield stable configurations with the exception of a $^4\Pi_u$ state, which corresponds to an electronic configuration of $(2\sigma_u^{-1}, 3\sigma_g^{-1}, 1\pi_g)$ in single particle notation [12,13]. The $(2\sigma_u^{-1}, 3\sigma_g^{-1})$ configuration is one of the stronger lines in the Auger decay of a N 1s core hole [14]. Therefore we expect that the corresponding line is also observed in the deexcitation spectrum of the N 1s → $1\pi_g$ core to bound state excitation. The measured ion kinetic energy is consistent with the $^4\Pi_u$ energy curve and a decay into N ($^4S°$) + $N^+$(3P), which is located 24.29 eV above the ground state of $N_2$ [6]. An alternate process yielding an $N^+$ ion with about 1 eV kinetic energy involves a curve crossing between a $^2\Sigma_u^-$ final state $(3\sigma_g^{-1}, 1\pi_u^{-1}, 1\pi_g)$ and the $^4\Pi_u$ state [13,15]. Again the $(3\sigma^{-1}, 1\pi_u^-)$ configuration is one of the stronger lines of the N-Auger spectrum [14]. All other electronic configurations of the Auger spectrum, which contribute to the peak in the kinetic energy range between 357 and 367 eV, are stable ionic configurations, when an extra $1\pi_g$ electron is added as in the deexcitation spectator final states. These states give rise to the peak around 375 eV in the deexcitation spectrum. In coincidence with these electronic configurations we observe $N_2^+$ in the ion mass spectrum.

In direct optical excitation the C-$^2\Sigma_u^+$ state of $N_2$ is excited at an energy of 25 eV [15,16]. The calculated potential energy curve is strongly bound and a curve crossing into the $^4\Pi_u$ state has been proposed as the pathway to the fragment ion configuration [6]. In our experiment we observe a transition directly into the $^4\Pi_u$ state. At the equilibrium internuclear distance of the core excited molecule the potential energy curve exhibits a shallow, about 1.3 eV deep, minimum. Depending on the vibrational state the molecule is in after the deexcitation, it is likely that the molecule overcomes this potential barrier and decays directly into N($^4S°$) and $N^+(^3P)$. The observed kinetic energy of the $N^+$ ions (~1 eV) is fully consistent with this process, since because of momentum conservation the neutral fragment has the same kinetic energy as the ionic one. The minimum of the $^4\Pi_u$ state is calculated to be about 1.5 eV above the energy of the fragments [6,13,15].

In the future we hope to be able to measure the coincidence spectra without the perturbation by the ion extraction field. This will result in better resolution in the electron channel. Together with more theoretical support in the interpretation of the deexcitation spectra, we will be able to individually resolve these states and determine their involvement in the production of ionic fragments.

These measurements were carried out at NSLS-Brookhaven, which is supported by the U.S. Department of Energy under contract DE-AC02-76CH00016. This work was also partially supported by NSF under grant No. NSF-DMR-8120331.

REFERENCES

1. W. Eberhardt, T. K. Sham, R. Carr, S. Krummacher, M. Strongin, S. L. Weng, D. Wesner, Phys. Rev. Lett. 50, 1038 (1983).
2. W. Eberhardt, C. T. Chen, W. K. Ford, E. W. Plummer, H. R. Moser, in "DIET II," Springer Series in Surf. Sci. Vol. 4, ed. by W. Brenig and D. Menzel, (Springer, Berlin 1985) p. 50.
3. W. Eberhardt, E. W. Plummer, In Whan Lyo, R. Reininger, R. Carr, W. K. Ford, D. Sondericker, Austral. J. of Physics 39, 633 (1986).
4. One exception to this has been observed to date. In HBr the creation of a Br 3d core hole was shown to directly lead to a dissociative state; P. Morin and I. Nenner, Phys. Rev. Lett. 56, 1913 (1986).
5. W. Eberhardt, E. W. Plummer, I. W. Lyo, R. Carr, W. K. Ford, Phys. Rev. Lett. 58, 207 (1987).
6. J. Eland in Electron Spectroscopy Vol. 3, ed. by C.R. Brundle and A. D. Baker, (Academic Press, London 1979) p. 231.
7. G. R. Wight, M. J. van der Wiel, C. E. Brion, J. Phys. B 9, 675 (1976).
8. P. Millie, I. Nenner, P. Archirel, P. Lablanquie, P. Fournier, J. H. D. Eland, J. Chem. Phys. 84, 1259 (1986).
9. W. Eberhardt, J. Stohr, C. Troxel, Jr., J. Feldhaus, E. W. Plummer, F. Sette, Rev. Sci. Instrum. 55, 354 (1984).
10. W. Eberhardt, E. W. Plummer, C. T. Chen, R. Carr, W. K. Ford, Nucl. Instrum. and Methods A246, 825 (1986).
11. W. Eberhardt, J. Stohr, J. Feldhaus, E. W. Plummer, F. Sette, Phys. Rev. Lett. 51, 2370 (1983).
12. F. R. Gilmore, J. Quant. Spectr. Rad. Transfer 5, 369 (1965).
13. A. L. Roche, H. Lefebvre-Brion, Chem. Phys. Lett. 32, 155 (1975).
14. H. Agren, J. Chem. Phys. 75, 1267 (1981).
15. P. Erman, Physica Scripta 14, 51 (1976).
16. J. L. Gardner, J. A. R. Samson, J. Chem. Phys. 62, 1451 (1975).

# Electron-Stimulated Desorption of $D_n^+$ Clusters from Condensed Deuterium Layers on Pd(111) Between 5 and 20 K

*R.H. Stulen*

Sandia National Laboratories, Livermore, CA 94550, USA

Electron-stimulated desorption (ESD) of ions from deuterium condensed on a Pd(111) surface has been measured at temperatures between 5 and 20 K. Exposure of a palladium sample to gas phase $D_2$ at surface temperatures, $T_s$, below 8 K results in the growth of solid deuterium multilayers. Bombardment of these layers by 100 eV electrons gives rise to the desorption of $D_n^+$ clusters which, except for n=2, are comprised of odd numbers of atoms with n as large as 21 observed. By examining the cluster size distribution as a function of the thickness of the solid deuterium layer we conclude that the larger clusters are produced when an ion, formed within the multilayer by electron excitation, collides with surrounding $D_2$ molecules in the condensed layer as it is ejected from the surface. The cluster size distribution also indicates that many of the larger clusters are produced with internal energies exceeding their binding energy and that these "hot" species dissociate into smaller fragments shortly upon leaving the surface. Examination of the temperature dependence of the various ion yields reveals that ion production correlates well with surface coverage in a straightforward manner. Finally, the dependence of the ESD signal upon the incident electron beam energy shows the existence of a threshold between 30 and 35 eV for the production of high kinetic energy $D^+$. This threshold is in the same region as the $D_2^+(2p\sigma_u)$ state which is known from gas phase experiments to dissociate to D(1s) and $D^+$.

## 1. Introduction

Studies of weakly bound, condensed gas layers have come under increasing interest within the field of electron and photon-stimulated desorption from surfaces [1]. Electronic excitations responsible for these phenomena also play a key role in the erosion of rare gas solids by fast ion beams [2]. At least two features of these condensed phase systems make them attractive for investigation. First, in contrast to atomic or molecular chemisorbed species, the excitations and dynamics for weakly bound, condensed gas layers are relatively straightforward [3] since they do not couple strongly to the underlying metal surface. Second, a great deal of detailed experimental and theoretical work has already been reported on the electronic structure of the corresponding gas phase species. This forms a strong basis of understanding for the excitations important to stimulated desorption phenomena from condensed layers.

Theoretically, the hydrogen isotopes are the simplest of the molecular species and certainly one of the earliest to have been studied in great detail. As a result, much is known about the physical and chemical properties of the molecule including its molecular structure, ro-vibrational states, excited electronic states and a host of thermodynamical data on its various phases: solid, liquid, and gas. For solid hydrogen, there has been one previous study, CLAMPITT and GOWLAND [4], of ESD from a sample at 3 K which demonstrated that large ion clusters could be formed by electron irradiation. In their work, they proposed that either $H^+$ or $H_3^+$ was the key nucleating species in the production of larger clusters. More recently, STULEN and KUBIAK [5] observed $D_n^+$ clusters from ESD of solid deuterium at temperatures near 10 K. There have also been a variety of gas phase studies using either electron impact[6-10], or intense photon beams from synchrotron radiation sources[11] to examine the excitations of $H_2$, $D_2$, and HD which lead to dissociation and ion production. Such studies are extremely helpful in extending our understanding of gas phase processes to related phenomena on surfaces.

The intent of the present work is to expand on Clampitt and Gowland's study in the thick molecular solid regime by examining electron beam energy effects, and to extend their results to the submonolayer regime where the $D_2$ molecule can be considered to be physisorbed to the underlying metallic substrate. Examination of the dependence of the cluster size distribution upon the thickness of the deuterium film should provide a well defined test of the original cluster production model. Such studies are not only intrinsically important in the characterization of ESD as a surface probe for physisorbed species, but also bear directly on a number of other interesting technological areas involving surfaces at or near liquid helium temperatures, in vacuum environments, while in the presence of electron, photon, or ion beams. These include, for example, cryopumping in radiation environments, compact superconducting electron storage rings for synchrotron radiation, liquid helium superconducting storage rings for high energy physics, and the use of frozen deuterium pellets in fusion applications.

## 2. Experiment

The experiments were performed in an ultra high vacuum chamber with a base pressure of ~$3 \times 10^{-11}$ Torr, equipped with an argon ion sputter gun and Auger electron spectrometer for surface preparation and characterization, a differentially pumped mass spectrometer for thermal desorption measurements of deuterium surface coverage, and a pulsed electron beam source with a time-of-flight (TOF) ion mass spectrometer for ESD. The palladium (111) crystal (~6 mm diameter by 1 mm thick) was mounted on a resistively heated holder which in turn was attached to a commercially available, continuous flow, liquid helium transfer line (Helitrans TM) equipped with a radiation shield. This arrangement allows the specimen to be heated to over 1000 K for cleaning purposes and cooled to 5 K for gas dosing and ESD experiments. The Pd(111) surface temperature was measured with a chromel-alumel thermocouple spot welded to the front face of the sample. This thermocouple was calibrated in-situ between 5 K and room temperature, by means of a solid state diode.

The time-of-flight ESD apparatus has been described previously[12]. Briefly, it consists of a pulsed electron beam which emits short bursts (~30-50 ns) of electrons at a repetition rate of ~ 100 kHz. During the 10µsec between pulses, any ions ejected from the surface are accelerated into a 10 cm long time-of-flight tube, biased negatively with respect to the sample by between 50 and 100 volts. At the end of the flight tube these ions are again accelerated, this time across a 1 mm gap to the potential, typically -1800 V, of a microchannel plate assembly. The time between the arrival of the incident electron beam pulse at the surface and the output of the microchannel plates gives the charge to mass ratio of the ejected particle. Except when noted otherwise, the incident electron beam was tuned to 100 eV in energy at an average incident current density small enough so as not to induce any significant change in adsorbate coverage during the course of the measurements.

In every case, exposure to deuterium followed the same general sequence: first the clean surface at $T_S$=6 K was dosed with ~ 25 Langmuirs (L) of $D_2$, sufficient to produce a surface saturated with atomic deuterium. This surface was warmed to ~ 30 K to desorb any weakly bound molecular $D_2$ that might be present, and then cooled quickly back to 6 K. The Pd(111) surface processed in this manner, denoted D/Pd(111), was subsequently used as the starting point for further gas dosing. At low exposures (<2.5 L), a weakly bound physisorbed molecular monolayer ($\theta$ = 1ML) is formed, and at high exposures (>2.5 L), multilayers of solid deuterium begin to grow in. In this paper, a monolayer of physisorbed molecular $D_2$, is defined as the maximum coverage prior to multilayer formation. This physisorbed layer has been carefully characterized and is the subject of a separate paper[13]. The results of that study are used here for the determination of coverage.

## 3. Results and Discussion

Three types of experiments are reported here: first, TOF measurements of the distribution of ions desorbed as a function of $D_2$ coverage (shown in Fig. 1), second, the dependence of the ESD signal on the energy of the incident electron beam, and third, the dependence of the $D^+$ ESD signal on surface temperature. All of the TOF data in Fig. 1 were taken at $T_s$ = 6 K with a 100 eV incident electron beam. Figure 1a shows results for coverages of

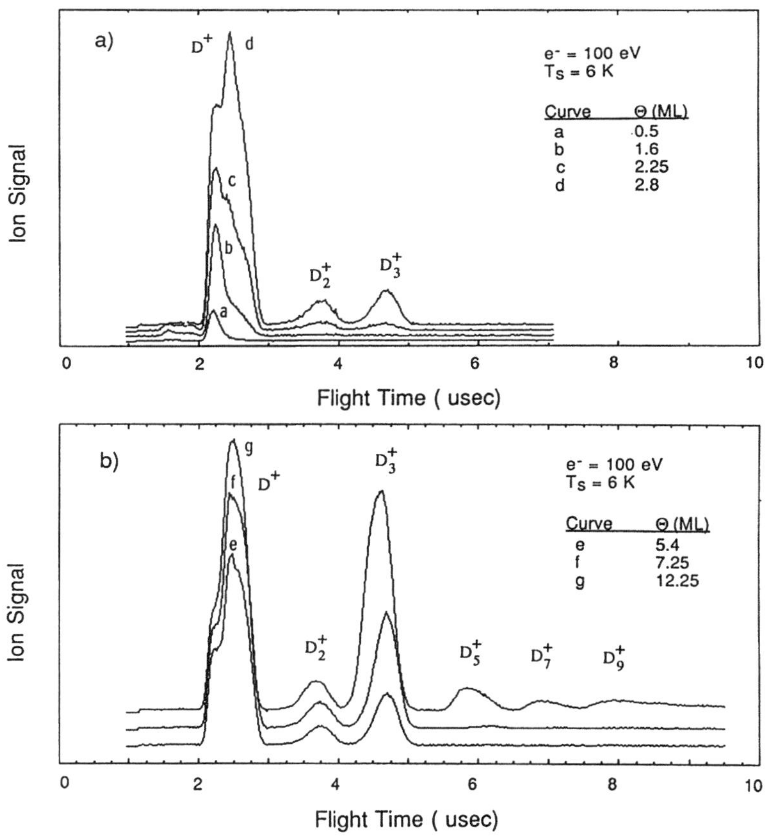


Figure 1 Time-of-flight spectra for various coverages of solid deuterium on a D/Pd(111) surface at $T_s$= 6 K. The curves are offset vertically for ease of viewing.

less than 3 monolayers, while Fig. 1b shows results for $\theta > 3$ ML. At $t < 2\mu sec$, some of the TOF spectra show a small peak due to hydrogen contamination from either the dosing gas or the ambient pressure in the chamber. This is a small effect and does not appear to influence our results. At t~2μsec, $D^+$ ions begin to appear and are followed at longer flight times by the larger deuterium ion clusters: the largest size shown in this series is visible at t = 8 μsec and corresponds to clusters of 9 atoms. At larger film thicknesses we have previously observed clusters as large as $D_{21}^+$ [5]. The rather broad width of the peaks reflects the distribution of initial kinetic energies of the desorbing ions as they are ejected from the surface and does not reflect any experimental limitations.

Curve a (Fig. 1a) is representative of the TOF spectra for physisorbed layers at coverages less than 1 ML. Only a single peak corresponding to $D^+$ is observed. Above 1 ML, two easily discernable qualitative changes occur in the spectra: first, above $\theta \approx 2$ ML larger clusters begin to appear, and second, structure starts to develop within the $D^+$ peak itself. This additional structure occurs at longer flight times, corresponds to production of slower ions, and will be referred to here as the low kinetic energy component of the mass 2 peak. Above coverages of approximately 3 ML it is the dominant component of the mass 2 peak. The situation may actually be somewhat more complicated in that the low energy portion appears to exhibit additional, subtle structure which is more apparent at certain coverages than others : for example, see curve c in Fig. 1a which shows a hint of three peaks in the $D^+$signal. For the purposes of this paper we will make no further distinctions between the low energy features.

The second clearly distinguishable change in the spectra is that as the film thickness increases, higher mass ion clusters emerge which correspond, except for the case of n = 2, to $D_n^+$ species with odd n. The coverage dependence is illustrated more clearly in Fig. 2 which is a plot of the integrated signal for each cluster size as a function of film thickness. These odd clusters have been studied theoretically [14] and are found to be stable, but exhibit decreasing dissociation energies with increasing size.

The fact that the larger species are seen only for thick multilayer films suggests that collisional effects may be important. It is interesting to note that there are changes,

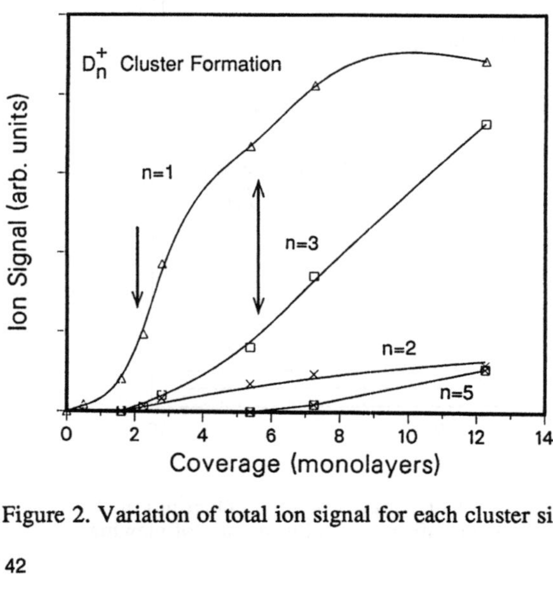


Figure 2. Variation of total ion signal for each cluster size as a function of coverage.

denoted by the arrows in Fig. 2, in the rate of production of $D_n^+$ at coverages corresponding to the onset for the next larger $D_{n+2}^+$ ion. For example, there is an increase in the $D^+$ signal at $\theta = 2$ ML which is coincident with the onset for $D_3^+$ production, and a similar change in $D_3^+$ when $D_5^+$ turns on. This behavior is representative of the decay of vibrationally hot clusters into smaller ones through the desorption of a $D_2$ molecule. From Fig. 1 it is apparent that the rise in the integrated $D^+$ signal, which is coincident with the $D_3^+$ onset (as seen in Fig. 2), is caused by the increase in the low kinetic energy mass 2 component, suggesting that it is the decay of vibrationally hot $D_3^+$ which leads to an increase in this portion of the $D^+$ signal. As we will show, the coincidence in the onset of $D_2^+$ and $D_3^+$ can also be explained by this mechanism.

From these results the following picture (Fig. 3) emerges: for $\theta < 1$ ML (Fig. 3a), excitation of the deuterium molecule by 100 eV electrons results in the production of a high kinetic energy $D^+$ ion through direct ionization of physisorbed $D_2$ and its subsequent dissociation. Collisions of the dissociating $D^+$ fragment with other physisorbed molecules within the same layer do not lead to ion cluster production presumably, in part, on account of rapid quenching by the substrate. Instead, the $D^+$ species observed in the TOF spectra (Fig. 1a, curve a) are only those with initial velocity directed away from the surface such that they have sufficient energy to escape and be detected. Once multilayer formation begins, secondary events involving collisions of the primary $D^+$ ion with adjacent $D_2$ along its original trajectory away from the surface, result in cluster formation, Fig. 3b. This collisional cascade process is capable of producing vibrationally hot clusters which can decay into smaller ones as they leave the surface. For example, the simultaneous appearance of $D_2^+$ and $D_3^+$ (Fig. 2) can be explained by a process that involves first the formation of a hot $D_3^+$ cluster, designated $D_3^{+**}$, followed by its fragmentation, either as $D_3^{+**} \longrightarrow D_2^+ + D$, or as $D_3^{+**} \longrightarrow D_2 + D^+$.

A key feature in the above description is the initial formation of an energetic $D^+$ ion. Fast ions have been observed and studied in great detail in electron impact studies of gas phase $H_2$, $D_2$, and HD. Three distinct processes have been found to occur in these previous studies[10]: first, direct ionization of $D_2$ to an excited state of $D_2^+$ whose potential energy is above its dissociation energy, second, autoionization of doubly excited states of neutral $D_2$, and third, direct ion pair ($D^+ + D^-$) production from doubly excited neutral states[10]. A few of the relevant excitation energies and dissociation limits, referenced to the zero-point level of ground state $H_2$, are summarized below (from SHARP [18]).

| Quantity | Energy |
|---|---|
| Zero point of ground state $H_2^+$ | 15.425 eV |
| Dissociation limit $H_2 \longrightarrow H^- + H^+$ | 17.321 eV |
| Dissociation limit of $H_2^+ \longrightarrow H(1s) + H^+$ | 18.075 eV |
| Dissociation limit of $H_2^+ \longrightarrow H(n,l) + H^+$ | 31.673 - ( 13.597/$n^2$ ) eV |
| Vertical excitation $H_2 \longrightarrow H_2^+ (2p\sigma_u)$ | 34.6 eV |

Previous electron impact experiments, CROWE and McCONKEY [6], on gas phase $D_2$ have shown that the $2p\sigma_u$ excited state of $D_2^+$ to be a major source of fast $D^+$ ions. This

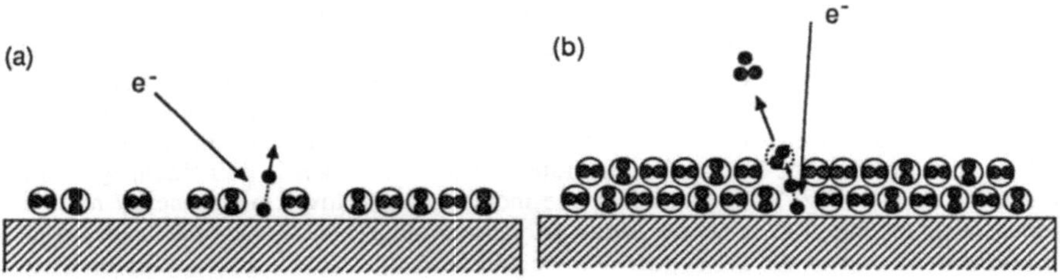


Figure 3. Model of ESD process for a) submonolayer coverages and b) multilayer coverages.

state lies approximately 34.5 eV above the ground state of the molecule and leads directly to a dissociation limit ( D(1s) + $D^+$) at 18.07 eV. Thus excitations to it can produce fragments with slightly greater than 8 eV kinetic energies each. In these previous studies[6,10] an 8 eV ion was in fact observed and exhibited an excitation threshold between 29 and 35 eV. Autoionization and pair production processes giving rise to low kinetic energy ions (<~5 eV) were also observed [8,10,11,16,17] with excitation thresholds below 30 eV.

For the ESD presented here we find that the $D^+(2p\sigma_u)$ state is also a major channel in the production of the most energetic ions. This can be deduced from the observation of an excitation threshold between 30 and 35 eV from the condensed layers, in good agreement with the gas phase results of CROWE and McCONKEY [6]. Figure 4 shows a sequence of TOF spectra for electron beam energies of 30 eV, 35 eV, and 40 eV. For each of these, the condensed layer was 3 monolayers thick, large enough to observe a few higher mass clusters. In the curve associated with 40 eV excitation, both high and low kinetic energy ions, denoted by A and B arrows, respectively, are clearly visible in the $D^+$ peak shape. At 30 eV, the high energy component has disappeared leaving only a residual B or low kinetic energy component. Thus we assign the structure labeled A to a direct dissociative ionization process analogous to the gas phase results. Precise measurements of the

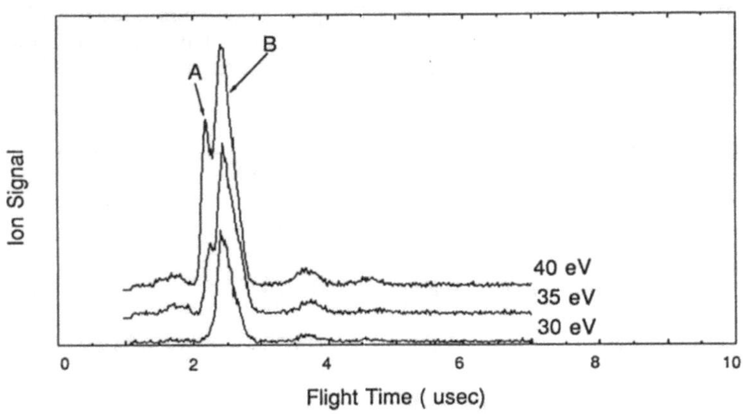


Figure 4. Dependence of the TOF curves on electron beam energy showing the existence of a threshold between 30 and 35 eV for the high kinetic energy portion of the $D^+$ peak (labeled A).

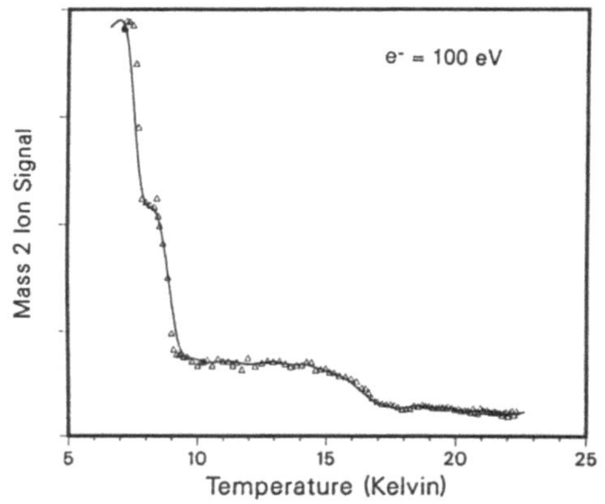


Figure 5. Temperature dependence of the $D^+$ signal with temperature as a ~50 ML thick film is heated from 7 K to 22 K at a rate of 0.1 K/sec and an e-beam energy of 100 eV.

kinetic energy distribution would be desirable to further confirm this analysis, and also to determine more about the nature of the low energy component but, unfortunately, because of the rather short flight time, and the uncertainties in the ion flight paths, this was not possible.

Finally, Fig. 5 shows the temperature dependence of the $D^+$ signal, measured by slowly warming (~ 1 K/sec) a thick film deposit ($\theta \approx 50$ ML) and integrating over all kinetic energies in the mass 2 TOF peak. The signal drops sharply as the temperature increases to 9 K, then remains fairly constant, and finally drops to a baseline at $T_s = 17$ K. To first order, this behavior follows directly from what is known about the desorption characteristics of these films: a) deuterium multilayer desorption begins at 8 K and is complete (for these films) by ~ 10 K, while b) the remaining physisorbed layer is completely desorbed upon reaching $T_s$ of 18 K. Therefore, the change in the signal with temperature reflects first, between 7 and 9 K, the transition of ESD from multilayers to ESD from a saturated physisorbed layer, and second, at 17 K, the desorption of the physisorbed $D_2$ layer. The shoulder between 8 and 9 K is most likely due to $D^+$ from electron beam induced dissociation of the gas phase $D_2$ produced during thermal desorption of the multilayers.

From the magnitude of the signal drop it can be seen that the $D^+$ production is roughly 8 times larger from the multilayers than from the physisorbed layer at $\theta \sim 1$ ML. The higher yield for the multilayers comes from a number of factors: a reduction in the reneutralization probability, a finite sampling depth larger than one layer deep, and from the decay of larger clusters which are only produced for $\theta > 1$. One estimate of the sampling depth can be determined by examining the saturation behavior of the A component of the $D^+$ peak as a function of coverage, Fig. 1. By $\theta \sim$ 5 ML it has to saturated. Since this feature comes from the direct dissociative ionization process for molecules excited both on the surface and in the layers underneath, we feel that the depth corresponding to 5 ML provides a reasonable estimate for the $D^+$ mean free path.

## 4. Conclusions

In general, the results presented here on condensed deuterium closely parallel previous electron impact studies in the gas phase. While the important excited states are the same as for low pressure gas phase work, the primary difference here lies in the observation of large ionic clusters. From a rather simple point of view, these results can be thought of as analogous to what one would observe in the gas phase at extreme pressures. The formation of these large deuterium ion clusters, analogous to what has been observed for solid $H_2$, is consistent with a mechanism involving collisional cascade events.

The initial step leading to cluster formation is the production of an energetic $D^+$ ion through direct dissociative ionization of the ground state $D_2$ molecule. This model is supported by the variation in ion yields as a function of layer thickness, which shows that $D_n^+$ clusters are formed only for coverages greater than one monolayer. At these coverages, energetic $D^+$ ions produced within the film, undergo collisions during ejection, resulting in the formation of larger clusters. The primary electron excitation initiating this process involves transitions to the $D_2^+(2p\sigma_u)$ excited state with a threshold between 30 and 35 eV consistent with previous gas phase experiments. Finally, the temperature dependence of the TOF ESD spectra exhibits structure at temperatures associated with both multilayer and monolayer desorption.

Acknowledgements

The author gratefully acknowledges the expert technical assistance of Steven J. Haney during the course of these measurements. This work has been supported by the U. S. Department of Energy under contract number DE-AC04-76DP00789.

References

1. R. A. Rosenberg, V. Rehn, A. K. Green, and P. R. LaRoe,in Desorption Induced by Electronic Transitions, DIET-I, edited by N. H. Tolk, M. M. Traum, J. C. Tully, and T. E. Madey, (Springer, 1983), p. 247; and Desorption Induced by Electronic Transitions DIET-II, edited by W. Brenig and D. Menzel, (Springer, 1985).
2. R. E. Johnson, W. L. Brown, Nucl. Instr. Meth. 198, 103 (1982).
3. N. Schwentner, E. E. Koch and J. Jortner, Rare Gas Solids Vol. III, edited by M. L. Klein and J. A. Venables ( Academic Press, New York,1985).
4. R. Clampitt and L. Gowland, Nature (London) 223 , 191 (1968).
5. R. H. Stulen and G. D. Kubiak, J. Vac. Sci. A 5 , 535 (1987).
6. A. Crowe and J. C. McConkey, Phys. Rev. Lett. 31 , 192 (1973).
7. C. Backx, G. R. Wight, and M. J. van der Wiel, J. Phys. B 9 ,315 (1976).
8. K. Kollman, J. Phys. B 11 , 339 (1978).
9. B. van Wingerden, Ph. E. van der Leeuw, F. J. de Heer, and M. J. van der Wiel, J. Phys. B 12, 1559 (1979).
10. M. Landau, R. I. Hall, and F. Pichou, J. Phys. B 14 , 1509 (1981).
11. M. Glass-Maujean, J. Chem. Phys. 85 ,4830 (1986).
12. J. O. Noell, C. F. Melius and R. H. Stulen, Surf. Sci. 157 , 119 (1985).
13. R. H. Stulen, to be published.
14. Y. Yamaguchi, J. F. Gaw, and H. F. Schaefer III, J. Chem. Phys. 78, 4074 (1983).
15. L. J. Kieffer and G. H. Dunn, Phys. Rev 158, 61 (1967).
16. S. Strathdee and R. Browning, J. Phys. B 12 , 1789 (1979).
17. S. L. Guberman, J. Chem. Phys. 78 , 1404 (1983).
18. T. E. Sharp, Atomic Data 2 , 119 (1971).

# Desorption Processes in Planetary Science

*R.E. Johnson*

Department of Nuclear Engineering and Engineering Physics, University of Virginia, Charlottesville, VA 22901, USA

Surfaces of materials in the solar system and the interstellar media, not protected by atmospheres, can be exposed to a variety of energetic particle radiations: cosmic rays, solar particles and photons, and plasma particles trapped in planetary magnetospheres. Of considerable interest in planetary science is the exchange of atoms and molecules between the gaseous and condensed phases. Hence our ability to describe desorption from surfaces, stimulated by electronic energy deposition in a variety of materials, is a limiting factor in interpreting many remote sensing and spacecraft data. For instance, line/band spectra can often be obtained on atoms/molecules in the gas phase in the interstellar medium or near remote objects in the solar system like Io (a satellite of Jupiter) or the planet Mercury. Further, the nature of the radiation environment may be known and, indeed, the local plasma may be ionized species desorbed from the surface. As reflectance spectra of the surface or, in the interstellar media, light extinction curves due to the presence of grains, give much less or, often, no information on the surface composition, a description of the desorption processes can be critical in placing constraints on the composition. Recently a number of review articles have appeared which summarize aspects of the importance of electronically induced desorption and sputtering in planetary science[1-6]. Therefore, I briefly list some areas of interest, giving reference to these article and some other primary or secondary references.

Free Na atoms were recently observed in the vapor phase over the daylight surface of Mercury[7], giving, hopefully, a first clue to the mineralogy of that surface. Attempts have been made to describe the total amount of Na observed in terms of photon desorption of Na from silicate surface[8]. The predominantly water ice surfaces of a number of the moons of Jupiter and Saturn show strong evidence of energetic ion bombardment[3,6]. Further, the types of ions observed in the local plasma indicate that they originate from the surface, most likely desorbed (sputtered) as neutrals, subsequently ionized and eventually re-impacting the surface[3,9]. The unique satellite of Jupiter, Io, has condensed $SO_2$ and $S_8$ deposited on the surface, with a local sulfur/oxygen plasma which originates from this material [3,10,11]. But also neutral sodium has been observed in an extended 'atmosphere' about Io[12], implying this species also is present on the surface and desorbed/sputtered by the local radiation field[3]. There is some concern that photon-stimulated desorption of water ice may be important in limiting the lifetime of icy ring particles at Saturn[13], or icy grains in the interstellar media. For the more distant solar system, additional more volatile species are likely to be condensed CO, $CH_4$, $NH_3$, etc. and there may be organics in these surfaces[14]. With the exception of the possible identification of $CH_4$ in the gas phase at Pluto, evidence for the other condensates is only indirect at present. The carbon-containing condensed gases subjected to electronic stimulation form new, more volatile molecules[15] (e.g. $H_2$ from $CH_4$) which are desorbed therefore, reflectance properties of such surfaces are affected[16]. They first become redder (i.e. enhanced absorption at short wavelengths) and then dark (non reflecting)[17]. Therefore, the dark rings at Uranus may be due to extensively exposed organic material in, possibly, an icy matrix. In fact 'dark' materials are ubiquitous in the outer solar system and recently Halley's nucleus was seen to be dark[18] suggesting it also has been processed by radiation[19].

The gas/solid relationship is very important in describing recent observations of grains and molecular vapor in the vicinity of Halley. That is, some species in the coma have spatial distributions that correlate strongly with the outflow of 'dust' grains from Halley[20], suggestive of stimulated desorption. Finally, the material in the solar system was condensed from material in an interstellar molecular cloud. In these clouds molecular species form by ion-molecule reactions, but grains are known to be present on which molecular hydrogen is formed and desorbed [21]. Therefore, it is likely that many of the large molecular species observed in these clouds also form on the grain surfaces and are desorbed. Although this is interesting in itself, a possibly more important problem is to deduce the grain composition as there have been suggestions that comets, which are debris from the formation of our sun, contain 'unaltered' interstellar material[22].

Acknowledgement

Work supported by the NSF Astronomy Division under grant Ast-85-11391.

References

1. W.L. Brown, L.J. Lanzerotti, R.E. Johnson: Science 218, 525 (1982).
2. R.E. Johnson, L.J. Lanzerotti, W.L., Brown: Nucl. Inst. Meth 198, 147(1982)
3. A.F. Cheng, P.K. Haff, R.E. Johnson, L.J. Lanzerotti: In Satellites, eds. J.A. Burns, M. Matthews (Univ. of Arizona, Press, Tucson, 1986) p. 403.
4. L.J. Lanzerotti, W.L. Brown, R.E. Johnson: Nucl. Inst. Meth. B14, 373(1986)
5. L.J. Lanzerotti, R.E. Johnson: In Ion Beam Modification of Insulators, eds. P. Mazzoldi, G. Arnold (Elsevier, The Netherlands, 1987) p. 631
6. A.F. Cheng, R.E. Johnson: In Origins and Evolution of Atmospheres, ed. S.K. Atreya (Univ. of Arizona Press, Tucson, 1987) in press.
7. A. Potter, T. Morgan: Science 229, 651 (1985). D. Hunten, D. Shemansky, T. Morgan: in Mercury (Univ. of Arizona Press, Tucson) in press.
8. M. McGrath, R.E. Johnson, L.J. Lanzerotti: Nature 323, 694(1986). A.F. Cheng, R.E. Johnson, L.J. Lanzerotti, S.M. Krimigis: Icarus in press (1987).
9. R.E. Johnson, L.J. Lanzerotti, W.L. Brown: Adv. Space Res. 4, 41(1984). R.E. Johnson, L.A. Barton, J.W. Boring, W.A. Jesser, W.L. Brown, L.J. Lanzerotti: In Ices in the Solar System, ed. J. Klinger et al. (Reidel, Dordrecht, 1985) p. 301. R.N. Clark, F.P. Fanale, M.J. Gaffey: In Satellites eds., J.A. Burns, M. Matthews (Univ. of Arizona Press, Tucson, 1986) p. 437.
10. D. Nash, M.H. Carr, J. Gradie, D.M. Hunten, C.F. Yoder: In Satellites, eds. J.A. Burns, M. Matthews (Univ. of Arizona Press, Tucson, 1986) p. 629.
11. J.W. Belcher: In Physics of the Jovian Magnetosphere eds. A.J. Dessler (Cambridge Univ. Press, N.Y. 1983) p. 68
12. R.A. Brown, C.B. Pilcher, D.F. Strobell: In Physics of the Jovian Magnetosphere ed. A. J. Dessler (Cambridge Univ. Press, N.Y., 1983) p. 197.

13. R.W. Carlson: Nature 238, 461 (1980). J.N. Cuzzi, J.J. Lessauer, L.W. Esposito, J.B. Holberg, E.A. Marout, G.L. Tyler, A. Boeschat: In Planetary Rings eds. R. Greenberg, A. Brahic (Univ. of Arizona Press, Tucson, 1984) p. 73.
14. D.J. Stevenson, A.W. Harris, J.I. Lunine: In Satellites eds. J.A. Burns, M. Matthews. (Univ. of Arizona Press, Tucson, 1986) p. 39.
15. W.L. Brown, G. Foti, L.J. Lanzerotti, J.E. Bower, R.E. Johnson: Nucl. Inst. Meth. B19/20, 899 (1987).
16. A.F. Cheng, L.J. Lanzerotti: J. Geophys. Res. 83, 2596 (1978).
17. G. Strazzulla, L. Calcagno, G. Foti, K.L. Seng: In Ices in the Solar System eds. J. Klinger et al. (D. Reidel, Dordrecht, 1983) p. 273
18. See e.g. Nature 321 (1986).
19. R.E. Johnson, J. Cooper, L.J. Lanzerotti, G. Strazzulla: Astron. Astro. in press (1987).
20. M.F. A'Hearn et al.: Astron Astro. in press (1987)
21. J.A. Nuth, R.E. Stencel: Interrelationships among Circumstellar, Interstellar, and Interplanetary Dust. (NASA Carf. Pub. 2403, Washington, D.C. 1986).
22. J.M. Greenberg: In Comets ed. L.L. Wilkening (Univ. of Arizona Press, Tucson, 1982) p. 131.

# Atom Selective Photochemistry with Monochromatic Soft X-Rays

*D.M. Hanson*

Department of Chemistry, State University of New York
Stony Brook, NY 11794-3400, USA

## 1. INTRODUCTION

While much detailed information has been obtained about the fragmentation of molecules following excitation or ionization of valence electrons, relatively little is known about corresponding phenomena involving core electrons. This paper reviews some recent research that we have done to learn about the chemical consequences of selective core electron excitation in molecules. Specifically, studies of nitrous oxide, acetone, and oxygen will be summarized, and a theory will be described that is directed at predicting which bonds in a polyatomic molecule rupture as a consequence of the multiple electron excitation accompanying decay of a core hole excited state.

One of the major problems in chemistry is to control, in a complex molecule, the atomic site at which chemical reactions occur. This research therefore was stimulated by studies of ion desorption from solid surfaces because these studies indicated that in many cases bond rupture and ion desorption was strongly correlated with the atomic site of the electronic excitation /1/.

"Selective excitation" refers to the fact that core electrons of different atoms can be excited selectively to different final state configurations. With tunable, monochromatic radiation from synchrotron sources, it is possible to excite selectively not only core electrons of different elements but also core electrons of the same element in different chemical environments in the molecule. This selectivity is possible because the differences in the binding energies of the core electrons are larger than the bandpass of modern soft x-ray monochromators. In addition, different types of core hole excited states can be prepared. The core electron can be promoted to an available valence orbital, a Rydberg orbital, a shape resonance, or the continuum. Also, multiple electron excited states result from shake-up and shake-off processes.

Since the x-ray emission rate is small compared to the Auger decay rate for the light elements in the periodic table, Auger decay generally is the most likely process to occur following core electron excitation or ionization in these elements. In Auger decay, one valence electron falls into the core hole and one or more other valence electrons are excited or ionized. For core hole configurations where the core electon is not ionized, this relaxation also is called autoionization or resonance Auger decay. This process generally depletes the bonding structure of the molecule, and, in the gas phase, fragmentation usually results. In special cases, fragmentation may occur prior to Auger decay /2/. Recent research has revealed that memory of the excitation selectivity is preserved in subsequent fragmentation events /3-8/. In order to understand the origin of this memory effect, it is necessary to answer two questions: What are the electronic processes associated with relaxation of the core hole excited state? and How are these relaxation processes related to the fragmentation channels? Information relevant to the first question can be obtained with the techniques of photoelectron spectroscopy. The techniques of mass spectroscopy can be used to provide information relevant to the second question. In addition, fluorescence spectroscopy can be used to detect neutral and ionic species produced in excited states. The detection of electrons, ions, and photons in coincidence serves to establish detailed cause and effect relationships in the relaxation scheme.

## 2. EXPERIMENTAL

The experiments were done at the National Synchrotron Light Source using the U15 beam line of the VUV storage ring. This beam line uses a toroidal grating monochromator, with a range of about 250 to 900 eV and a resolution of about 300. The monochromator is protected from contamination by a thin aluminum window barrier placed in the beam line, thereby reducing the vacuum and differential pumping requirements considerably.

The apparatus, consisting of a coincidence time-of-flight (TOF) mass spectrometer and retarding field electron energy analyzer, has been described previously /4/. The sample was introduced into the ionization chamber as a supersonic beam through a 0.1 mm diameter nozzle and a 0.5 mm diameter skimmer. The background pressure in the ionization chamber during the experiments was around $5 \times 10^{-7}$ Torr. The molecular beam crosses the soft x-ray beam at right angles between a pair of grids that are biased to establish a gradient across the interaction region of 240 V/cm. Electrons, accelerated at right angles to the plane of the beams, pass through the retarding field analyzer and are detected by a channel electron multiplier array (CEMA). Ions, accelerated in the opposite direction, pass through the TOF mass analyzer and are detected by another CEMA. Mass spectra are obtained by using the energy discriminated electrons as trigger events for a 100 MHz signal averager. The resolution of the electron analyzer, though not sufficient to obtain electron energy spectra, is adequate for distinguishing between Auger and low energy electrons. The linewidths in the mass spectra are determined by the kinetic energy released in the fragmentation. In other experiments, luminescence was detected by using a gas cell at pressures of 0.1 Torr with a cooled photomultiplier tube and 0.25 m monochromator.

## 3. RESULTS AND DISCUSSION

### 3.1 Nitrous Oxide

The electron and ion yield spectra of nitrous oxide in the regions of the nitrogen and oxygen K edges have intense sharp peaks corresponding to excitation of a 1s electron of the terminal nitrogen (401 eV), the central nitrogen (405 eV), and oxygen (536 eV) to the lowest unoccupied molecular orbital /4/. By positioning the monochromator on these features, it is possible to select the atomic site of the core hole and to obtain mass spectra characterizing the resulting chemical relaxation. Two statements summarize the observations. The fragmentation pattern resulting from the core hole on the central nitrogen is similar to that resulting from the core hole on the oxygen and different from that resulting from the core hole on the terminal nitrogen. The ratio of $NO^+$ to $N_2^+$ is greater than 1 when the core hole is on either the terminal nitrogen or oxygen.

Several existing concepts are inconsistent with these observations. The fragmentation is not determined by the excitation energy since this is similar for the core hole positioned on the nitrogen atoms and considerably more energy is required to excite an oxygen 1s electron. The simple idea of a "Coulomb explosion" /9/ would predict a high yield of atomic ions and little if any specificity in the fragmentation pattern. The idea of valence hole correlation and localization /10/ would predict bond rupture around the atomic site of the core hole. Consequently, the yield of atomic ions should be largest when the core hole is on the central nitrogen, and the ratio of $NO^+$ to $N_2^+$ should be less than one when the core hole is on the oxygen.

The changes in the mass spectra with the atomic site of excitation and the similarity of the mass spectra with the core hole on the central nitrogen or the oxygen can be understood qualitatively in terms of the atomic populations and the overlap populations of the valence molecular orbitals /4/. The atomic populations identify the valence orbitals that participate in the Auger decay process, and the overlap populations reveal the contributions these orbitals make to the bonding structure of the molecule. The differences in the mass spectra obtained with excitation of the terminal nitrogen and excitation of the central nitrogen can be attributed to the larger atomic populations of the bonding orbitals and the smaller atomic populations of the antibonding orbitals on the central nitrogen. Consequently Auger decay depletes the bonding orbitals and enhances fragmentation when the core hole is on the central nitrogen. When the core hole is on the oxygen atom, the observed ion yields resemble the case of the central nitrogen core hole because of the similarity in the atomic populations for those two atoms.

Auger decay involving the 2 pi orbital may be responsible for the observed decrease in the ratio of $NO^+$ to $N_2^+$ as the site of the core hole is shifted from the terminal to the central nitrogen. This orbital is N-N bonding and N-O antibonding. Holes in this orbital therefore should weaken the N-N bond and stabilize the N-O bond. Auger decay is more likely to create holes in this orbital when the initial core hole is on the terminal rather than the central nitrogen because the atomic population is larger for the terminal nitrogen. Consequently the ratio of $NO^+$ to $N_2^+$ should be larger for the case of the terminal nitrogen core hole.

## 3.2 Acetone

Ion and electron yield spectra were taken scanning photon energies from 250 to 600 eV /5/. The oxygen pi resonance appears as a strong peak at 531 eV followed by an intense shape resonance peak at 545 eV. Electron energy spectra, obtained by scanning the potential applied to the center grid of the retarding potential analyzer, reveal, after correcting for the transmission efficiency, that about equal numbers of low energy and Auger electrons are produced following core electron excitation, even to the pi resonance, as was the case for nitrous oxide /4/. It is thought that these electrons originate from double Auger decay or autoionization processes following Auger decay. Above the ionization threshold, direct photoelectrons also contribute to the low energy electron yield. Mass spectra were obtained with excitation at 526 eV, just below the pi resonance, at 531 eV, on the pi resonance, at 535 eV, in the valley between the pi resonance and the shape resonance, at 545 eV, on the shape resonance, and at 597 eV, well above the oxygen K ionization edge. Studies of acetone fragmentation with excitation in the region of the carbon K edge have been reported previously /3/.

The mass spectra consist predominantly of protons and five skeletal groups. These groups are $CH_n^+$, $C_2H_n^+$, $C_3H_n^+$, $CH_3CO^+$, and the singly charged parent. A peak at m/q = 29 could arise from either the doubly charged parent or $C_2H_5^+$. The most prominent peaks correspond to $H^+$, $CH_3CO^+$, and $CH_3^+$. The appearance of the $C_3$ and $C_2$ groups suggests that neutral atomic oxygen and possibly carbon monoxide are produced since neither oxygen or carbon monoxide ions are abundant in any of the mass spectra. It is apparent that methyl ion production is strongly associated with excitation of either the pi resonance or the shape resonance. This observation suggests that an important process following oxygen core excitation is carbon - carbon bond rupture. A detailed analysis reveals that the increase in methyl ion production associated with the resonances simply tracks the change in the core electron excitation cross section and that no special phenomenon occurs at these points /5/. The increase in the yield of $CH_3^+$ is for oxygen excitation is somewhat counterintuitive because the bond rupture, though site specific, is not adjacent to the site of the initially created core hole. The concepts of population analysis, as used for nitrous oxide, may explain this result, but the data necessary for the analysis are not available in the literature.

The previous studies at the carbon K edge /3/ associated $C^+$ and $O^+$ formation with excitation of the carbon 1s electrons at the pi resonance and also noted the production of $CH_3CO^+$ and $CH_3^+$ above the C 1s ionization edge. Neither carbon nor oxygen atomic ions are significant products of oxygen 1s excitations. Again, although the chemistry is site specific, the simple idea of bond scission around the atomic site of the core hole is not adequate to explain the observations.

### 3.3 Oxygen

Fluorescence in the visible and ultraviolet regions of the spectrum provides a convenient means for detecting decay products, both neutral and ionic, produced in excited electronic states. Fluorescence spectra with resolved rotational and vibrational structure provide information about the energy spacing between electronic states, about the structure and bonding properties of these states, and about the populations of rotational and vibrational levels, which can characterize the populating mechanisms. The production of the doubly charged molecular cation by Auger decay following core electron ionization is of particular interest because little is known about the properties of these ions and because the fluorescence decay of the Auger final state essentially allows Auger spectroscopy to be done with the high resolution of an optical spectrometer.

The first observation of dispersed luminescence following core electron excitation by monochromatic synchrotron radiation was reported for the case of oxygen /11/. Two series of bands were observed. The relative intensity of these two series varies with changes in the excitation energy in the region of the oxygen K edge. The longer wavelength series matches the known luminescence of $O_2^+$. The shorter wavelength series with peaks at 2.97, 2.80, and 2.64 eV does not match any known luminescence from this ion or other possible atomic or molecular species. It was proposed that the shorter wavelength bands are due to $O_2^{2+}$ and are the analog of the first positive system of $N_2$, which is isoelectonic to $O_2^{2+}$.

Double zeta Gaussian type molecular orbital, configuration interaction calculations for $O_2^{2+}$ have been completed /12/. These calculations are entirely consistent with the above assignment of the shorter wavelength bands. Both states involved in the transition have a minimum in the potential energy as a function of internuclear separation. At the position of the minimum for the upper state, the electronic energy spacing of the two states is 2.52 eV, and at the position of the minimum for the lower state, the energy difference is 2.92 eV. The relative positions and depths of the two potential curves appear to be consistent with the observation of only three bands in the fluorescence spectrum.

Recently we have completed studies of the fragmentation of oxygen following core electron excitation and ionization. We have observed that the relative yield of $O^+$ and $O^{2+}$ ions varies significantly with photon energy through the resonances and into the continuum /13/, as does the relative intensity of two series of fluorescence bands /14/. Such strong variations have not been observed by us or reported in the literature for other systems.

## 4. TRANSITION STRESS THEORY

The transition stress theory that we are developing is directed at analyzing the forces acting on the atomic nuclei of a molecule when one or more electrons are excited or ionized and the electronic configuration changes from that for ground state. The objective is to have a relatively simple method for predicting which bonds in a polyatomic molecule break as a result of the multiple electron changes in the electronic configuration that result from core electron excitation or ionization and subsequent Auger decay. The theory is being developed and tested first for diatomic molecules /15-17/.

The orbital stress for an electron is defined as the average magnitude of the forces exerted by that electron on the two nuclei in a diatomic molecule. When the electonic configuration changes and an electron is promoted from one orbital, i, to another, f, a transition stress on the molecule is produced. The transition stress is the difference between the orbital stress of orbital f and that of orbital i. The transition stresses are additive for multiple electron excitation. Both the nuclear separation and the electronic wavefunctions change to remove this stress. These changes are called the nuclear relaxation and electronic relaxation, respectively. Transition stresses for many electronic states of several diatomic molecules and ions were calculated. It was found that the ratio of the nuclear relaxation to the electronic relaxation is essentially constant for any molecule, independent of the electronic configuration or state. In some cases this ratio is the same for ionic states of the molecule as well. The bond lengths and vibrational force constants of excited states are related to the transition stress, and the rotational constants are linear functions of the transition stress /15-17/. The theoretical relationships and calculated values are in good agreement with experimental values and with the results of ab initio calculations.

The transition stress theory is based on concepts of one electron molecular orbital theory. The transition stresses are calculated for the orbitals and virtual orbitals of the molecule in its lowest energy configuration. With this information, the bond lengths, rotational constants, and vibrational frequencies of excited states of the molecule and its ions can be predicted successfully. This success indicates that the theory can account, with reasonable accuracy, for the changes in the forces exerted on the nuclei when the electronic configuration changes. The formalism does not yet apply to states where configuration interaction is significant.

## 5. ACKNOWLEDGMENTS

The work reviewed in this paper depended upon the contributions of my colleagues, research associates, and graduate students, as documented by the references. This successful collaboration is much appreciated. The research was supported by the National Science Foundation under grants CHE-8215061 and CHE-8703340.

## 6. REFERENCES


1. (a) D.M. Hanson, R. Stockbauer, T.E. Madey: J. Chem. Phys. 77, 1569 (1982); (b) R. Stockbauer, E. Bertel, T.E. Madey: J. Chem. Phys. 76, 5639 (1982); (c) J.A. Kelber, M.L. Knotek: Surf. Sci. 121, L499 (1982); (d) J.A. Kelber, M.L. Knotek: J. Vac. Sci. Technol. A1, 1149 (1983); (e) R.L. Benbow, M.R. Thuler, Z. Hurych, Phys. Rev. Lett. 49, 1264 (1982); (f) M.L. Knotek, V.O. Jones, V. Rehn: Phys. Rev. Lett. 43,300 (1979); (g) R. Jaeger, J. Stohr, J. Feldhaus, S. Brennan, D. Menzel: Phys. Rev. B23, 2102 (1981); (h) R. Jaeger, J. Stohr, R. Treichler, K. Baberschke: Phys. Rev. Lett. 47, 1300 (1981)
2. P. Morin, I. Nenner: Phys. Rev. Lett. 56, 1913 (1986)
3. W. Eberhardt, T.K. Sham, R. Carr, S. Krummacher, M. Strongin, S.L. Weng, D. Wesner: Phys. Rev. Lett. 50,1038 (1983)
4. J. Murakami, M.C. Nelson, S.L. Anderson, D.M. Hanson: J. Chem. Phys. 85, 5755 (1986)
5. M.C. Nelson, J. Murakami, S.L. Anderson, D.M. Hanson: J. Chem. Phys. 86, 4442 (1987)
6. K. Muller - Dethlefs, M. Sander, L.A. Chewter, E.W. Schlag: J. Phys. Chem. 88, 6098 (1984)
7. (a) P. Morin, G.G.B. de Souza, I. Nenner, P. Lablanquie: Phys. Rev. Lett. 56,131 (1986); (b) G. Dujardin, I. Nenner: Chemical Physics 105, 291 (1986)
8. (a) R.G. Hayes: J. Chem. Phys. 86, 1683 (1987); (b) R.G. Hayes: J. Phys. Chem., in press.
9. (a) T.A. Carlson, R.M. White: J. Chem. Phys. 44, 4510 (1966); (b) T.A. Carlson, M.O. Krause: J. Chem. Phys. 56, 3206 (1972)
10. (a) D.R. Jennison, J.A. Kelber, R.R. Rye: Phys. Rev. B25, 1384 (1982); (b) P.J. Feibelman: Surf. Sci. 102, L51 (1981); (c) D.E. Ramaker, C.T. White, J.S. Murday: J. Vac. Sci. Technol. 18, 748 (1981)
11. K. Tohji, D.M. Hanson, B.X. Yang: J. Chem. Phys. 85, 7492 (1986)
12. H. Yang, J.L. Whitten, D.M. Hanson: to be published
13. D. Lapiano, C.I. Ma, K. Wu, D.M. Hanson: to be published
14. K.Tohji, D.M. Hanson, B.X. Yang: to be published
15. Y. Zhang, D.M. Hanson: Chem. Phys. Lett. 127, 33 (1986)
16. Y. Zhang, D.M. Hanson: J. Chem. Phys. 86, 347 (1987)
17. Y. Zhang, D.M. Hanson: J. Chem. Phys. 86, 666 (1987)

# Part III

# Neutrals and Negatives

# Photon Stimulated Desorption of Neutrals by Synchrotron Radiation from Chemisorbed, Physisorbed, and Condensed Layers

*P. Feulner, S. Auer, T. Müller, A. Puschmann, and D. Menzel*

Physik Department E20, Technische Universität München,
D-8046 Garching, Fed. Rep. of Germany

## 1. Introduction

The intention of this contribution is to show that photon stimulated desorption investigations with synchrotron light can increase the understanding of the DIET process, at least for simple adsorbates like diatomic molecules or rare gases on metal surfaces and particularly if desorption of neutral particles is monitored as well. Generally, the desorption rate $dN_k/dt$ for a species k (e.g. $CO^*$, $CO^+$, $O^+$, $O^{++}$, etc.) can be written as /1/

$$dN_k/dt = i(\hbar\omega) \cdot \sigma_j(\hbar\omega) \cdot P_{escape}(k,j), \tag{1}$$

where $i(\hbar\omega)$, $\sigma$ and P denote the photon flux, the cross section for a primary excitation of type j and the probability that this excitation j leads to desorption of a particle k. Understanding DIET quantitatively requires the ability to assign primary electronic excitations and to calculate the interdependent dynamics of the evolution of the electronic excitation and the nuclear motion of the adsorbate or its fragments /1,2/. In this way, escape probabilities and branching ratios can be obtained. Experimental data are necessary, in particular on energies and symmetries of primary and secondary thresholds and resonant maxima of the PSD yield vs. photon energy, for the assignment of the primary electronic excitations; and also branching ratios, total desorption cross sections as function of coverage, angular distributions, and kinetic as well as internal energies of the desorbing particles, for the evaluation of P. First calculations for physisorbed species and chemisorbed diatomic molecules based on ESD data were obtained. They are presented at this conference /3-6/. Using photons instead of electrons for stimulation makes the assignment of the primary electronic excitations much easier, because ionization thresholds and excitations to bound or resonant final states can be discerned much better; conclusions on the symmetry of the states involved can be drawn from polarization dependent measurements with synchrotron light /7/. To obtain total desorption cross section values, the investigation of DIET of neutrals is necessary for those systems for which these dominate. This is the case for valence excitations in most cases /1/ (see however /8/). Then PSD of neutrals (n-PSD) may also be important for technical reasons, e.g. for the design of vacuum equipment for synchrotrons or storage rings /2/.

## 2. Experimental

A poor signal/background ratio is the main obstacle for n-PSD measurements with synchrotron radiation. Without a wiggler or undulator, light intensities of $\leq 10^{12}$ photons/s are available. Cross sections for desorption are of the order of $10^{-19}$ $cm^2$ for chemisorbed adsorbates (see below) or even less, resulting in total desorption yields of $\approx 10^8$ particles/s from a monolayer (ML). In an ordinary UHV chamber with a typical pumping speed of 500 l/s, this flux leads to a pressure rise $\Delta p_{PSD}$ of only $10^{-14}$mbar, which is not detectable without sophisticated equipment. Indirect monitoring of neutral PSD from measurements of the decreasing surface concentration under photon irradiation with, e.g., AES or PES, is ruled out even more clearly. Even for optimum UHV conditions, readsorption rates two orders of magnitude higher than the PSD rate are expected. Direct detection is necessary, therefore.

We present a simple detector with sufficient sensitivity for n-PSD work. It is similar to one used for measurements of angular distributions of n-ESD described in detail at DIET II /9/. A high efficiency electron impact ionizer is mounted on a Balzers QMG 311 quadrupole mass spectrometer (QMG). The ionizer is enclosed in an auxiliary glass volume of ≈60 ml with an aperture of slightly less than sample size in its front, see Fig.1. For measuring n-PSD, the sample is positioned close (≈2 mm) to this aperture and irradiated from the side. A platinum shield behind the aperture prevents stray light from entering the mass filter. PSD ions are suppressed by a retard potential of ≈100 eV applied between sample and QMS. The light from the monochromator is modulated at a frequency of 20 Hz by a magnetically driven chopper. This frequency is chosen for optimum signal/background ratio according to the characteristic pumping speed of the auxiliary volume. The signal from the multiplier of the QMS is processed with a lock-in amplifier. Data are accumulated with a minicomputer. As mentioned, $\Delta p_{PSD}$ will always be much smaller than the background pressure $p_{res}$, especially for masses normally abundant in the residual atmosphere of an UHV system. The noise of the PSD signals obtained with this device is then $\approx\sqrt{p_{res}}$. Therefore, isotopes not present in the residual gas of the UHV system are dosed for ultimate sensitivity, like $C^{18}O$ instead of $C^{16}O$. In this way, less than $10^8$ particles/s can be detected for CO and rare gases. Main advantages of this detector are its simplicity and its high sensitivity, which can be improved even further by lowering $p_{res}$. Cross section estimates from comparisons of PSD yields to TPD spectra taken in situ from layers of well known coverage are possible. A serious drawback compared to detection methods based on excitation of the desorbing particles by laser light is the missing of any selectivity for internal excitations. For highly reactive particles, like reactive atoms or fragments of molecules which might adsorb on the metal parts of the ionizer, a decreased detection sensitivity is expected.

Though polarization dependent measurements are possible in principle, only $A_z$ polarization could be realized with the set-up used in this work. For ionic PSD, a rotatable QMS described in /7/ was used. Nearly pure $A_z$ (light vector 6° from the surface normal) and mixed polarization (the angle between surface normal and E-vector was 60°, resulting in a $A_{xy}/A_z$ ratio of 1/0.6) could be realized with this apparatus.

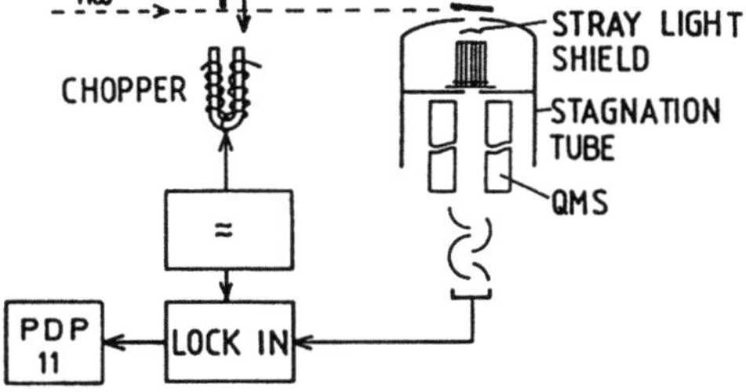


Fig.1 Detector for neutral PSD

## 3. Results and Interpretation

### 3.1 Chemisorbed CO/Ru(001)

DIET of molecular and fragment ions off CO/Ru(001) has been studied with PSD and ESD for valence /10,11,12/ and core excitations /7/, whereas DIET of neutrals has so far only been investigated with ESD (see /9/ and references cited therein). With ESD, primary thresholds of 14 and 28 eV electron energy in vacuum (see below) are found for $O^+$ and $CO^+$ desorption, respectively /12/. For $CO^\circ$, a strong secondary threshold is obtained at 14 eV too. A much smaller desorption signal is observed for electron energies below that value. Due to the smallness of the signal, this primary threshold is difficult to discern. It is estimated to be around 10 eV /12/ (For CO/W(100), a threshold at 4.5 eV with a maximum at 8 eV has been observed, see /1/ and references cited therein). In ESD, the $CO^\circ$-yield increases nearly linear if the kinetic energy of the stimulating electrons is raised from 14 to 50 eV /12/; no influence of C1s or O1s core ionization on the $CO^\circ$ signal is observed /12/. For valence excitations, ionic and neutral yields are maximal at a CO coverage of 1/3 ML. At that coverage, a $(\sqrt{3}x\sqrt{3})R30^\circ$ is observed in LEED /2/. In PSD, resonances at 41 and 50 eV were found for $O^+$/10/ (see however below). For core excitations, ionic PSD by complex multi-hole, multi-electron excitations are obtained, in particular for fragment ions and multiply charged fragment ions /7/. The purpose of our recent investigation was to clarify if multiple-hole, multiple-electron excitations are important for n-PSD in the valence region, to assign the excitations leading to DIET for this energy range, and to evaluate absolute desorption cross section data as function of photon energy.

We consider the energy range 12 to 28 eV first. These measurements were performed at the 1m SEYA monochromator at BESSY, Berlin. No $O^+$ emission is obtained for this energy range in agreement with ESD results /12/. For $CO^+$, primary thresholds of 18 eV are obtained for $A_z$ and $A_x$ polarization, see Fig.2. For saturated layers, the $CO^+$ signal is lower by a factor of 4.5 than for the $\sqrt{3}$-layers. Due to the low photon flux at this beam line, the s/b ratio for $CO^\circ$-PSD is very poor, but an increased yield around 18 eV as for the ionic species is clearly discernible (for $CO^\circ$, only data for $\sqrt{3}$ layers were taken; desorption due to the low threshold excitation channel found in

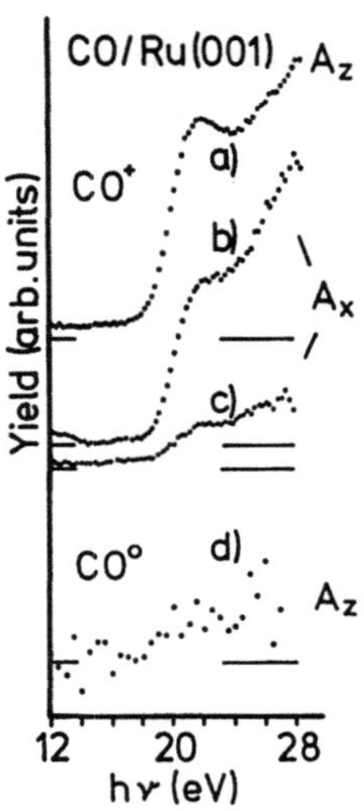


Fig.2. $CO^+$ and $CO^\circ$ signals off $\sqrt{3}$ (a,b,d) and saturated layers (c) of CO/Ru(001), for $A_z$ and mixed (mainly $A_{xy}$) polarization

ESD could not be detected). This energy is in very good agreement with total energies for $5\sigma^{-2}$ final states from AES measurements /13/. It has been shown for NO/Pt(111) that ionization of the $5\sigma$ level and subsequent screening of this hole via charge transfer into the $2\pi$ level of the NO makes the molecule behave like weakly chemisorbed $O_2$, which can desorb from the surface /5,6/. We think that a similar mechanism could lead to desorption of CO after a $5\sigma^{-2}$ excitation, although the $5\sigma$ orbital is more strongly involved in bonding to the substrate for CO than for NO, and we assign the threshold at 18 eV to the $5\sigma^{-2}$ state. The narrow maximum at 22 eV which is more pronounced for $A_z$ than for $A_x$ polarization must be an excitation to a bound resonance below the vacuum level. A possible candidate with correct symmetry is the $5\sigma^{-2}2\pi_a^{+2}$ excitation. The discrepancy concerning the energies for the thresholds found in ESD /12/ and PSD is due to the fact that in /12/ for the stimulating electrons final states at the vacuum level were assumed. If we consider these final states to be close to the Fermi level (and the data corroborate this) good agreement of ESD and PSD is obtained.

PSD off CO/Ru(001) for 25 to 120 eV will be considered next. These data have been published /14/ and will be described only briefly therefore. The measurements were performed at the TGM 1 beamline at BESSY. A primary threshold for $O^+$ and secondary thresholds for $CO^{\bullet}$ and $CO^+$ are obtained at 35 eV for $A_z$ light, see Fig.3a (the $O^+$-signal at lower energies is entirely due to second order light). The edge-jump, i.e. the yield ratio for 30 and 43 eV, is 1:3 and 1:4 for $CO^{\bullet}$ and $CO^+$, respectively. Narrow resonant peaks at 43 and 47 eV for $CO^{\bullet}$, $CO^+$ and $O^+$, and at 51 eV for the ionic species $O^+$ and $CO^+$ only, are superimposed on a broad maximum centered around 45 eV. Further maxima occur for $A_z$ polarization at 63 eV ($CO^{\bullet}$, $CO^+$ and $O^+$) and at 102 eV ($O^+$ only), see Fig.3b. For $A_x$ polarization, resonant structures are far less pronounced (Fig.3b). Here, the PSD yields follow the sample current in shape, indicating dominant contributions from XESD by photo or Auger

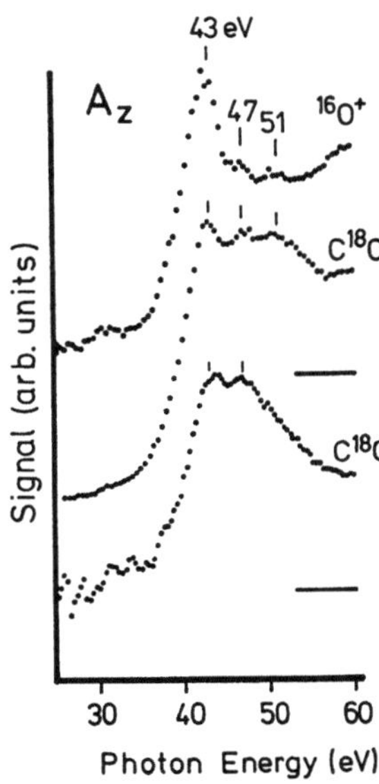


Fig.3a. Ionic and neutral PSD yield from √3 layers of CO/Ru(001); $A_z$ polarization only (after/14/)

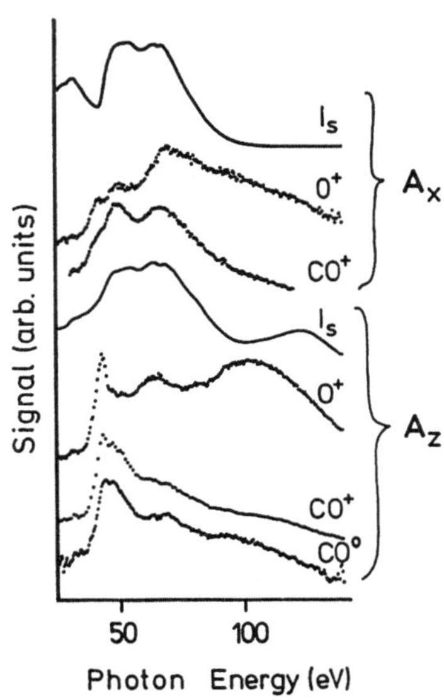


Fig.3b. PSD yield and sample current from √3 layers of $C^{16}O$/Ru(001); polarization as indicated (after/14/)

electrons. For $A_z$ light however, this desorption mechanism is clearly ruled out by the different energy dependences of sample current and PSD yield. From imperfect parts of the Ru(001) sample, in particular $O^+$ signals larger by a factor of ≈3 than from the well ordered surface are obtained /15/; the polarization dependence observed for the clean sample (see Fig.3b) is missing and the resonances at 43 and 51 eV are slightly shifted in energy to 41 and 50 eV (the peak at 47 eV is not observed at this surface), ruling out DIET via Ru 4p core hole excitation (the BE's of the Ru $4p_{1/2}$ and $4p_{3/2}$ levels are 46.5 and 43.2 eV, respectively). The observed maxima must reflect different electronic excitations of the CO molecule itself, as already concluded in /10/. The evolution of these excitations leads to different branching ratios as mentioned above; the dependence of the ratio of the PSD peaks at 43 and 47 eV on isotopic mass, see Figs. 3a and 3b, strengthens this conclusion. The $CO^{\bullet}$ desorption cross section is estimated to be $10^{-19}<\sigma_{des}<10^{-18}cm^2$ from TPD traces taken in-situ (the accuracy of $\sigma_{des}$ is limited by the uncertainty of the photon flux).

In order to assign these primary excitations, comparison with gas phase data were made /14/. For $O^+$-emission from CO gas, however, only a broad maximum at 5o eV assigned to $3\sigma$ ionization is obtained. Additional structures due to ionization of multielectron states, which arise from CI and are seen in UPS of inner valence levels for CO gas (outer valence orbitals are mimicking the $3\sigma$ orbital) can be discerned only after differentiation of the $O^+$-yield vs. photon energy. Shape resonances ≈18 eV in width are observed for these multielectron states about 12 eV above ionization thresholds (see references cited in /14/). The main shape of the yield between 35 and 55 eV in our spectra may have the same origin. The peaks at 43, 47 and 51 eV are too narrow for shape resonances. Transitions to bound resonances as in the case of the 22 eV peak mentioned above are more likely. Tentatively, the maxima at 43, 47 and 51 eV have been assigned to the $3\sigma^{-1}5\sigma^{-1}2\pi_a{}^{+2}$, $3\sigma^{-1}4\sigma^{-1}2\pi_a{}^{+2}$ and $3\sigma^{-1}5\sigma^{-1}2\pi_b{}^{-1}2\pi_a{}^{+3}$states (see /14/ for details). Curve crossing during the evolution along the the repulsive curve must lead to the branching ratios observed. The high desorption cross section compared to the primary excitation cross section indicates that the probability for localisation at least of part of the excitation must be high /1,2,14/. As for core ionization /7/, the small cross section for these complex excitations appear to be compensated by enhanced localisation /14/. On the other hand, excitation must be removed from the $3\sigma$ during evolution because otherwise more desorption of fragments would be expected. The breaking of the bond to the surface is probably due to the additional charge in the $2\pi_a$, which is antibonding for it. A possible assignment for the broad peak at 62 eV is the $3\sigma^{-2}2\pi_a{}^{+2}$ excitation. The even broader peak at 102 eV seen only for $O^+$ may be due to the double shape resonance $3\sigma^{-2}6\sigma^{+2}$. The $2\pi$ is not involved, which could explain the missing of molecular desorption from this state.

Coverage dependent PSD yields are obtained for these inner valence excitations as well. Contrasting the findings for $5\sigma^{-2}$ ionization, a decrease of the $CO^+$-signal by a factor of ≈ 0.5 is found at saturation coverage compared to a $\sqrt{3}$-layer /16/ in agreement with ESD results obtained with electron energies of ≈100 eV /11/. This finding bears on the question whether different primary excitations are monitored with ESD and PSD. In section 1, it was shown that identical <u>thresholds</u> are obtained if the energy scale for the electrons is aligned properly. Here we show that also similar <u>coverage dependences</u> are obtained in ESD and PSD. In Fig.4, ESD-yield ratios for $CO^{\bullet}$ signals from $\sqrt{3}$ layers and layers of 1/2 ML coverage obtained for different electron energies are shown. For 25 eV electron energy, i.e. below the $3\sigma$ ionization threshold, this ratio is lower than for faster electrons by a factor of 1.5, in qualitative agreement with the PSD findings. The

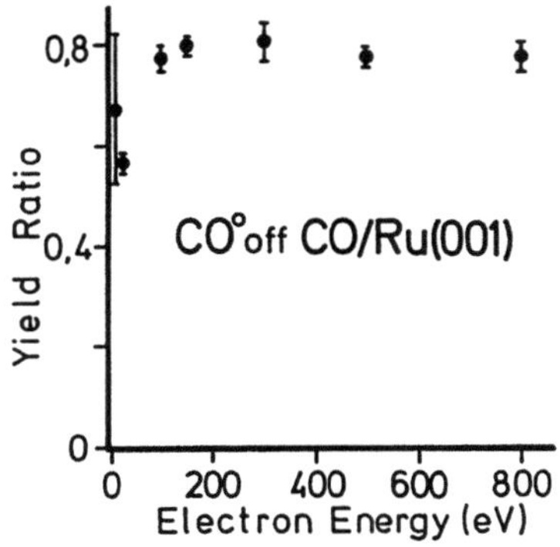


Fig.4. Ratio of $CO^{\bullet}$ ESD signals from 1/3 and 1/2 ML coverage layers of CO/Ru(001) vs. electron energy

linear increase of ESD yield vs. electron energy found in /12/ is obviously due to superposition of different yield curves related to the different primary excitations obtained in PSD which cannot be distinguished in ESD experiments. Interestingly, for electron energies below the $5\sigma^{-2}$ ionization threshold the n-ESD yield seems to depend less strongly on coverage, as for the $3\sigma$ related excitations (see Fig.4; the large error bar is due to the small signal below the $5\sigma^{-2}$ threshold).

In conclusion, we have shown that at least tentative assignments of the primary excitations for CO/Ru(001) in the energy range 12 to 120 eV from PSD data are possible. Roughly, three regimes with different coverage dependences, different branching ratios and localisation probabilities for the electronic excitation, and different kinetic energies of the desorbing particles /16/ related either to excitations of outer valence levels only, to $3\sigma$ ionization, or to $3\sigma$ double ionization can be discerned. Especially for the $3\sigma^{-1}$ states, high localisation probabilities were found. We should mention, however, that there is no simple correlation of localisation probability and coupling to the substrate. For $N_2$/Ru(001) which is weakly chemisorbed, only $N^+$PSD is observed and DIET by $3\sigma^{-2}$ excitation is prevalent /8/; primary core excitation is very effective and shows strong contributions of complex excitations /12/. Systematic investigations of different adsorbates on different substrates are necessary. In view of the complexity of the excitations, our assignments are tentative and theoretical investigations should further clarify the situation.

### 3.2 Argon and Krypton on Ru(001)

Diet of rare gases from mono- and submonolayers was first observed in LEED studies of these layers /17/. ESD of physisorbed adsorbates from metals has since been investigated by several groups (see /3/ and references cited therein). The results have been discussed in terms of the so called ANTONIEWICZ mechanism /18/ (see references cited in /3/). According to this model, the primary process is ionization of the adatom. Because the ionic potential well is deeper than that for the neutral, electronic exitation is transferred into nuclear motion towards the surface. Resonant tunneling from the surface brings the particle back to the repulsive branch of the neutral state. Depending on its kinetic and potential energy at this time, either desorption or recapture occurs /18/. A different mechanism for DIET of rare gases is presented at this conference /4/. This model assumes the excited rare gas atom to behave like an alkali atom and predicts different kinetic desorption energies than the ANTONIEWICZ mechanism, if the known escape probabilites and their dependence on sample temperature are used. DIET from condensed layers of rare gases as model substances for wide gap insulators was a focal point of the discussions at DIET II. Desorption has been linked to formation of either ions or excitons, which are self-trapped as ionimers or excimers, respectively, after a random walk through the solid /19/. The bond distance of these dimers is much smaller than the lattice constant of the rare gas solids (RGS). When these dimers decay radiatively (ionimers have to capture an electron first), evolution of the system along the repulsive branch of the potential curve releases enough energy for desorption, if deexcitation occurs near the surface /19/. Another mechanism for DIET especially for the light rare gases Ne and Ar, has recently been proposed by COLETTI et al. /20/. In this model, DIET occurs via self-trapping of atomic excitons at the surface of the RGS. The resulting lattice stress can relax through expulsion of the excited atoms; luminescence light due to the deexcitation of the free atom is indeed observed /20/( see also /21/).

Figure 5 shows neutral PSD signals from Ar and Kr multilayers of different thickness. These data were recorded at the 1m-Seya beamline at BESSY (see /22/ for details). Narrow maxima are obtained for the excitonic regime for Ar and for Kr. For photon energies higher than the band gap of the RGS, DIET by ionimer formation is obtained /19/. Due to the large diffusion length for ions, this mechanism prevails for thick layers /2,19/. If the photon energy exceeds the sum of gap energy and exciton excitation energy, two step processes (i.e. exciton excitation by photoelectrons) become possible, see Fig. 5. The high resolution data in Fig.6 exhibit considerable structure which can be assigned by comparison with absorption spectra from SAILE et al. /23/. Surface and bulk excitons of $1^{st}$ and $2^{nd}$ order and their splitting can be discerned. Excitons of higher order are observed for thick layers only, in agreement with their large diameters. Small deviations in energy are due to different sample temperatures in both cases /22/. Differences in relative peak intensities must be ascribed to excitation dependent coupling to desorption. Generally, the mechanisms proposed for DIET from RGSs are confirmed by these results. We mention that data of G.ZIMMERER for these layers are in perfect agreement with our results /21/.

For Ar monolayers, DIET after ionization is prevalent, see Fig.7. Compared to gas phase ionization yields, final state effects modulate the desorption signal from the surface. Contributions from excitonic excitations are obtained as well (but not for Kr monolayers, where stronger coupling of the surface exciton excitation to the substrate is expected /24/). Surprisingly, identical excitation energies for the surface exciton are found for mono-

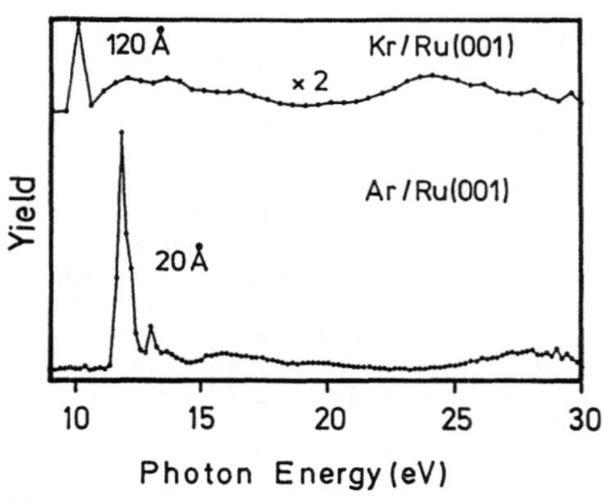


Fig.5. Wide range PSD spectra from Argon and Krypton multilayers of different thickness on Ru(001)

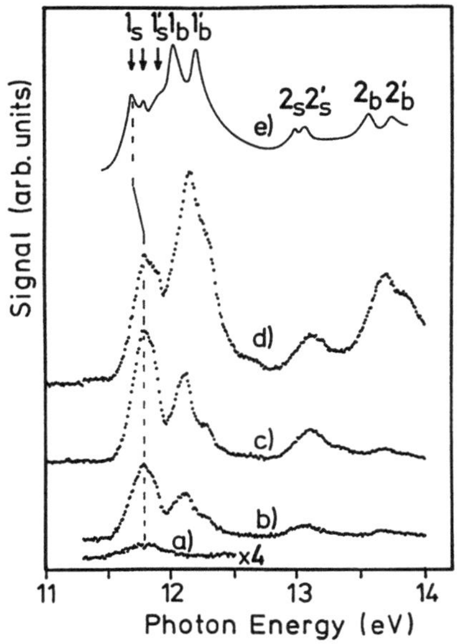


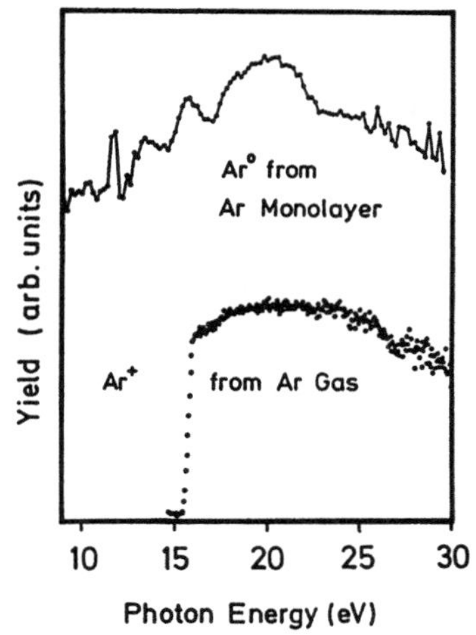


Fig.7. PSD from Ar monolayers compared to $Ar^+$ ion yield from Argon gas

Fig.6. "High" resolution PSD from Ar monolayers (a) and multilayers (b: 15Å, c: 25 Å, d: 60 Å) compared to optical absorption of a 23 Å layer (e; from /23/) (after /22/)

and multilayers, see Fig. 6. An explanation for this could be that the excitation energy depends mainly on next neighbour distances which could be similar in both cases. The exact desorption mechanism for PSD from RG monolayers is still in question, see above. An interesting point is if the desorption process after excitonic excitation differs from that after ionization. Autoionization by resonant tunneling to the metal and subsequent PSD as for primary ionic excitation or DIET after formation of an intermediate, strongly bound complex of the excitonically excited atom and the surface (e.g. due to chemical interaction) which might result in a movement towards the surface as for ions, or mixtures of both processes seem possible. Measurements of kinetic energy distributions of the desorbing particles are desirable to discriminate these pathways /3,4/.

## 4. Acknowledgement

We gratefully acknowledge the cooperation of A.Cassuto, D.Coulman, W.Riedl, R.Treichler, and W.Wurth, and helpful discussions with them and with W.Brenig, A.R.Burns, Z.W.Gortel, H.J.Kreuzer, D.R.Jennison, E.B.Stechel and G.Zimmerer. The help of W.Braun and H.Petersen of BESSY was decisive for the success of the synchrotron experiments. This work has been supported by the Deutsche Forschungsgemeinschaft through Sonderforschungsbereich 128 and by the German Ministry of Research and Technology, BMFT, under Grants 05 237 MZ and 05 366 CAB.

## 5. References

/1/ see, e.g., *Desorption Induced by Electronic Transitions*, DIET-I, edited by N.H.Tolk, M.M.Traum, J.C.Tully and T.E.Madey, (Springer, Berlin 1983)

/2/ D.Menzel, this volume

/3/ Z.W.Gortel, this volume; Z.W.Gortel, H.J.Kreuzer, P.Feulner and D.Menzel, Phys. Rev. B35 (1987) 8951;

/4/ D.R.Jennison, this volume

/5/ A.R.Burns, this volume; A.R.Burns, Phys.Rev.Lett. 55 (1985) 525

/6/ E.B.Stechel, this volume; A.R.Burns, D.R.Jennison and E.B.Stechel, Phys. Rev. Lett. 58 (1987) 250

/7/ R.Treichler, W.Riedl, W.Wurth, P.Feulner and D.Menzel, Phys.Rev.Lett. 54 (1984) 462

/8/ P.Feulner, D.Coulman, A.Puschmann and D.Menzel, in preparation

/9/ P.Feulner, in: *Desorption Induced by Electronic Transitions*, DIET II, edited by W.Brenig and D,Menzel (Springer, Berlin 1985), p.142

/10/ T.E.Madey, R.Stockbauer, S.A.Flodström, J.F.van der Veen, F.J.Himpsel and D.E.Eastman, Phy. Rev. B23 (81) 6847

/11/ T.E.Madey, Surface Sci.79 (1979)575; P. Feulner, H.A. Engelhardt and D. Menzel: Appl. Phys. 15 (1978) 355

/12/ P.Feulner, R.Treichler and D.Menzel: Phys. Rev. B24 (1981) 7427

/13/ W.Wurth, Ph.D. thesis, Technische Universität München 1987

/14/ S.Auer, P.Feulner and D.Menzel, Phys. Rev. B35 (1987) 7752

/15/ P. Feulner, S. Auer, W. Riedl, A. Cassuto and D. Menzel, Surface Sci. 182 (1987) L221

/16/ S.Auer, P.Feulner, T.Müller, A.Puschmann and D.Menzel, in preparation

/17/ H.H.Farrell, M.Strongin, and J.M.Dickey, Phys.Rev. B6 (1972) 4703; J.Unguris, L.W.Bruch, E.R.Moog and M.B.Webb, Surface Sci. 87 (1979) 415; 109 (1981) 522

/18/ P.Antoniewicz, Phys.Rev. B21 (1980) 3811

/19/ in: *Desorption Induced by Electronic Transitions*, DIET II, edited by W.Brenig and D.Menzel (Springer, Berlin 1985): J.Schou, O.Ellegaard, P.Borgesen and H.Sorensen, p. 170; H.Haberland, p. 177; R.Pedrys, D.J.Oostra and A.E.de Vries, p. 190; W.L.Brown, C.T.Reimann and R.E.Johnson, p.199

/20/ F. Coletti, J.M. Debever and G. Zimmerer, J. Physique Lett. 45 (1984) L467

/21/ G. Zimmerer, this volume

/22/ P. Feulner, T. Müller, A. Puschmann and D. Menzel, Phys. Rev. Lett., in press

/23/ V.Saile, M.Skibowski, W.Steinmann, P.Gürteler, E.E.Koch and A.Kozevnikov, Phys.Rev.Lett. 37 (1976) 305

/24/ M.Tsukada and W.Brenig, Surface Sci. 151 (1985) 503

# Rovibrational Laser Spectroscopy of ESD Neutrals from Chemisorbed Species

*A.R. Burns, E.B. Stechel, and D.R. Jennison*

Sandia National Laboratories, Albuquerque, NM 87185, USA

## 1. Introduction

Ever since the introduction of intense tunable laser radiation, the fields of molecular spectroscopy and dynamics have expanded tremendously. Indeed, with each technical improvement in laser resolution and output power, new applications have emerged with correspondingly increased levels of sophistication and information. In the field of electron-stimulated desorption (ESD), not only has the utilization of lasers resulted in the detection of neutral species [1-5], which were previously difficult to observe without specially designed analyzers [6], but also the quantum-specific nature of the resonant laser interaction with the neutrals has yielded valuable information concerning internal energies [2-5]. In this article, we will discuss some of the experimental methods in the application of laser resonance-ionization spectroscopy (RIS) to the study of the ESD of neutrals, in particular, chemisorbed NO and CO desorption from Pt(111). We will also show how the detailed information obtained in these experiments has identified a new desorption mechanism.

State-selectivity and sensitivity are the primary advantages of RIS. The pulsed output of the high intensity laser also selects, via the time-of-flight (TOF) of the neutrals from the surface, a range of translational energies of interest. Thus one has the rotational, vibrational, and translational detection specificity required for a detailed comparison to the multidimensional dynamics calculations discussed in this volume [7]. The high sensitivity of ion detection is well suited for measuring desorption thresholds, which are required in the determination of excitation mechanisms. It is also important at low adsorbate coverages, where yields reveal probabilities of desorption and excitation localization [7,8].

## 2. Resonance ionization spectroscopy (RIS) of NO and CO

As shown in Fig. 1, the first step in any RIS scheme is the resonant transition between bound electronic states. In the second step, the absorption of another photon induces photoionization of the excited-state population. Once ionized, the desorbed molecules can be detected with the same high efficiencies found in conventional ion mass spectrometry. As in the case of CO, some neutral species have high-lying excited states and/or ionization potentials which require multiphoton transitions. Rotational structure in the two bound electronic states of the first step can be resolved as a function of rotational quantum-number J ($\Delta J = -1,0,+1$) [9] for each vibrational level $\nu$.

There are many possible electronic transitions in NO and CO that can be resonantly excited. The following are transitions that are reasonably strong and are accessible to the usual tuning range of commercial lasers (215-700 nm). The molecule NO is routinely photoionized in the (1+1)-photon scheme involving the $A^2\Sigma^+(\nu') \leftarrow X^2\Pi_{1/2,3/2}(\nu'')$ transition [9], where $\nu''$ is

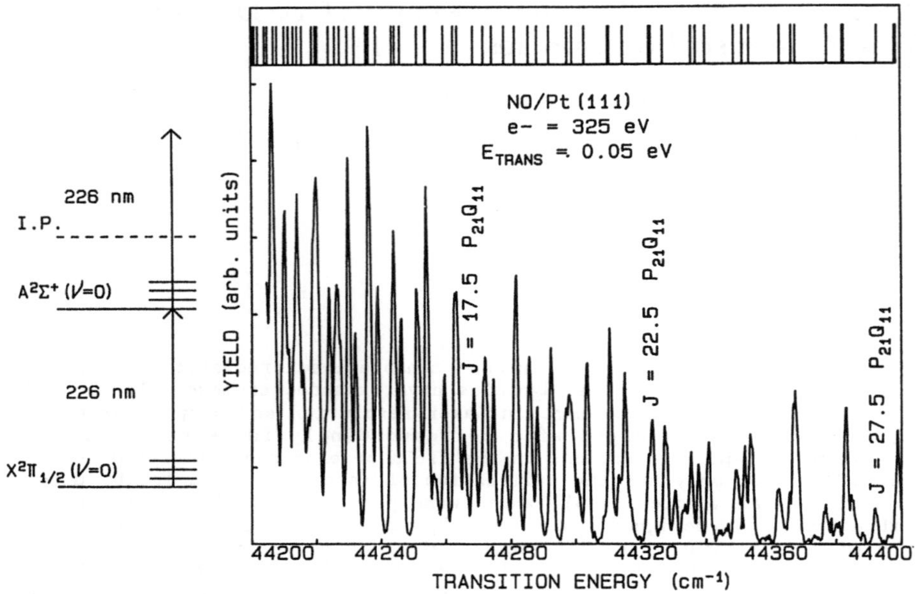


Fig. 1 (Left): RIS scheme for NO. (Right): Rotational spectrum of NO.

the vibrational level to be probed, and $\nu'$ is selected to maximize the signal, given that the Franck-Condon factors are well known [10]. A part of the $\nu$=0 rotational spectrum of this transition for desorbed NO is shown in Fig. 1. Ground-state CO is detected in a (2+2)-photon scheme involving the $A^1\Pi(\nu=1) \leftarrow X^1\Sigma^+(\nu=0)$ transition [11]; and metastable $a^3\Pi$ $CO^*$ is detected in a (1+1)-photon scheme involving the transition $b^3\Sigma^+(\nu=1) \leftarrow a^3\Pi(\nu=2)$ [9].

## 3. Experimental Apparatus

A TOF laser ionization spectrometer that incorporates most of the capabilities required for the analysis of neutrals desorbed by electron impact is shown in Fig. 2. The basic features are similar in concept to those used in the RIS detection of sputtered atoms [12], and include the following: (1) A 4 nsec pulse-width tunable laser beam that is focussed by a cylindrical lens into a ribbon-like ionization region (~1 $cm^2$ square, 0.01 cm thick), parallel to the surface at a fixed distance (0.5 cm); (2) a pulsed electron gun which is oriented at 45° with respect to the surface; (3) grids for acceleration of the photoionized species into a 10 cm flight tube; and (4) a multichannel plate (MCP) array. The wide diameter (1.0 inch) of the detector results in a high ion collection efficiency. Overall, the sensitivity of the apparatus approaches neutral densities of $10^6$ $cm^{-3}$. This density in the laser beam corresponds to an ESD cross-section of ~$10^{-19}$ $cm^2$, assuming that the velocity of the neutrals is ~$10^5$ cm/sec.

The computer control of the experimental parameters and data acquisition is summarized in Fig. 3. In the TOF measurement of translational energies, the time delay between the electron pulse and the laser pulse is produced by a programmed delay generator. As will be shown below, the delay times can be programmed to collect the data in a step-wise fashion that is linear in energy. In the threshold experiments, the voltages which maintain the energy and focus of the electron gun are scanned together in order to maintain a constant spot size (2 mm) on the surface for the electron energy

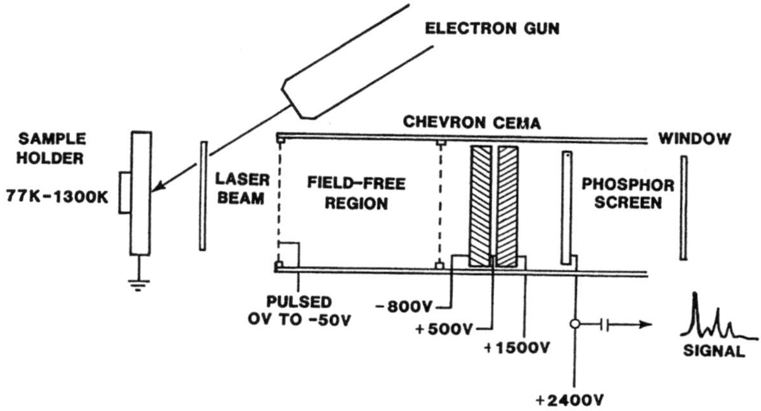


Fig. 2 Experimental geometry for RIS detection of ESD neutrals.

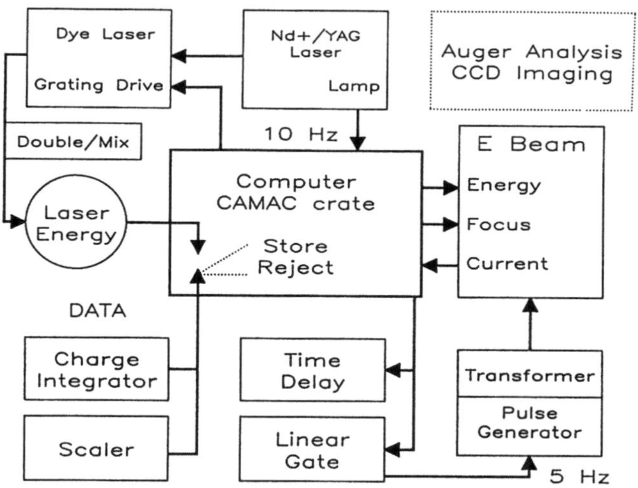


Fig. 3 Computer control of desorption experiments

range 20-1000 eV. Below 20 eV, the spot size begins to increase due to the difficulty of focussing low-energy electrons; however, the large focal region of the laser insures that no signal is lost. The beam current is monitored at the sample surface, while the spot size is monitored with a Faraday cup which can be rotated into the sample plane. When rotational spectra are required, the wavelength of the dye laser is scanned by stepping the grating drive.

The tunable uv laser energy used in the NO experiments discussed below is generated with the use of two angle-tuned harmonic crystals. The $Nd^+$:YAG-pumped dye laser light (~50 mJ) is first doubled to produce 8-12 mJ of 280-340 nm light, and is then mixed with the 1064-nm fundamental of the $Nd^+$:YAG laser to produce ~1-2 mJ of 220-260-nm light. For the CO experiments, only the doubled dye laser light was needed. In any case, the harmonics are separated by a Pellin-Broca prism which causes the beam to move as the laser is tuned. Thus the uv beam that goes into the vacuum chamber must pass through other prisms to counteract the beam movement.

In all of the computer-controlled experiments discussed above, the uv laser energy is integrated and digitized along with the ion signal. The non-linear nature of the RIS technique precludes a simple normalization of the data with laser energy [13]. One way of insuring a consistent signal level during data collection is to reject any data point that is obtained when the laser energy deviates by more that a few percent (e.g. 10%) from a preset level. Other methods require extensive signal calibration procedures that can be time consuming without the use of high-speed computers [13].

There are two background signals that must be minimized and accounted for: (1) the ions coming from stimulated processes on the surface; and (2) the residual gas pressure in the chamber (e.g., NO). Ions leaving the surface coincident with the neutrals can be swept out of the laser ionization region by a 100-200 nsec HV pulse applied to the front grid of the spectrometer during the time delay between the electron beam pulse and the laser pulse. In this way, the region can be kept field free until after the electron beam pulse is fired. Another long, lower-voltage, pulse is applied to the same grid just before the laser pulse in order to have an acceleration field for the laser-ionized species. The gas background is important only for ground-state, $\nu$=0 signals. In these cases, the electron beam pulse is disabled by a linear gate every other "shot." The laser signal obtained at this time in then subtracted from the enabled shot and stored in memory. The gas background is easily minimized ($<10^{-10}$ torr) by the use of a dosing tube in front of the sample when the clean sample is exposed to the gas.

## 4. Translational Energies

The translational energy distributions for desorbed NO, CO and CO* are shown in Fig. 4. These are really TOF distributions plotted on an energy scale. They are obtained by varying the time delay between a 250 nsec electron beam pulse and the 4 nsec laser pulse, with the laser 0.5 cm away from the surface. The relative intensity of the signal as the delay time is varied is a complicated function of the pulse widths, the velocities, and the angular spread of the neutrals. When comparing the data to theoretical calculations [7], it is more reasonable to convolute the theory to correspond to the data rather than to deconvolute the data.

The $\nu$=0 and 1 vibrational levels of NO have TOF distributions peaked at ~0.05 eV. As the laser is tuned to the $\nu$=2 and 3 levels, a second desorption channel peaked at ~0.3 eV is observed. This is the first time that vibrational specificity has been used to distinguish two ESD channels. In contrast, CO and CO* display no low energy channel, but instead exhibit a single peak at ~0.5 and ~1.0 eV, respectively. This indicates a significant difference in the dynamics of CO and NO desorption from Pt(111) [7].

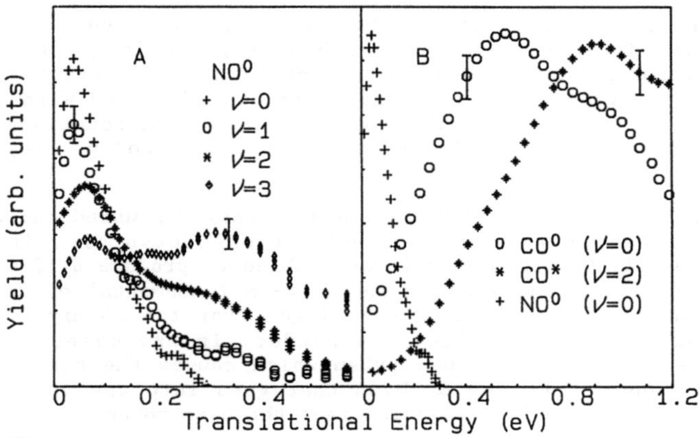


Fig. 4 TOF distributions for NO and CO desorbed from Pt(111) by 315 eV electrons. In (A), the NO vibrational levels are not relative (see text). In (B), the CO ground-state and metastable $a^3\Pi$ CO* distributions are compared with NO. The CO vibrational populations are unknown. From [4].

## 5. Thresholds

The thresholds of the NO and CO desorption channels discussed above are shown in Fig. 5. Threshold data require an increase in sensitivity due to the reduced excitation cross-sections and beam currents at low electron energies. Thus the electron beam pulse width is increased to ~800 nsec, resulting in a larger "window" ( e.g., 0.045-0.055 eV), of selected translational energies.

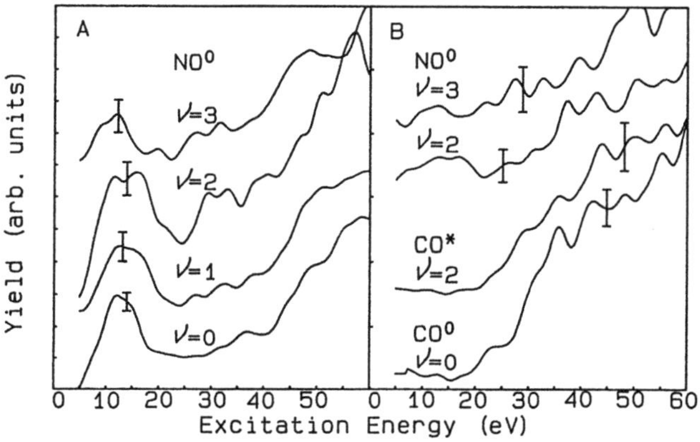


Fig. 5 (A) Threshold region for "slow" NO ($E_{trans}$ = 0.045-0.055 eV). (B) Threshold region for "fast" NO ($E_{trans}$ = 0.3-0.5 eV), ground-state CO ($E_{trans}$ = 0.28-0.46 eV), and metastable CO* ($E_{trans}$ =0.91-2.5 eV). The excitation energy is relative to the Fermi Level. From [4].

The ~6 eV threshold, together with the broad resonance at ~13-14 eV, suggests that the "low" energy NO channel is due to the creation of a single hole, either in the $1\pi$ or the $5\sigma$ levels. Lifetime calculations presented elsewhere [8] indicate that the $5\sigma$ hole is much longer lived and thus has a much greater probability for causing desorption. The higher thresholds (~20 eV) for the "high" energy NO and CO channels suggest multiple hole formation.

## 6. Internal Energy Distributions for "Low" Energy NO Channel

Relative vibrational populations are obtained by summing over the rotational distributions for each level. For the NO $\nu$=0,1,2,and 3 levels, at $E_{trans}$ = 0.045-0.055 eV and 325 eV excitation energy, the results are 1.00, 58, 0.26, 0.43, respectively. These non-Boltzmann populations reflect the $2\pi$ screening of the $5\sigma$ hole in the excited molecule [7,8]. When plotted in the usual manner, ln(I(J)/(2J+1)) vs. rotational energy, the rotational "temperatures" for each of these vibrational levels were 543, 581, 642, and 481 K (± 5%), respectively. It will be shown elsewhere [7] that the rotational energy of the desorbed NO molecule is derived in large part from the zero-point energy of the rotationally-hindered ground-state molecule.

## 7. Summary

We have discussed the application of laser RIS to the problem of understanding the ESD of neutral molecules. With appropriate laser wavelengths and fluences, most atoms and molecules of importance to the understanding of ESD can be analyzed with the techniques presented here.

We have obtained detailed data on both the translational and internal energies, as well as the state-selective thresholds, for NO and CO desorption from Pt(111). The interpretation of the results is discussed in detail elsewhere in this volume [7,8], but may be briefly summarized as follows.

The $5\sigma$ hole that is created in the NO and CO molecules is immediately screened by an extra electron donated from the metal in the $2\pi$ orbital of the excited molecule. The presence of two $2\pi$ electrons in NO weakens the surface bond and frees the hindered rotational motion. The unhindered rotational motion allows the excited NO to access weakly bound portions of the ground-state, resulting in desorption with the "low" 0.05 eV translational energy and the observed rotational distribution. The same physics is not observed for the CO molecule because the presence of only one $2\pi$ electron does not free the hindered rotor. This greatly reduces the probability of CO desorption via the "low" energy channel.

Acknowledgments

This work was performed at Sandia National Laboratories, supported by the U.S. Department of Energy under Contract DC-AC04-76DP00789.

References

1. N. H. Tolk, et al., in Desorption Induced by Electronic Transitions, DIET II, edited by W. Brenig and D. Menzel (Springer-Verlag, New York, 1985) pp. 152-159.
2. A. R. Burns, Phys. Rev. Lett. 55, 525 (1985).
3. A. R. Burns, J. Vac. Sci. Technol. A4, 1499 (1986)
4. A. R. Burns, E. B. Stechel, and D. R. Jennison, Phys. Rev. Lett. 58, 250 (1987).
5. A. R. Burns, D. R. Jennison, and E. B. Stechel, J. Vac. Sci. Technol. A xx, xxx (1987).
6. P. Feulner, in Ref. 1, pp. 142-151.
7. E. B. Stechel, D. R. Jennison, and A. R. Burns, this volume.
8. D. R. Jennison, E. B. Stechel, and A. R. Burns, this volume.
9. Gerhard Herzberg, Spectra of Diatomic Molecules 2nd ed., (Van Nostrand Reinhold Co., New York, 1950), p. 264, 558.
10. Spectroscopic Data, v. 1, edited by S. N. Suchard (IFI/Plenum, New York, 1975), pp. 308-347 (CO) and pp. 720-743 (NO).
11. Wanda R. Ferrell, C. H. Chen, M. G. Payne, and R. D. Willis, Chem. Phys. Lett. 97, 460 (1983).
12. Paul H. Kobrin, G. Alan Schick, James P. Baxter, and Nicholas Winograd, Rev. Sci. Instrum. 57, 1354 (1986).
13. D. C. Jacobs, R. J. Madix, and R. N. Zare, J. Chem. Phys. 85, 5469 (1986).

# Electron Stimulated Desorption of Excited Hydrogen Atoms from an Alkali Promoted Surface

*P.D. Johnson*[1], *X. Pan*[1], *J. Tranquada*[1], *S.L. Hulbert*[2], *and E. Johnson*[2]

[1]Physics Department, Brookhaven National Laboratory
Upton, NY 11973, USA
[2]National Synchrotron Light Source,
Brookhaven National Laboratory, Upton, NY 11973, USA

We observe directly the radiation emitted from excited hydrogen atoms following their electron stimulated desorption from an alkali-promoted Ni(001) surface. We observe Lyman α, β, γ radiation. The line width and energy of the emitted radiation indicate that the hydrogen atoms are well removed from the metal surface at the point of deexcitation. By comparing the desorption cross-section for these states with the desorption cross-section of hydrogen ions we find that the excited hydrogen atoms are emitted directly from the surface rather than resulting from the reneutralization of desorbed ions. The intensity of Lyman α radiation is over an order of magnitude greater than the intensity of Lyman β and γ radiation suggesting either a higher probability for excitation into the hydrogen 2p state initially or a higher probability of resonance ionization of the higher excited states.

## 1. Introduction

The desorption of ions and neutral atoms from solid surfaces through electron- or photon-induced electronic transitions remains a subject of considerable interest [1]. Because of their ease of detection most experimental and theoretical studies have concentrated on the desorption of ions.

Menzel and Gomer [2] and independently Redhead [3] (MGR) have suggested that ion desorption occurs via a one-electron transition to a "repulsive" final state. The excited atom subsequently moves away from the surface with kinetic energy equal to the potential drop on the repulsive curve. On the other hand, Knotek and Feibelman [4] have suggested that the Auger decay of appropriate core holes producing multiple-hole final states may also lead to desorbed ions through coulomb repulsion. Of course for both mechanisms the possibility of observing such final states is influenced or determined by the probability of survival in these states.

However, whilst most studies are of desorbed ions it is known that most electronically desorbed particles are in a neutral state when far removed from the surface. There have been several studies of these neutral particles both in the ground state [5] and in the various excited states [6]. Detection techniques have included both time-of-flight measurements [7] and optical techniques using laser excitation [6]. An 0.3 m spectrometer has also been used to detect the Balmer radiation emitted from hydrogen atoms desorbed from alkali-halide surfaces [6].

In this paper we report the observation of Lyman radiation emitted from excited hydrogen atoms desorbed from an alkali-promoted Ni(001) surface. Because of their simplicity a study of the desorption of the excited states of hydrogen may serve as a model for other systems. Using a 1.0 m spectrometer we are able to measure the intensity of radiation as a function of incident electron energy. We find that the different excited states show different desorption thresholds and further preliminary ion desorption measurements from the same surface indicate that the excited atoms do not necessarily result from the reneutralization of emitted ions. We find that the relative intensity of Lyman $\alpha$ to the higher excited states, Lyman $\beta$ and Lyman $\gamma$, suggests that either the 2p state is preferentially filled in the initial excitation or the higher excited states have a lower survival rate.

## 2. Experimental

The experiments reported in this paper were carried out in a typical surface science chamber having a background pressure of the order of 2 - 3 x $10^{-10}$ $\tau$. LEED and Auger were used to characterize the surfaces under investigation. The photons emitted from the desorbed excited atoms were detected with a grating spectrograph originally designed for Inverse Photoemission studies [8]. The low energy electron source used for these desorption studies has been described in detail elsewhere [9]. Desorbed ions were detected using a time-of-flight detector that has been described in detail elsewhere in these proceedings [10].

Following cleaning cycles of argon bombardment and annealing, a Ni(001) crystal was exposed to either potassium or sodium evaporated from SAES getters. The coverage of alkali atoms ranged from 0·25 to 0·5 monolayers. The sample then received a typical exposure to hydrogen of approximately 5L at room temperature. We would emphasize, however, that it was not necessary to subsequently expose the sample to hydrogen to allow the observation of the Lyman series. Indeed it would appear that the residual hydrogen pressure in the chamber above ($\leq$ 1 x $10^{-10}$ $\tau$) was sufficient to allow this observation. Indeed such an observation suggests that the presence of the alkali metal increases the sticking coefficient of hydrogen.

## 3. Experimental Results

Figure 1 shows a spectrum of the radiation emitted from Ni(001) surface following the evaporation of sodium with no subsequent exposure to hydrogen. Sharp lines corresponding to Lyman $\alpha$, $\beta$ and at this particular energy (30 eV), a very small $\gamma$ radiation are observed. The width of these lines is determined solely by the wavelength resolution of the spectrograph and there is no indication, within that resolution, of any broadening due to the proximity of the surface. Thus a picture emerges of excited hydrogen atoms leaving the surface and subsequently making radiative transitions. However the fact that there is no observable shift in the measured energies of these discrete lines from the standard Lyman series suggests that the radiative transitions must occur within a very short distance (< 1 mm) from the source point of the spectrograph. Transitions occurring outside of this distance would result in an overall shift of the image and a concomitant loss of energy calibration.

In Fig. 2 we show a representative plot of the intensity of these lines as a function of the incident electron beam energy. We observe that each line shows a discrete threshold separated from the other lines. We note that these thresholds move to lower energies if the surface is

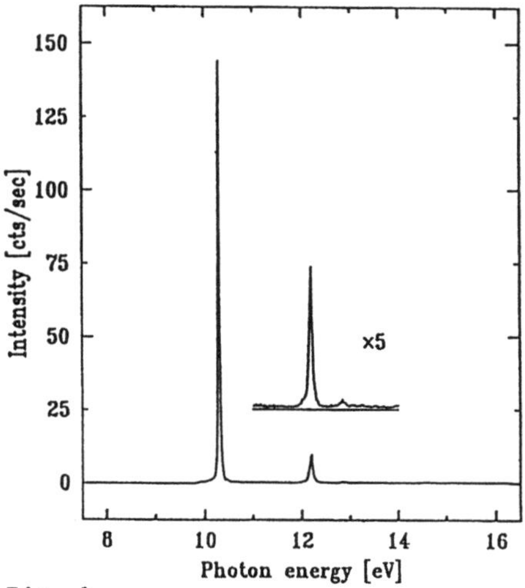


Fig. 1

Lyman α (10·2 eV) β (12·1 eV) and γ (12·75 eV) emitted from Ni(001) following evaporation of sodium. The incident electron beam energy is 30 eV with respect to the Fermi level $E_F$.

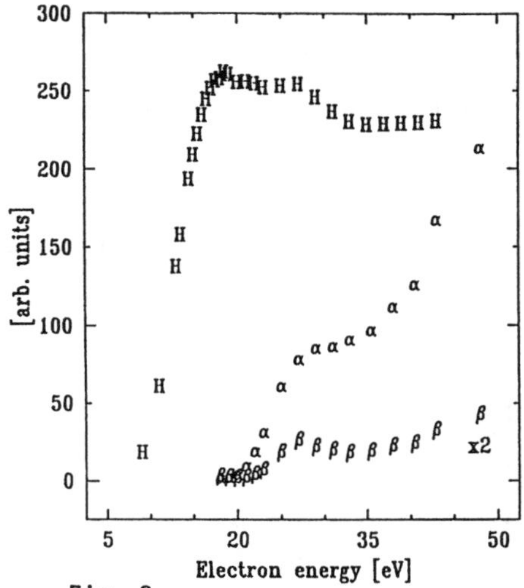


Fig. 2

The intensity of hydrogen ions (H) and Lyman α (α) and Lyman (β) emitted as a function of incident electron beam energy with respect to the Fermi level. The intensity of ions and excited states have not been normalized to each other.

exposed to hydrogen. Both the α and the β state show an increase in intensity above 30 eV, the threshold for Na 2p excitation. This would indicate the possibility of a Knotek-Feibelman-type mechanism contributing to the desorption signal. Very little difference is observed in either Fig. 1 or 2 if the surface is predosed with potassium rather than sodium.

We show in Fig. 2 the desorbed hydrogen ion yield from the same surface as a function of incident electron beam energy. The threshold for the desorption of ions from this surface, 10 eV, is considerably lower than was previously measured for a clean Ni(111) surface, 23 eV.

## 4. Discussion

The thresholds for the observation of the excited states following exposure to hydrogen, 14.0 eV for Lyman α and 20 eV for Lyman β, clearly differ from the equivalent thresholds for gas phase hydrogen, confirming that the excitation mechanism occurs at the surface. However the further observation that these thresholds differ for different excited states suggests that the desorption channels are not related. Thus we believe the excited states do not arise through the reneutralization of desorbed ions. Conversely our measurement of a lower threshold energy for the desorbed ions suggests that the latter do not originate through the resonance ionisation of desorbed excited states. However, even though the different channels, ions and excited atoms, appear to exhibit their own desorption thresholds we anticipate that, if the excited level, (2p, 3p, etc.) is appropriately located with respect to the Fermi level, there will be resonance transitions between the two. The observation that the excited states exhibit different desorption thresholds leads naturally to a picture of one-electron transitions to repulsive final states, i.e., a model of the type suggested by Menzel, Gomer and Redhead [2,3]. However at higher energies it would appear that desorption is also possible via a Knotek-Feibelman-type mechanism following excitation of the alkali atom core levels.

Figure 2 shows that above the threshold energies the intensity of each excited state rises up to some plateau or equilibrium level. It is of interest to ask how these equilibrium values relate to the transition rates calculated for transitions from the respective excited states to the 1s ground state. Bethe and Salpeter [11] in their classic work have tabulated the intensities to be expected from these transitions. In particular they assume that one electron arrives in the excited state, $|n,l\rangle$ per unit of time (for the benefit of our discussion either through the initial excitation or through cascade processes from higher levels) and thus the intensity I of a given transition is proportional to the product of the transition rate for that transition and the lifetime of the excited state. If $A_{n\ell}^{n'\ell'}$ is the transition rate between the state $|n\ell\rangle$ and the state $|n'\ell'\rangle$ then

$$I \alpha \frac{A_{n\ell}^{n'\ell'}}{\sum_{n'\ell'} A_{n\ell}^{n'\ell'}}$$

where $1/\sum_{n'\ell'} A_{n\ell}^{n'\ell'}$ is the lifetime of the state $|n\ell\rangle$.

For equal occupation rates of the excited 2p 3p and 4p states the different line in our measured Lyman series should have the ratio 1·18: 1·03: 1·00 for α: β: γ respectively. Clearly the spectrum in Figure 1 does not reflect such ratios for the different lines. We therefore conclude that the higher intensity of Lyman α radiation in that spectrum is either the result of a higher initial excitation probability for that excited state or a higher survival rate for the hydrogen in that excited state.

As the initial excitation mechanism is not known it is not possible to say whether excitation to the 2p state would dominate over excitation to the higher excited states. However we note in passing that dipole excitation very definitely favors transitions to the lower excited states [11]. Following desorption the principle non-radiative mechanisms reducing the numbers of atoms in the excited states are Auger deexcitation and Resonance ionisation [12]. It seems unlikely that the Auger mechanism would necessarily favor the survival of atoms in the 2p state and we therefore concentrate our discussion on the Resonance mechanisms.

Resonance ionisation of an excited state requires the excited electron to be at a level above the Fermi level. This will always be true of the higher excited states but not necessarily true of the 2p state particularly in the presence of the alkali atoms. It is well established that saturation coverage of the alkali metals reduces the work function by approximately 3·0 eV. Thus at large distances from the surface the 2p state will be below the Fermi level. However we note that near the surface all the excited levels will be shifted to lower binding energies through relaxation effects by an amount approximately equal to the image shift. Indeed with such a shift the higher excited states will be in the continuum when close to the surface. We are currently investigating the role of the Fermi level and the Resonance processes in greater detail.

## 5. Conclusions

Our studies of the desorption of excited hydrogen atoms from an alkali promoted surface show that desorption occurs independently of the mechanism leading to the production of positive ions. It would appear that desorption is possible both via one - electron transitions characteristic of the MGR model and also Knotek-Feibelman transitions following ionisation of the alkali atom core levels. Lyman α is the dominant line in the deexcitation spectrum but it is uncertain whether this reflects preferential excitation into the 2p state initially or a higher survival rate for this state. Finally we again note that in the presence of the alkali atoms the hydrogen atoms appear to have a higher sticking coefficient than on the clean surface.

## 6. References


1. These published proceedings represent the third volume in a series on Desorption Induced by Electronic Transitions (DIET).
2. D. Menzel and R. Gomer, J. Chem. Phys. 41, 3311 (1964).
3. P. A. Redhead, Can. J. Phys. 42, 886 (1964).
4. M. L. Knotek and P. J. Feibelman, Phys. Rev. Lett. 40, 964 (1978).
5. P. Feulner, W. Riedl and D. Menzel, Phys. Rev. Lett. 50, 986 (1983).
6. N. H. Tolk, R. F. Haglund, Jr., M. H. Mendenhall, E. Taglauer and N. G. Stoffel: in DIET II, edited by W. Brenig and D. Menzel, Springer-Verlag, Berlin, 1985.
7. H. Overeijnder, R. R. Tol and A. E. deVries, Surf. Sci. 90, 265 (1979).
8. P. D. Johnson, S. L. Hulbert, R. F. Garrett and M. R. Howells, Rev. Sci. Inst. 57, 1324 (1986).
9. N. G. Stoffel and P. D. Johnson, Nucl. Inst. and Meth. A 234, 230 (1985).
10. E. D. Johnson, R. F. Garrett, M. L. Knotek and F. Sette, in these proceedings.
11. H. A. Bethe and E. E. Salpeter, "Quantum Mechanics of One and Two-Electron Atoms", Springer-Verlag, Berlin (1957).
12. H. D. Hagstrum, Phys. Rev. 96, 336 (1954).



Acknowledgement

Work supported by the Division of Materials Sciences, U.S. Department of Energy, under Contract No. DE-AC02-76CH00016.

# Dissociative Attachment in ESD from Condensed Molecules

*L. Sanche*

MRC Group in Radiation Sciences, Faculty of Medicine, University of Sherbrooke, Quebec, Canada J1H 5N4

## 1. INTRODUCTION

In recent years, electron stimulated desorption (ESD) of negative ions has been observed for a number of adsorbate/substrate systems and mechanisms have been proposed to explain the origin of the signal [1]. ESD experiments performed with physisorbed molecules have been particularly helpful in demonstrating [2,3] the presence of a dissociative attachment (DA) mechanism responsible for the production of negative ions below the threshold of dipolar dissociation. The DA process constitutes a particular channel for the decay of transient negative ions formed at or near solid surfaces. For such an ion formed by a low-energy electron (1-30 eV) impinging on a physisorbed molecule AB, the main decay channels may be represented by reactions (1) to (3):

$$e^- + AB \rightarrow AB^- \rightarrow AB + e^- \quad (1)$$
$$\rightarrow AB^* + e^- \quad (2)$$
$$\rightarrow A + B^- \quad , \quad (3)$$

(1) and (2) being the elastic and inelastic channels, and (3) the DA channel. Experimentally, the $B^-$ signal is recorded as a function of the energy of electrons impinging on molecules AB condensed on a clean polycrystalline metal substrate held at cryogenic temperatures. The ESD yield of these anions and their kinetic energy (KE) distributions can be measured for submonolayer, single layer and multilayer coverages of the substrate. It is also possible to deposit spacer rare gas layers between the substrate and the molecules under investigation to reduce the effects of the metal on the molecular process.

The aim of this chapter is to provide a comprehensive summary of the most recent ESD results on DA from condensed (i.e., physisorbed) molecules. These results are discussed with particular emphasis on the microscopic description of the sequence of steps leading to anion emergence in vacuum.

## 2. EXPERIMENT

The ESD apparatus [4] consists esentially of an electron gun, a mass spectrometer (MS) and a closed-cycle refrigerated cryostat. These components are housed in an ultra-high-vacuum (UHV) system reaching working pressures in the $10^{-11}$ torr range. The condensed films are grown on a polycrystalline platinum ribbon press fitted to the cold tip of the cryostat and isolated electrically from this latter by a ceramic sheet. With this arrangement, it is possible to measure the electron current at the substrate and clean the ribbon by resistive heating. Gases or vapors initially prepared in a UHV manifold are introduced into the system through a valve connected to a tube having an opening located in front of the metal substrate.

The electron beam is incident at 40° from the target surface. It has an energy resolution of 0.3 eV full width at half maximum (FWHM) and produces at the sample a spot of about a 2 mm$^2$ area above 2 eV. A portion of the negative ions desorbed from the film are focused, by ion lens at the entrance of a quadrupole MS whose axis is oriented at 70° with respect to the surface. Three grids of about 60% transparency are placed in front of the MS in order to analyse the ion energies by the retarding potential method. The signal is stored in a computer as a function of electron energy or as a function of the retarding potential. The first energy derivative of this latter recording provides the energy distribution of the anions. The apparatus can therefore be operated in two modes: the ion-yield mode in which mass selected and energy discriminated anions are measured as a function of electron energy and the ion-energy mode in which anions of a selected mass produced at a fixed electron energy are measured as a function of the ion energy. An energy resolution of 0.5 eV FWHM can be achieved in the ion-energy mode but it is not possible to determine the absolute energies of the ions.

The absolute electron energy, the film thickness and charge accumulation in a condensed film can be determined by measuring at the metal substrate the electron energy dependence of the current intensity transmitted through the film. The low-energy electron transmission (LEET) spectrum thus obtained contains several features which can be related to the band structure of the film and electronic energy losses [5,6]. At low coverages, LEET spectra produce oscillatory structures which are related to the number of layers forming the film via the quantum interference of the electron wave between the substrate-film and film-vacuum boundaries [7]. These quantum size effects (QSE) can be utilized to calibrate a precise amount of gas leaked into vacuum near the substrate in terms of a given number of condensed layers. In turn, this calibration can serve to estimate the thickness at higher coverages or in materials where QSE are not observed. The lowest-energy feature in LEET is due to electrons having just enough energy to enter the film and defines, for a low-intensity monochromatic electron beam, the zero of electron energy with respect to the vacuum level [8]. Any shift in the energy of this feature during an experiment is due to charge accumulation in the film [8].

In the present experiments, the absolute energy scale for impinging electrons was calibrated on transmission features whose energies are known within ±0.1 eV from high resolution LEET spectra and also by measuring the onset of electron transmission through the films. The overall accuracy varies from ±0.3 eV to ±0.5 eV depending on the incident current (1-10 nA) and charge accumulation in the film during the experiment. The films were grown at 20 K. Their thickness could be estimated within 10% to a factor of two depending on the growth properties of the condensed layers.

## 3. RESULTS

The electron-energy dependence of the $O^-$ ion yields desorbing from a 3-layer $O_2$ [9], a 6-layer CO [2] and a 4-layer NO [3] film is shown in Figs. 1 to 3, respectively.

Each structure in these curves is due to the formation of a transient anion (i.e., $O_2^-$, $CO^-$ and $N_2O_2^-$, respectively) occurring at a particular electron energy. These transitory states decay into their respective dissociation limits (i.e., $O^-$ + O, $O^-$ +C and $O^-$ + $N_2O$). In the top curve of Fig. 2, the 14-eV peak arises from $CO^-$ decay into the limit $C^-$ + O. Since NO dimerizes when condensed, $N_2O_2^-$ is considered to be the anion responsible for the four features in the bottom curves of Fig. 3. The upper curve in Fig. 3 is the gas-phase DA cross section [10]. In all three figures the rising background on the high-energy side is due to multiple scattering and dipolar dissociation (i.e., $e^- + AB \rightarrow AB^* + e^- \rightarrow A^+ + B^- + e^-$). When compared to the gas-phase total DA yield [10,11] the curves in Figs. 1 to 3 exhibit more features.

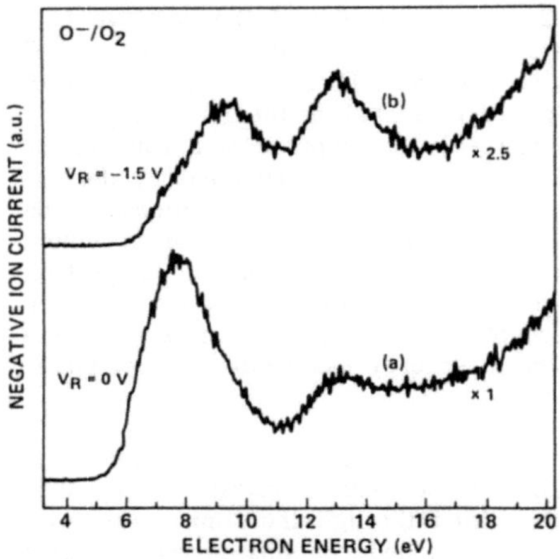


Fig. 1

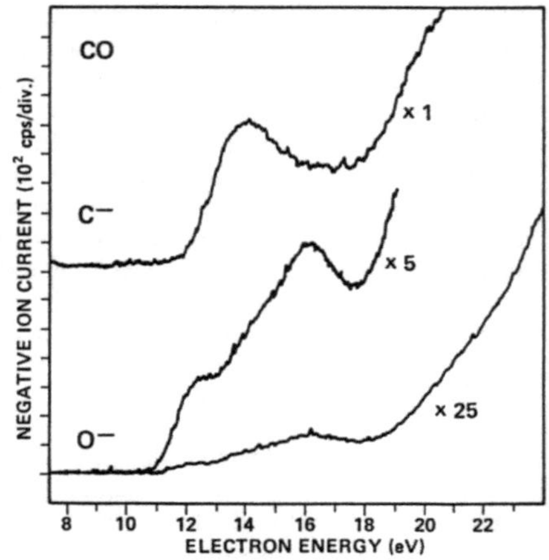


Fig. 2

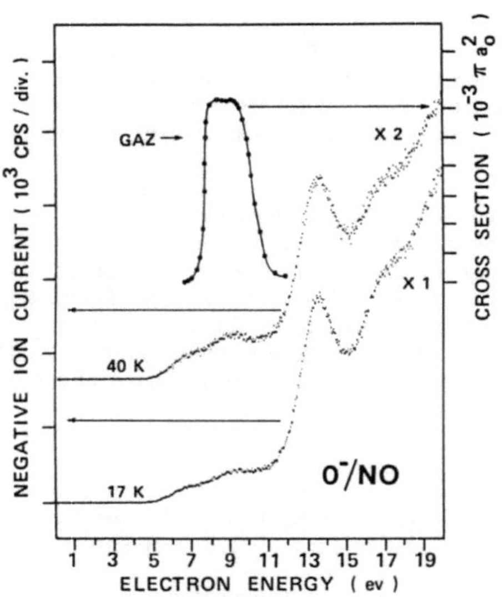


Fig. 3

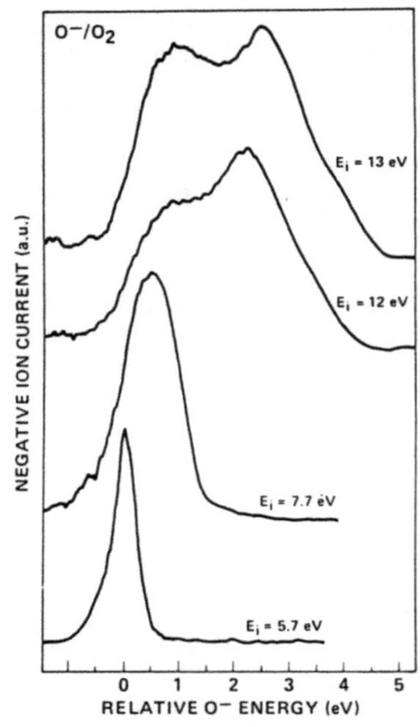


Fig. 4

The lower curve in Fig. 1 was recorded [9] with a potential $V_R = 0$ volt on the retarding grids whereas in the upper one this potential was set at -1.5 eV to repel low-energy ions. The difference between these two curves indicates that the 13-eV peak contains more high-energy ions than the one at 8 eV. This latter has been ascribed [2] to the DA reaction

$$e + O_2,\ {}^3\Sigma_g^- \rightarrow O_2^-,\ {}^2\Pi_u \rightarrow O^-({}^2P) + O({}^3P)\,.$$

The 13-eV peak arises from the decay of $O_2^-$ states $^2\Sigma_g^+$ and/or $^2\Sigma_u^+$ which are forbidden by symmetry rules in the gas phase. The origin of these $O_2^-$ states has been postulated from the analysis of the kinetic energy distribution of $O^-$ ions for specific electron-impact energies $E_i$ [9] shown in Fig. 4. No specific assignments have been made yet of the $CO^-$ and $N_2O_2^-$ states leading to DA.

The effect on the $O^-$ yield of isolating the $O_2$ molecule from the metal substrate by 1.3, 2.2 and 4.3 monolayers (ML) of Argon, respectively [12], is shown in the lowest three curves of Fig. 5. For all curves but the upper one, which represents the gas-phase cross section [10] for $O^-$ production, the $O_2$ coverage is 0.1 ML. The curves labelled 0.37 ML and 0.7 ML are the $O^-$ ESD yields when these amounts of Ar are coadsorbed with 0.1 ML of $O_2$. In Fig. 6, the electron energy dependence of $O^-$ and $OH^-$ signals recorded on a 3-layer $O_2$ film covered with a single $C_nH_{2n+2}$ (n = 1,2,4,5,6) layer are compared. The similitude of all curves indicates [13] that a portion of the $O^-$ ions created at the $O_2$ surface by DA reacts with saturated hydrocarbons via $O^- + C_nH_{2n+2} \rightarrow OH^- + C_nH_{2n+1}$.

ESD yields of $Cl^-$ for one ML and five ML of condensed $Cl_2$ [4] are shown at the bottom and top of Fig. 7, respectively. The large peak around 5 eV arises from decay of a $^2\Pi_u^{-*}$ core-excited $Cl_2^{-*}$ resonant state [4]. A broad peak due to multiple inelastic scattering is seen in the upper curve near 11 eV. A small "hump" appears on the left-hand side of the 5-eV peak when the electron gun is adjusted to focus at lower energy (inset, Fig. 7). This feature has been ascribed [4] to the formation of a $^2\Pi_g$ $Cl_2^{-*}$ state. These assignments have been made by comparison with gas-phase

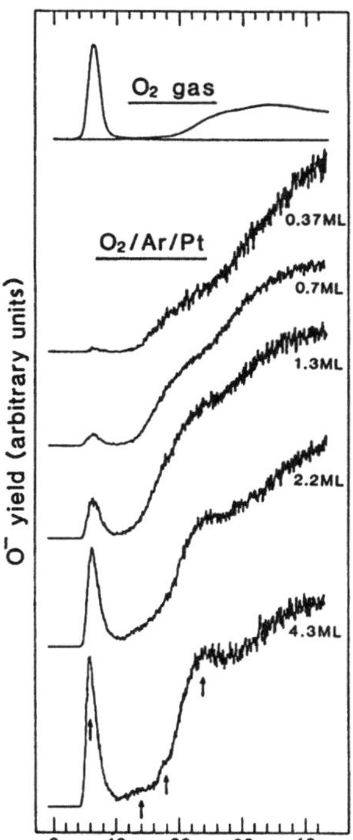


Fig. 5

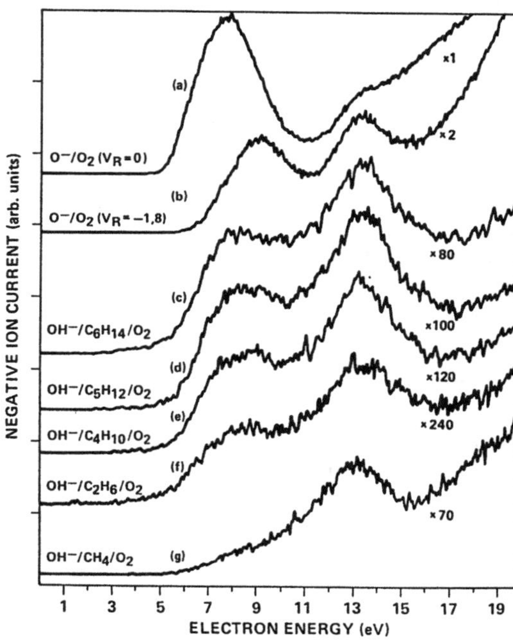


Fig. 6

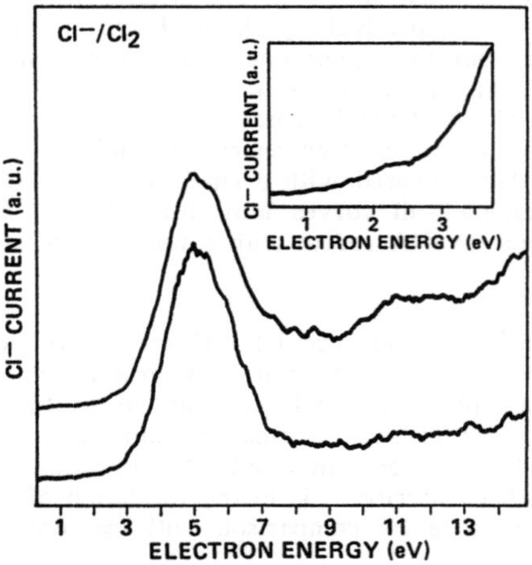


Fig. 7

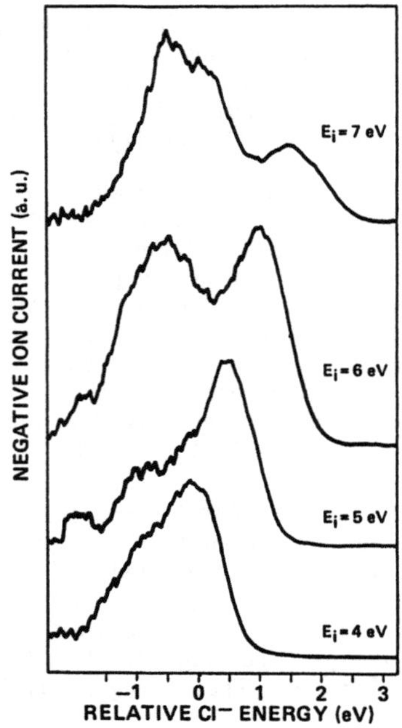


Fig. 8

data [14] and on the basis of the $Cl^-$ kinetic-energy distributions shown in Fig. 8 [4] from which the dissociation limits of the fragments can be known.

## 4. DISCUSSION

DA processes observed in ESD from molecules condensed on metal surfaces can be understood by considering the possible sequence of interactions that leads to stable anion emergence into vacuum: (1) the elastic and inelastic scattering of the electron before its capture at a molecular site; (2) the formation of a transient anion; (3) the decay of the anion by DA; and (4) the modification of the energy and momentum of the dissociated stable anion as it recedes from its site of formation. It is also instructive to consider the simple laws of energy and momentum conservation that govern DA processes in a free molecule. According to these laws, the dissociation energy $E_D$ is shared between the fragments of a diatomic dissociating anion $AB^-$ as

$$E_A = \frac{m_B}{m_A+m_B} E_D \text{ and } E_B = \frac{m_A}{m_B+m_A} E_D ,$$

where $E_A$, $E_B$ and $m_A$, $m_B$ are the kinetic energies and masses of fragments A and $B^-$ at infinite separation, respectively. Deviations from this energy partition in ESD indicates interaction with the solid in one or more of steps (1) to (4). If such interactions are absent in the condensed phase these laws will be conserved as can be shown by plotting the position of the high-energy peak in Fig. 8 as a function of incident electron energy $E_i$. This plot yields a straight line with slope 1/2 indicating that the dissociation energy is equally shared between Cl and $Cl^-$ [4]. Thus, $Cl^-$ ions forming the high-energy peak in Fig. 8 have no interaction with the $Cl_2$ lattice or the metal substrate and their formation only involves steps (2) and (3). However, the structure which occurs on the low-energy side of this peak is due to electrons which have in step (1) produced electronically excited $Cl_2$ states before forming the $Cl_2^{-*}$ $^2\Pi_u$ state [4]. In the analysis of the $O^-$ kinetic-energy distribution from condensed $O_2$ of Fig. 4, plotting the position of the low-energy peak as a function of $E_i$ yields a straight line with a slope smaller than 1/2. This result and the broadening of the peak can be explained by postulating that during step (1) electrons lose energy to vibrational excitation [9]. Multiple elastic scattering in step (1) can also influence DA and, when coherent, enhance its intensity up to an order of magnitude [15].

Steps (2) and (3) rely on intrinsic properties of the free electron-molecule system, but a number of factors at and near the surface of solids can modify these properties. Among these factors we find the image charge and charge exchange induced by the metal or neighbor molecules and the perturbation of the molecular symmetry of the anion. The effect of these factors induced by the metal substrate is dramatically illustrated in Fig. 5 by the increase of two orders of magnitude in the DA signal as the distance of the $O_2^-$ $^2\Pi_u$ anion from the metal is increased by spacer Ar layers [12]. In this example, the image force contributes to charge exchange with the metal and engenders a lowering of the anion symmetry. The lower symmetry, charge exchange, as well as any increase in the number of decay channels, tend to reduce the lifetime of transitory anions. Since the magnitude of the DA process depends exponentially on this lifetime [16], the magnitude of DA is expected to be weaker in the condensed phase. Even if this expectation is generally verified, many of the condensed systems investigated (e.g., Figs. 1, 2 and 3) exhibit more anions states than observed in the gas-phase total ionisation yields. This surprising result involves essentially step (2) where due to perturbation of the isolated electron-molecule potential and orientation of the molecule in space [17] the probability of anion formation can be considerably modified. Such a perturbation can even cause symmetry forbidden states to become accessible [9]. In fact, from the analysis of the data in Fig. 4 [9], it can be deduced that the 13-eV peak in Fig. 1 contains contributions from the lowest-energy $O^-(^2P) + O(^3P)$ limit which, at 13 eV in the Franck-Condon region, correlates only with sigma "plus" symmetry forbidden states.

Finally, during step (4) post-dissociation interaction (PDI) of the stable anion with the metal and/or neighbor molecules reduces the magnitude of the ESD signal. Considering that energy losses by the receding anion are expected to be relatively small [18], PDI effects may be considered to arise from the pull on the anion by the image and polarization forces and from reactive scattering. The first mechanism can be illustrated by coadsorbing rare gas atoms with other molecules on a metal surface. As shown in the curves labelled 0.37 and 0.7 ML in Fig. 5, screening of the metal image force by Ar atoms increases the number of desorbing $O^-$ ions (i.e., less $O^-$ ions are drawn back to the metal). Even in the simplest case of a film composed of diatomic molecules AB reactive PDI of fragment $B^-$ gives rise to a large number of possible product channels, namely

$$\begin{aligned} B^- + AB \rightarrow AB_2^- &\rightarrow AB_2 + e^- \\ &\rightarrow AB + B^- \\ &\rightarrow \text{different products} . \end{aligned}$$

These reactions may or may not involve the intermediate state $AB_2^-$. When they do, the momentum and energy are shared between the constituents of $AB_2^-$. This energy partition may involve internal vibrational energy deposition in diatomic fragments. Such effects modify the kinetic energy distribution of the $B^-$ ions.

The occurrence of PDI can be demonstrated by monitoring DA-induced products of reactions between different molecules coadsorbed on a metal surface. The formation of $OH^-$ from a multilayer $O_2$ surface covered with $C_nH_{2n+2}$ molecules is shown as an example in Fig. 6. This results stands as clear-cut evidence [13] of the occurrence of post-dissociation reactive scattering.

## ACKNOWLEDGMENTS

The author is indebted to R. Azria, L. Parenteau and D. Hunting for their critical evaluation of the present article. This work was sponsored by the Medical Research Council of Canada.

## REFERENCES

1. For a review of the processes leading to negative ion formation in ESD see E. Bauer (Chapt. 3.1) and D. Lichtman (Chapt. 3.2) in Desorption Induced by Electronic Transitions, N.H. Tolk, M.M. Traum, J.C. Tully and T.E. Madey, eds (Springer, New York, 1983).

2. L. Sanche: Phys. Rev. Lett. 53 , 1638 (1984).

3. L. Sanche and L. Parenteau: J. Vac. Sci. Technol. A 4, 1240 (1986).

4. R. Azria, L. Parenteau, L. Sanche: J. Chem. Phys. (in press).

5. G. Bader, G. Perluzzo, L.G. Caron, L. Sanche: Phys. Rev. B30, 78 (1984).

6. R.M. Marsolais, M. Michaud, L. Sanche: Phys. Rev. A35, 607 (1987); and papers cited therein.

7. G. Perluzzo, L. Sanche, C. Gaubert, R. Baudoing: Phys. Rev. B30, 4292 (1984); G. Perluzzo, G. Bader, L.G. Caron, L.Sanche: Phys Rev. Lett. 55, 545 (1985).

8. L. Sanche, G. Bader, L.G. Caron: J. Chem. Phys. 76, 4016 (1982).

9. R. Azria, L. Parenteau, L. Sanche: to be published.

10. D. Rapp and D.D. Briglia: J. Chem. Phys. 43, 1480 (1965).

11. A. Stamatovic and G.J. Schulz, J. Chem. Phys. 53, 2663 (1970).

12. H. Sambe, D.E. Ramaker, L. Parenteau, L. Sanche: to be published.

13. L. Sanche and L. Parenteau: to be published.

14. R. Azria, R. Abouaf and D.T. Billy: J. Phys. B15, L569 (1982) and papers cited therein.

15. H. Sambe, D.E. Ramaker, L. Parenteau, L. Sanche: to be published.

16. J.N. Bardsley, A. Herzenberg, F. Mandl: Proc. Phys. Soc., London 89, 321 (1966).

17. J.W. Davenport, W. Ho, J.R. Schrieffer, Phys. Rev. B17, 3115 (1978).

18. F. Linder: Private communication.

# Image Charge, Coherent Scattering, and Symmetry Effects in Electron Stimulated Desorption: $O^-$ from $O_2$ Condensed on Metals

*D.E. Ramaker and H. Sambe*

Department of Chemistry, George Washington University
Washington, DC20052, USA

## 1. Introduction

Considerable progress has been made towards our understanding of electron/photon stimulated desorption (ESD/PSD) [1]. Identification of the excited states responsible for desorption reveals that they usually are many particle excitations (i.e. n holes (h)-n electrons (e), such 2h, 2h-1e etc.) and arise from widely different excitation mechanisms [2]. Furthermore, it is known that the lifetime of these localized excitations, which determine the probability for desorption, is very much altered by covalent interactions between the active site and its immediate environment [3]. In this work we shall examine "environmental" effects, which do not arise from alteration of the excited-state lifetime, but rather from: (1) the image charge induced in the metal substrate, (2) adsorbate-adsorbate interactions which reduce the symmetry and hence cause certain selection rules to breakdown, and (3) coherent scattering of electrons in the substrate.

We review previously reported [4] ESD yields of $O^-$ from $O_2$ condensed on polycrystalline Pt in the electron energy range 0 to 40 eV. Of the various species which may desorb ($O^-$, O, and $O^+$), $O^-$ desorption is the most convenient to study here for three reasons: 1) the $O^-$ desorption yields appear to be dominated by resonant ESD processes, thus the $O^-$ yields are easier to interpret quantitatively than the non-resonant ESD $O^+$ yields; 2) the ESD $O^-$ yields are about as straightforward to interpret as PSD $O^+$ yields, but they do not require use of a synchrotron for measurement; and 3) the O yields are more difficult to observe experimentally, and they may have contributions from both resonant and nonresonant mechanisms making them more complex to interpret [2,5].

The apparatus utilized to obtain the experimental data described here has been described previously [6], and it is also fully described in the article by Sanche in this volume. The incident angle of the electron beam is 20° from the surface, and the $O^-$ ions are measured by a quadrapole mass spectrometer positioned at 70° from the surface. The film thicknesses are estimated to within 20% by a method described previously [6]. A rare gas (RG) is condensed near its sublimation temperature on a clean Pt ribbon, and the $O_2$ gas is condensed onto the RG film at 17 K. Electron Transmission Spectra (ETS) data [7] reveal that the RG films are essentially ordered, since Bragg reflection (BR) minima are observed around 5 and 9 eV for Ar, 4, 7, and 9 eV for Kr, and 2.5, 5, and 7.5 for Xe. In contrast, Xe films prepared well below the sublimation temperature do not exhibit BR minimum in ETS [7]. Data also indicate that the Ar films seem to be more ordered than the Kr and Xe films [8].

The three different "environmental" effects under study in this review, can be isolated in the $O^-$ yields from $O_2$/RG/Pt system. The effects of ion neutralization are minimized by inserting a rare gas (Ar, Kr, or Xe) layer between the Pt metal and the $O_2$ layer. The RG layer also serves as a spacer layer so that the distance between the ion and the metal surface can be varied, thus varying the effect of the image charge force on the desorbing ions [4]. The image charge effects are best studied for $O_2$(0.16 ML)/RG(1–6 ML)/Pt. Adsorbate-adsorbate interactions become important at larger $O_2$ coverages when symmetry breakdown effects become visible. Finally, in the range $O_2$(0.16 ML)/RG(6–35 ML)/Pt, coherent scattering effects dominate.

Figure 1 compares the electron-energy dependence of the ESD $O^-$ yield from $O_2$(.1 ML)/Ar(1–32 ML)/Pt as a function of incident electron energy. The electron impact $O^-$ dissociation of $O_2$ gas is well understood [9]. In the energy range from 4.4 to 10 eV, the $O^-$ ions are produced via the transient $O_2^-(^2\Pi_u)$ anion in a process referred to as dissociative attachment (DA). Above 17 eV, the $O^-$ ions are produced predominantly via an excited neutral $O_2^*$ molecule in a process referred to as dipolar dissociation (DD).

## 2. Image Charge Effects

Fig. 2 shows the Ar thickness dependence of the $O^-$ yield from $O_2$/Ar/Pt with a constant (0.1 ML) $O_2$ coverage. The sharp maximum probably corresponds to the case when the Ar plus $O_2$ coverage form one ML to within experimental

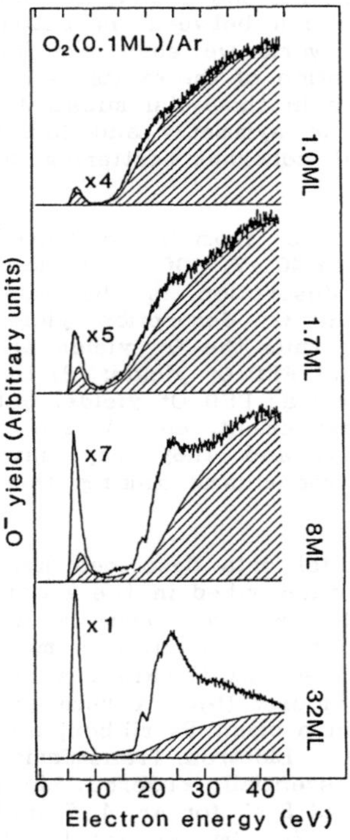


Fig. 1 Comparison of the ESD $O^-$ yield curves obtained from $O_2$/Ar/Pt with a constant (0.1 ML) $O_2$ coverage and variable (1–32 ML) Ar thicknesses [4]. The shaded areas indicate the estimated contributions due to the direct process.

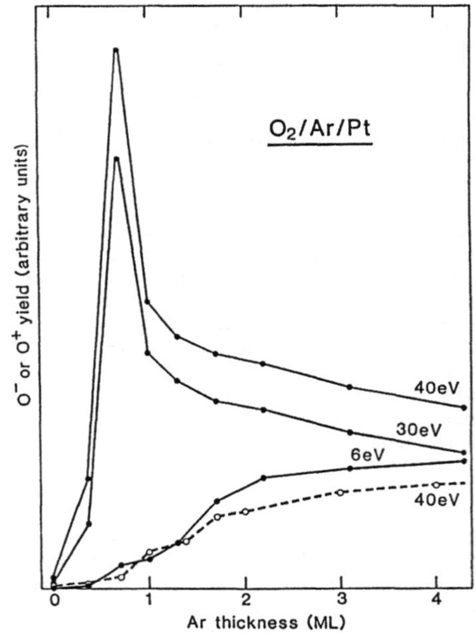


Fig. 2. Ar thickness dependence of the $O^-$ yields produced by ESD from the $O_2$/Ar/Pt samples with a constant (0.1 ML) $O_2$ coverage [4]. The incident electron energies are as indicated. Also shown (dotted curve) is the $O^+$ yield produced by ESD with incident energy of 40 eV on $O_2$(0.16 ML)/Ar(variable)/ Pt [4].

uncertainty. Fig. 2 also shows the $O^+$ yield from $O_2$(0.16 ML)/Ar(variable)/Pt samples produced by 40-eV electron impact. In $O_2$ gas, $O^+$ formation by 40-eV electron impact arises predominantly via an $O_2^+$ intermediate state [10]. Therefore, we expect that both the $O^+$ yield and the DA contribution to the $O^-$ yield are produced via charged intermediate states, $O_2^+$ and $O_2^-$, respectively. Fig. 2 shows that the image charge has a similar effect on these charged states and their dissociated ions.

Four effects that the image charge has on the desorption yield can be summarized using the semi-quantitative, adiabatic potential energy curves for $O_2$ depicted in Fig. 3 [4]. The solid curves give the ground state of $O_2$, the $^2\Pi_u$ $O_2^-$ state and its partner dissociating to the lowest $O^-$+O limit, and a $O_2^*$ state and its partner dissociating to the lowest $O^+$+$O^-$ limit [4]. The two $O_2^-$ and two $O_2^*$ states each form a pair of states which at large separation have wavefunctions approaching $[\psi(O_a^-)\psi(O_b) \pm \psi(O_a)\psi(O_b^-)]$ and $[\psi(O_a^+)\psi(O_b^-) \pm \psi(O_a^-)\psi(O_b^+)]$ respectively, where $\psi(O_a)$ denotes the wavefunction for an isolated O atom at site "a". The dotted curves give the corresponding potential energy curves in the presence of the image charge. The symbol $O^+$/M in Fig. 3, for example, means that the $O^+$ ion is near a metal surface. The image charge due to the Pt lowers the energies of the ionic species such as $O^+$/M, $O^-$/M, and $O_2^-$/M, while the energies of the neutral species such as O/M, $O_2^*$/M and $O_2$/M remain the same. The magnitude of this lowering depends on the distance between the ion and the metal surface, and hence on the thickness of the rare gas spacer layer. The curves in Fig 3 are drawn assuming the maximum energy lowering, 2.5 eV [4].

The first effect of the image charge involves the energy separation of the O + $O^-$ dissociation limit into the O/M+$O^-$ and $O^-$/M+O limits. Consequently, to desorb $O^-$ ions, the lower $O_2^-$/M intermediate state must undergo a non-radiative transition into its upper partner state. The mechanism for this "charge transfer" transition, $O^-$/M+O to O/M+$O^-$, is similar to that for symmetric charge transfer in atom-atom collisions of identical nuclei [11] and occurs because of the breakdown of the adiabatic approximation. The probability for this transition is 1/2 when the separation of the dissociation limits vanishes, and decreases quickly as the separation increases. In

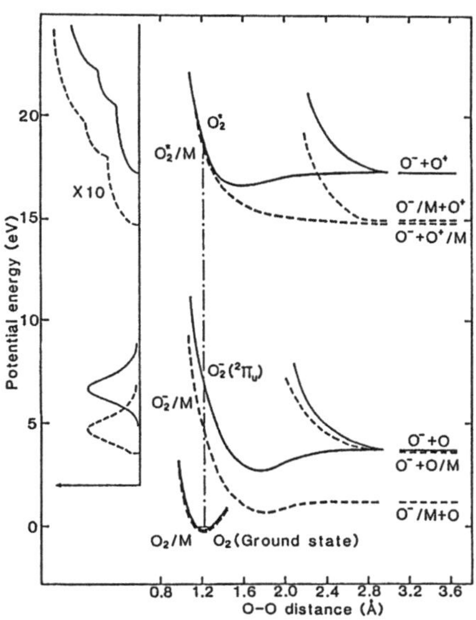


Fig. 3 Potential energy curves of $O_2$ gas (solid curves), and the corresponding potential energy curves under the influence of an image charge (dotted curves). Spectra on the left side show schematically the relative number of vibrational and/or electronic states which have sufficient energy to yield $O^-$ ions, for $O_2$ gas (solid lines) and for $O_2$/M (dotted lines) [4].

contrast, the separation between the $O^-/M+O^+$ and $O^-+O^+/M$ dissociation limits is negligibly small. Nevertheless, based on the relative atomic sizes of $O^+$ and $O^-$, we expect that the $O^-+O^+/M$ limit is slightly lower than the $O^-/M+O^+$ limit. This means that the DA contribution should increase and the DD contribution should decrease slightly as the thickness of the rare gas layer increases.

The second effect of the image charge involves the kinetic energy of the escaping $O^-$ ions. As can be seen easily in Fig. 3, in comparison with the corresponding $O^-$ kinetic energy from $O_2$ gas, the kinetic energy of the $O^-$ ions via $O_2^-/M$ is decreased while that via $O_2^*/M$ is increased. The slower escaping ions have a greater chance to be recaptured or neutralized, thus decreasing the DA $O^-$ yield, while the faster $O^-$ ions have a greater chance for escape, thus increasing the DD $O^-$ yield. However, this mechanism probably causes a relatively small effect.

The third effect of the image charge involves the number of vibrational and/or electronic intermediate states (i.e., amount of configuration space) which have sufficient energy to yield $O^-$ ions. The curves on the left side of Fig. 3 schematically indicate the configuration space for production of $O^-$ from $O_2$ gas (solid lines) and from $O_2/M$ (dotted lines). These plots indicate that the configuration space for $O_2^-/M$ decreases slightly while that for the $O_2^*/M$ state increases significantly compared to that for $O_2$.

Finally, the fourth effect of the image charge involves the quenching rate of the intermediate state. Near a metal surface, some of the $O_2^-$ or $O_2^*$ states may be neutralized or de-excited before the desorption process can get underway. The image charge influences these destruction rates of the $O_2^-$ intermediate state by attracting the ion towards the metal surface. The inward motion and subsequent neutralization are similar to the first two steps of the Antoniewicz "bounce" mechanism [12], which is known to be active for neutral desorption from physisorbed systems [13]. Again in contrast, the de-excitation rate of the $O_2^*$ intermediate state is not affected by the image charge. Thus, intermediate state quenching has a greater effect on the DA process than on the DD process.

All four image charge mechanisms predict a depletion of the DA process relative to the DD process. This is consistent with the two-orders of magnitude reduction seen in the $O^-$(via DA)/$O^-$(via DD) ratio for the $O_2$(0.37 ML)/Ar/Pt system compared with that of $O_2$ gas. As the thickness of the rare gas spacer layer decreases from 4 to 1 ML, the image charge force increases. Accordingly, with decreasing rare gas thickness, the $O^-$ yield via the DA process should decrease and that via the DD process should increase. This is consistent with Fig. 2. As indicated above, the dissociation branching effect for the DA process, and the configuration space effect for the DD process, are primarily responsible for this behavior.

## 3. Symmetry Effects

Figure 4 compares the $O^-$ yield from $O_2$ gas with that from $O_2$(3 ML)/Pt at 17 K [14] and $O_2$ (1 ML)/W at 300 K [15] in the electron energy range 2–20 eV, i.e. in the range where the DA contribution dominates. The data for $O_2$/Pt is also shown with retarding potentials, $V_R$, of −1.5 and −1.8 eV. These later data provide a measure of the kinetic energy of the desorbing $O^-$, since only ions in excess of $V_R$ can be collected. Note that application of the retarding potentials for $O_2$/Pt eliminates the 7 ev peak, making it comparable to that for $O_2$/W. The data also reveal the presence of peaks around 9 and 13 eV in the $O^-$ yields from the chemisorbed systems, which are absent for $O_2$.

A theoretical analysis [16] indicates that the 9 and 13 eV features arise from the $^2\Sigma_g^+$ and $^2\Sigma_u^+$ resonances which have excitation energies in this

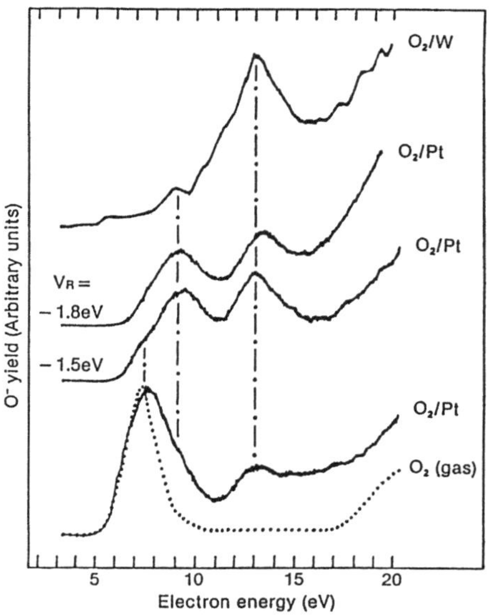


Fig. 4 Comparison of the $O^-$ yields from $O_2$ gas with that from $O_2$(3 ML)/Pt at 17 K [14] (with retarding potentials of $V_R$ = 0, -1.5, and -1.8 eV), and from $O_2$(1 ML)/W at 300 K [15].

region. These two $\Sigma^+$ resonances do not appear in the yield from $O_2$ because of a selection rule which does not allow $\Sigma^- \rightarrow \Sigma^+$ transitions from the $^3\Sigma_g^-$ ground state of $O_2$ in the gas phase, or even at low coverages on the RG/Pt substrate [16]. However at higher coverages, $O_2$–$O_2$ interactions on the surface, and perhaps even dimer formation, causes this selection rule to break down [16].

Image charge effects are responsible for the remaining differences seen in Fig. 4. The principal $^2\Pi_u$ DA contribution at 7 eV is absent in the yield from $O_2$/W primarily because of the branching ratio mechanism described above. It is present in the yield from $O_2$/Pt at the 3 ML $O_2$ thickness because the image charge effect is already decreased. The retarding potentials of 1.5 and 1.8 eV eliminate this contribution because the $^2\Pi_u$ resonance apparently produces $O^-$ of kinetic energy less than these energies while the $O^-$ from the two $^2\Sigma^+$ resonances have larger kinetic energies [14]. Although the image charge via the branching ratio mechanism essentially eliminates the $^2\Pi_u$ contribution, it increases the two $^2\Sigma^+$ contributions because these two resonances correlate to the upper O/M + $O^-$ dissociation limits, while their partner states (the $^2\Sigma_u^+$ and $^2\Sigma_g^+$) correlate to the lower O + $O^-$/M dissociation limits [16]. Thus the image charge effect can increase or decrease the DA yield depending on whether the excited resonance correlates to the upper or lower limit.

The data and interpretation above provides an answer to the long standing question concerning the active sites for $O_2$/W; i.e. are they atomic or molecular sites [17]. The presence of the two $^2\Sigma^+$ resonance contributions and the similarity of the DD contributions for $O_2$/W and $O_2$ gas clearly indicate that the minority molecular sites are the active ones in this case. However, the $O^-$ yields from $O_2$(1 ML)/Mo at 300 K [15] suggests that both the molecular and the atomic sites are active, because spectral features in the total yield align with those from $O_2$/W and from the $OH^*$ yield from $OH/TiO_2$ [5]. Portions of the $OH^*$ yield have been shown to arise from the DA mechanism, and therefore must involve breaking O–Ti bonds, i.e. the $OH^*$ spectral lineshape should mimic the $O^-$ spectral lineshape from atomic O sites on Pt [5]. A least squares fit of the $O^-$ from $O_2$/W and $OH^*$ from $OH/TiO_2$ spectral lineshapes to the $O^-$ yield from $O_2$/Mo suggest that about half of the $O^-$ yield from $O_2$/Mo arises from the minority molecular sites and half from the predominant atomic sites [16].

## 4. Coherent Scattering Effects

Fig. 1 shows that for $O^-$ from $O_2$/Ar(1-32 ML)/Pt, the three features at 6, 18, and 24 eV grow faster than the remainder of the spectrum, and at large thicknesses (>20 ML) dominate the spectrum [4]. The $O^-$ yields at these three energies are plotted as a function of the Ar thickness in Fig. 5. The spectral shape of the 6-eV feature depends on the RG, the thickness and structural order of the RG layer, and the $O_2$ coverage. It is broad and symmetric at small RG thicknesses but sharp, asymmetric, and shifted to lower energy at larger thicknesses. It is much narrower at smaller $O_2$ coverages than at larger coverages (Fig. 6). The dependence on the RG and on the structural order of the RG can also be seen in Fig. 6 [4].

The experimental results described above can be explained in terms of three processes which we call the direct (D), elastic-indirect (EID), and inelastic-indirect (IID) processes [4]. In the D process, an incident electron collides with an $O_2$ on the surface and produces an $O^-$ directly. In the EID process, an incident electron at an energy below the first electronic excitation energy (first $E_{ex}$) of the RG, passes through the $O_2$ layer without loss of energy, undergoes quasi-elastic multiple scattering in the RG, and returns to the surface, where it collides with an $O_2$ and produces $O^-$ as in the D process. The IID process is identical to the EID, except that the electron upon initial

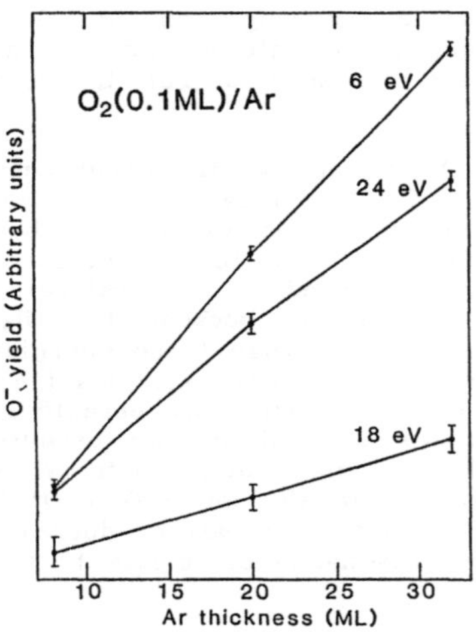


Fig. 5 Ar thickness dependence of the ESD $O^-$ yields from $O_2$(0.1 ML)/Ar/Pt [4]. The incident electron energies are as indicated.

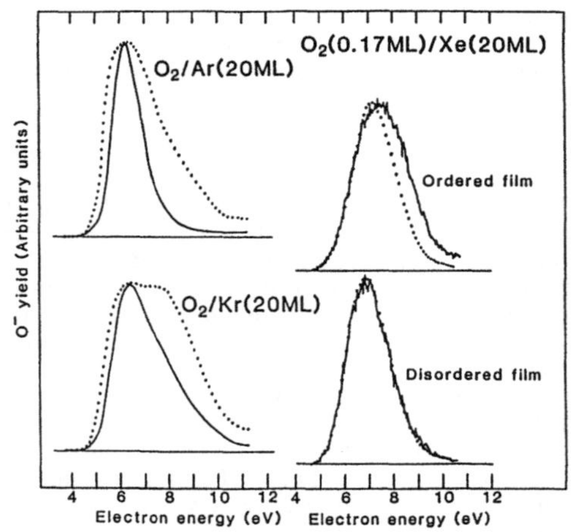


Fig. 6 LEFT) Comparison of the ESD $O^-$ yield curves obtained with a constant rare gas (Ar or Kr) thickness (20 ML) but with different $O_2$ coverages, 0.03 ML (solid curves) and 1.0 ML (dotted curves)[4]. RIGHT) ESD $O^-$ yield curves obtained from $O_2$(0.1 ML)/Xe(20 ML)/Pt but prepared at 40 K (top) and 17 K (bottom). Also shown are ESD $O^-$ yield curves from $O_2$ gas (dotted curves) [4].

entry into the RG film suffers loss of energy by electronically exciting the RG. At electron energies above the first $E_{ex}$, the inelastic mean free path in RG films is very short (≈ 10 A° [4]).

The $O^-$ yield via these processes are proportional to the following expressions [4]:

$$
\begin{aligned}
D &\propto \Theta\, \sigma(E) \\
EID &\propto [1-\Theta\, \sigma_{BI}(E)]\; P_B(RG,\tau,E)\; \Theta\, \sigma(E), \\
IID &\propto [1-\Theta\, \sigma_{BI}(E)]\; P_{ex}(RG,E_{ex},E)\; P_B(RG,\tau,E-E_{ex})\; \Theta\, \sigma(E-E_{ex}).
\end{aligned}
\qquad (1)
$$

Here, E denotes the incident electron energy, $\Theta$ the $O_2$ coverage, and $\sigma(E)$ the $O^-$ yield cross section from $O_2$ on the RG film. $\sigma(E)$ is assumed to be independent of the RG thickness beyond 7 ML. The $[1-\Theta\, \sigma_{BI}(E)]$ factor gives the probability of passing through the $O_2$ layer without loss of energy, where $\sigma_{BI}(E)$ is equal to a sum of the elastic backscattering and the total inelastic cross sections of $O_2$. $P_B(RG,\tau,E)$ denotes the probability of the electron returning to the surface, which depends on the RG, the RG thickness ($\tau$), and the electron energy in the film. $P_{ex}(RG,E_{ex},E)$ denotes the electronic excitation probability of the RG film by electron impact at energy E. Since $\sigma_{BI}(E)$ and $P_{ex}(E)$ are slowly varying functions of E [4], the E dependence of the D, EID, and IID contributions are primarily determined by $\sigma(E)$ and $P_B(E)$. Because of the short inelastic mean free path, the $O^-$ yield, Y(RG,t,E), below and above the first $E_{ex}$ are given by Y = D + EID and Y = D + IID, respectively.

Figure 5 shows that the $O^-$ yields at 6, 18, and 24 eV all increase as the Ar layer thickness increases [4]. Equation (1) indicates that this relationship exists because of the two indirect processes. The 6, 18 and 24 eV features observed for $O_2$(0.1 ML)/Ar(32 ML) arise predominantly from the EID, IID, and IID processes respectively. The spectral shape of $\sigma(E)$, or the D contribution, is expected to be similar to the $O^-$ yield from $O_2$ gas except for the relative magnitude of the DA and DD contributions. Polarization of the thick RG layer reduces the $O^-$ yield via the DA process, but enhances that via the the DD process, compared with that for $O_2$ gas . Our estimates of the D contribution, indicated in Fig. 1 by the shaded areas, are based on the above considerations. The intensity ratio EID/D for $O_2$(0.1 ML)/Ar(32 ML) is about 70 at the peak energy of the 6-eV feature which means that $P_B$ is around 90. For thicker and more ordered Ar films, we expect even larger values of $P_B$. Since the maximum $P_B$ obtainable by incoherent multiple scattering is around 2, coherent scattering must be playing a dominant role.

The spectral shape of the 6-eV feature for the different RG films can be correlated with the BR minima observed in ETS data. The 6-eV line shape from a disordered Xe film is virtually identical with that from $O_2$ gas (Fig. 6). Since $\sigma(E)$ is virtually identical in the 5-10 eV range with that from $O_2$ gas, $P_B(E)$ for the disordered film must be nearly invariant with E over this same energy range. However, $P_B(E)$ for the ordered film must vary over this energy range, since the line shape of the ordered Xe film differs from that for $O_2$ gas (Fig. 6). The 6-eV line shape for Ar also depends on the $O_2$ coverage so that at the lower coverage (0.03 ML), it is narrower than $\sigma(E)$ and its peak is shifted to lower energy (Fig. 1 and 6). All these suggest that $P_B(E)$ for Ar is strongly enhanced near the first BR minimum around 5 eV. The peak for Kr at the lower $O_2$ coverage is also shifted to lower energy (Fig. 6); however, the line shape is wider than $\sigma(E)$ apparently because two BR energies (4 and 7 eV) partially overlap with $\sigma(E)$. The $O^-$ yield is enhanced for Ar more strongly than for Kr, because the first BR minimum for Ar is closer to the peak of $\sigma(E)$, and perhaps also because the Ar film is more ordered. This analysis indicates that $P_B(E)$ is enhanced near BR energies, especially strongly near the first BR minimum [4].

It is well known that for a perfect crystal the electronic wavefunctions just above and below the energy band gaps, which arise from the BR's, have standing-wave character [18]. We conclude that this standing-wave character is responsible for the large enhancement of $P_B$ near the BR energies. The increase of $P_B$ with RG thickness is also consistent with this since the standing-wave character should increase with RG thickness. However, this raises an interesting question. How does a relatively small increase in the amplitude of the electronic wavefunction (i.e. at most a factor of 2 due to the standing-wave character) cause such a dramatic increase in the $O^-$ desorption yield (a factor of $10^2$)? Clearly, either some quantum mechanical matrix element effect causes the $O_2 \rightarrow O_2^-$ cross-section to increase non-linearly with standing-wave character, or the $O_2 \rightarrow O^-$ desorption branching ratio increases with standing-wave character. These possibilities must still be investigated.

## 5. Summary

In this work we have observed the following environmental effects: 1) the image charge causes desorption yields arising from neutral intermediates to increase, while those arising from ionic intermediates may decrease or increase depending on the dissociation limit of the ionic intermediate state, 2) $O_2$-$O_2$ interaction effects at the surface reduce the local symmetry, cause a breakdown in the $\Sigma^- \nrightarrow \Sigma^+$ selection rule, and thus introduce additional dissociative attachment contributions in the $O^-$ yield, and 3) coherent scattering in the rare gas may enhance the $O^-$ yield up to two orders of magnitude near the Bragg reflection energies. The observed image charge and adsorbate-adsorbate interaction effects should be generally observed in other systems. The coherent scattering effects will be large only for low energy desorption thresholds when the inelastic mean free path of the electrons is long. This explains why $X^+$ desorption, which generally has higher energy thresholds, may be less affected by coherent scattering and hence has not exhibited these large effects to date.

## Acknowledgements

We have greatly benefited from discussions with Leon Sanche. This work was supported in part by the Office of Naval Research.

## References

1. a) Desorption Induced by Electronic Transitions DIET I, ed. by N.H. Tolk, M.M. Traum, J.C. Tully, and T.E. Madey, Springer Ser. Chem. Phys., Vol. 24 (Springer, Berlin, Heidelberg 1983); b) Desorption Induced by Electronic Transitions DIET II, ed. by W. Brenig and D. Menzel, Springer Ser. Surf. Sci., Vol. 4 (Springer, Berlin, Heidelberg 1985).
2. D.E. Ramaker: p. 70 in Ref. 1a and p. 10 in Ref. 1b above.
3. D.E. Ramaker: J. Vac. Soc. Technol. A1, 1137 (1983).
4. H. Sambe, D.E. Ramaker, L. Parenteau, and L. Sanche: Phys. Rev. Lett. 59, 236 (1987); 59, 505 (1987).
5. F.L. Hutson, D.E. Ramaker, V.M. Bermudez, and M.A. Hoffbauer: J. Vac. Sci. Technol. A3, 1675 (1985).
6. L. Sanche and L. Parenteau: J. Vac. Sci. Technol. A4, 1240 (1986).
7. G. Bader et al.: Phys. Rev. B30, 78 (1984); B26, 6019 (1982).
8. G. Perluzzo et al.: Phys. Rev. Lett. 55, 545 (1985).
9. G.J. Schulz: Rev. Mod. Phys. 45, 423 (1973); R. Locht and J. Momigny: Int. J. Mass Spectrom. Ion Phys. 7, 121 (1971); D. Rapp and D.D. Briglia: J. Chem. Phys. 43, 1480 (1965).
10. R. Locht and J. Schopman: Int. J. Mass Spectrom. Ion Phys. 15, 361 (1974); Chem. Phys. Lett. 26, 596 (1974).
11. S. Geltman: Topics in Atomic Collision Theory (Academic Press, New York, 1969).

12. P.R. Antoniewicz: Phys. Rev. B21, 3811 (1980).
13. P. Feulner et al.: Phys. Rev. Lett. 53, 671 (1984); Q.J. Zhang, R. Gomer, and D.R. Bowman: Surf. Sci. 129, 535 (1983).
14. R. Azria, L. Parenteau, and L. Sanche: to be published.
15. Z.X. Liu and D. Lichtman: Surf. Sci. 114, 287 (1982).
16. H. Sambe and D.E. Ramaker: to be published.
17. E. Bauer: p. 104 in Ref. 1a.
18. C. Kittel: "Introduction to Solid State Physics", 6th ed. (John Wiley & Sons, NY, 1986), Chap. 7.

# Techniques for the Detection of Photodesorbed Negative Ions

*C.E. Young*[1], *E.L. Schweitzer*[1], *M.J. Pellin*[1], *D.M. Gruen*[1,*], *Z. Hurych*[2,***], *P. Soukiassian*[2,**], *M.H. Bakshi*[2], *and A.S. Bommannavar*[2]

[1]Materials Science and Chemistry Divisions, Argonne National Laboratory, Argonne, IL 60439, USA
[2]Department of Physics, Northern Illinois University, De Kalb, IL 60115, USA

## 1. Introduction

The negative ion channel in stimulated desorption has been the subject of very few studies. The existing experimental data, consisting entirely of electron-stimulated desorption (ESD) experiments, have been summarized in a recent review by KNOTEK [1]. Systems studied have included $H^-$ from H/W [2], $O^-$ from CO/W [3], $O^-$ from O/W [4,5], $O^-$ from O/Mo [5], and $O^-$ from Mo(100) [6] (the other substrates being polycrystalline). For the surfaces investigated above, the ion yield for negatives was one to two orders of magnitude smaller than that for positives. In some cases, both positive and negative ions could be assigned to the same chemical state on the surface through coverage dependence of the signal, for example [2]. There is also a report of ESD of $H^-$ from potassium-dosed polycrystalline Mo in an investigation relating to the conversion of proton beams to negative ions at surfaces [7]. The associated application is in the production of intense neutral beams for fusion experiments. It is also noteworthy that negative ions could have deleterious effects on the performance of forthcoming synchrotron radiation sources utilizing positron beams because of trapping and scattering effects.

This paper reports the direct observation of $H^-$ ions released from a Cs-dosed W(100) crystal by photon-stimulated desorption (PSD).This study utilized the 3m toroidal grating monochromator beamline at the University of Wisconsin-Madison Synchrotron Radiation Center. The main technical problem to be overcome in such experiments is the large background from photoemitted electrons which dominate the weak anion signal by many orders of magnitude. The solution ultimately employed utilized both magnetic suppression of photoelectrons and time-of-flight (TOF) mass separation. No internal modifications to the basic cylindrical mirror analyzer (CMA) [8] were required. We are not aware of any previous reports of the detection of negative ions released from surfaces via photon bombardment, with the exception of high flux laser experiments, in which plasma formation is involved in the ionization process [9].

---

*Work supported by the U.S. Department of Energy, BES-Materials Sciences, under Contract No. W-31-109-ENG-38.
**On leave from SPAS-Commissariat a l'Energie Atomique, CEN Saclay, 91191, Gif-s-Yvette, France.
***Work supported by NSF Grant No. DMR-8405-173.

## 2. Experimental

The separation of electrons from negative ions by the application of a magnetic field is a well-known technique based on the expression for the Larmor radius R of a particle of mass m in a magnetic field B:

$$R_{[cm]} = 144.0 \{m_{[u]}E_{[eV]}\}^{1/2} / B_{[Gauss]}, \quad (1)$$

where E is the component of particle kinetic energy perpendicular to the direction of B. Such a momentum analysis is effective in the separation of electrons from ions at a comparable energy, the electron orbits being much more strongly perturbed. In the case of stimulated desorption, where the ions have characteristic kinetic energies of a few eV, interference from photoemitted electrons of high energy can still occur, as was found to be the case in the present work. Instruments employing a magnetic mass spectrometer have the intrinsic characteristic of removal of electrons from the negative signal [10,11]. More flexibly, apparatus applying magnetic discrimination just prior to the detector has been used to provide the capability to monitor electrons or anions selectively [12].

In the present instance, a coil of 12.5 mm radius consisting of 5 turns of 0.8 mm dia. copper wire was placed near the sample, as shown in Fig. 1. The coil was axially aligned with the CMA and placed near its entrance aperture. This orientation preserves the conical symmetry of the transmitted trajectories. The axial magnetic field of a loop falls off as $1/r^3$ asymptotically, r being the axial distance from the coil center. The computed range (FWHM) of $|B_{axial}|$ is 19 mm., giving selective electron deflection principally near the CMA entrance. This design avoids the necessity of modifying the CMA, but has the drawback of failing to supress electrons produced away from the vicinity of the sample.

Typically a coil current of ~5 A was employed, giving a maximum on-axis magnetic field of 12 Gauss, and corresponding Larmor radii, at E = 1 eV, of 3 mm for electrons and 11 cm at 1 atomic mass unit. An experimental check with desorbed $H^+$ showed little influence of the magnetic field at these levels. The electron suppression achieved in the region of low kinetic energy proved to be about two orders of magnitude. This was significant for the reduction of channeltron fatigue [13] when operating at the high gains desirable for the detection of negative ions. CMA energy scans showed a rise in transmitted signal due to energetic electrons beginning at ~ 8 eV. The presence of significant residual electron signal below this energy (apparently mainly due to electrons released in the CMA) required the use of the background-subtraction method described next.

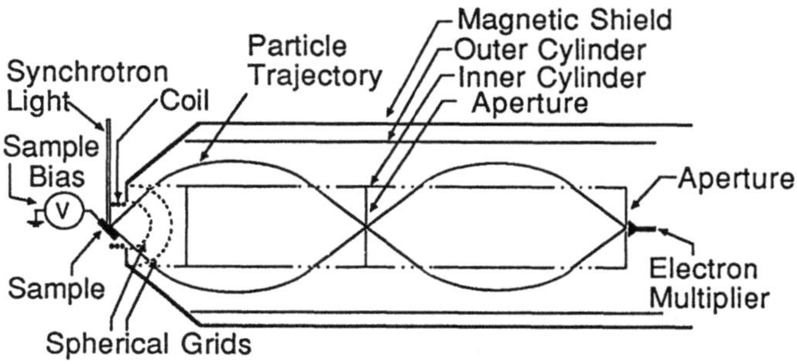


Fig. 1. Magnetic coil and sample configuration for negative ion detection

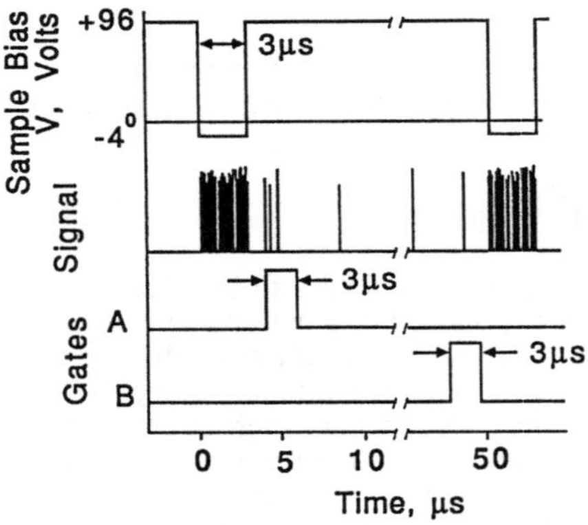


Fig. 2. The pulse sequence employed for the special TOF mode employed. A photoelectron burst, partially suppressed magnetically, is detected during the interval of negative sample bias. The strong retarding potential applied over most of the cycle results in a low-pass TOF mass spectrometer (see text).

Operation in the mode illustrated in Fig. 2 provided electrostatic suppression of sample photoelectrons during most of the TOF cycle. The flight time for unit mass desorbed at low energy was ~ 4 μs in this instrument [14] (CMA pass energy: 60 eV). The target bias was therefore reset to a value sufficient to retard all sample photoelectrons at the incident photon energy prior to the arrival of any $H^-$. At the start of each flight sequence a small negative potential was applied to the target for a few μs, sufficient for low energy $H^-$ to reach the CMA entrance aperture. Heavier anions at similar energy are prevented from reaching the analyzer entrance by the re-application of the strong retarding potential. A relatively long signal gate (A in Fig. 2) was used to collect the weak ion + background electron signal and a second gate (B), in a region where ion signal could not be present, allowed background subtraction. In this method, a higher mass of interest could be selected for observation by lengthening the sample negative pulse duration and adjusting gate A (the lower masses then being obscured by target electrons). However, due to the $\sqrt{m}$ dependence of flight time, the $H^-$ case is the most favorable in terms of separation from heavier species.

## 3. Results and Discussion

Experiments were performed in $10^{-11}$ Torr vacuum on a W(100) crystal, cleaned and dosed with 1 monolayer of Cs as described previously [15], then exposed to 5 Langmuirs of hydrogen. A brief description of the principal results follows; a more extensive discussion is given elsewhere [16]. Figure 3a shows the kinetic energy distribution of $H^-$ ions obtained by scanning the energy-transmission window of the CMA (pass energy: 60eV), at a fixed incident photon energy of 50 eV. The distribution is seen to be characterized by small values of the kinetic energy values for the peak (~ 1 eV) and FWHM (~ 3 eV). In Fig. 3b is shown the dependence on incident photon energy of the $H^-$ signal, taken with the analyzer energy window at the maximum signal position of Fig. 3a. Within the noise limits, no dependence on the metal core levels could be observed. No corresponding ESD data for hydrogen on tungsten was reported [2]. A further observation of the present work was that the $H^-$ signal was subject to decay with a half-life of ~ 2 minutes at ~ 1 x $10^{-10}$ Torr chamber pressure, indicating the ease with which the active desorption sites could be poisoned. Of course, there may have remained other negative ion desorption processes at lower rates, which were hidden by the residual photoelectron arrival rate (typically 1 to 10 per second) in the present apparatus. This situation can be substantially ameliorated by relocating the electron suppression to a position just prior to the detector [12].

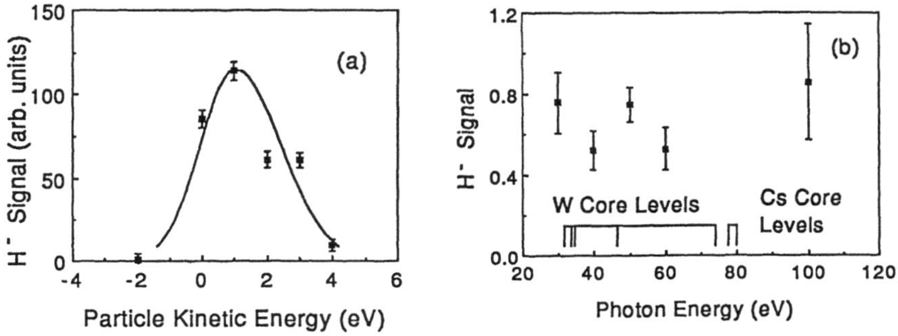


Fig. 3. (a) Kinetic energy distribution of $H^-$ ions desorbed at a fixed incident photon energy of 50 eV; (b) dependence of the H- signal on the energy of the desorbing photon

## References


1. M. L. Knotek: Rep. Prog. Phys. 47, 1499 (1984)
2. J. L. Hock, J. H. Craig, and D. Lichtman: Surf. Sci. Lett. 85, L218 (1979)
3. J. L. Hock, J. H. Craig, and D. Lichtman: Surf. Sci. 87, 31 (1979)
4. J. L. Hock, J. H. Craig, and D. Lichtman: Surf. Sci. 85, 101 (1979)
5. Z. X. Liu and D. Lichtman: Surf. Sci. 114, 287 (1982)
6. M. L. Yu: J. Vac. Sci. Technol. 16, 518 (1979); Phys. Rev. B19, 5995 (1979)
7. A. Kh. Ayukhanov and E. Turmashev: Sov. Phys. Tech. Phys. 22, 1289 (1977)
8. Physical Electronics Corp., Model 15 - 250 (bandpass = 1.2% of pass energy)
9. R. D. Knight, R. A. Walch, S. C. Foster, T. A. Miller, S. L. Mullen, and A. G. Marshall: Chem. Phys. Lett. 129, 331 (1986)
10. J. L. Hock and D. Lichtman: Surf. Sci. 77, L184 (1978)
11. R. Middleton: Nucl. Instr. Meth. 144, 373 (1977)
12. R. I. Hall, I. Cadez, C. Schermann, and M. Tronc: Phys Rev. A15, 599 (1977)
13. L. A. Frank, N. K. Henderson, and R. L. Swisher: Rev. Sci. Instr. 40, 685 (1969)
14. M. M. Traum and D. P. Woodruff: J. Vac. Sci. Technol. 17, 1202 (1980)
15. P. Soukiassian, R. Riwan, J. Lecante, E. Wimmer, S. R. Chubb, and A. J. Freeman: Phys. Rev. B31, 4911 (1985)
16. Z. Hurych, P. Soukiassian, M. H. Bakshi, A. S. Bommannavar, C. E. Young, E. L. Schweitzer, M. J. Pellin, and D. M. Gruen: Phys. Rev. Lett., submitted

# Part IV

# Electron Stimulated Desorption Ion Angular Distribution (ESDIAD)

# Observation of Molecular Rotors on Surfaces by ESDIAD: Studies of $PF_3$ and $NH_3$ Chemisorption on Ni Surfaces

*J.T. Yates, Jr.*[1], *M.D. Alvey*[1], *M.J. Dresser*[1], *A.-M. Lanzillotto*[1], and *K.J. Uram*[2]

[1]Department of Chemistry, University of Pittsburgh, Pittsburgh, PA 15260, USA
[2]IBM, Thomas J. Watson Research Center, Yorktown Heights, NY 10598, USA

## 1. Introduction

We demonstrate in this paper that ESDIAD (electron stimulated desorption ion angular distribution) can give detailed information about the dynamical behavior of chemisorbed species and about the intermolecular forces which modify the dynamical motions of these species. Traditionally, spectroscopic methods have been used to study the dynamical behavior of molecules in gases, liquids, and in the adsorbed layer. In studies of the adsorbed molecular rotors $H_2$, HD, and $D_2$, EELS has been used to observe transitions between essentially free rotational states present on Cu(100) [1]. For CO hydrogen-bonded to OH groups on $SiO_2$ surfaces, rotational wings have been observed by IR [2]. Recently, the axial rotation of adsorbed methoxy species ($CH_3O$) on $Al_2O_3$ surfaces has also been studied in detail by IR [3]. These spectroscopic methods detect transitions between librational/rotational levels, but do not give information about the directionality of the surface forces which determine the barrier for hindered rotational motion. To provide this type of geometrical information, the ESDIAD method, by virtue of its ability to image chemical bond directions, has been employed here to observe hindered rotation in a chemisorbed layer. In addition, the freezing of the rotational motion of adsorbed molecules through the action of intermolecular forces between neighbor molecules has been directly observed using ESDIAD.

In this study, the digital ESDIAD technique was used to determine the P-F bond directions in chemisorbed $PF_3$ molecules on Ni(111) at low temperatures, and to observe the effect of increasing temperature on the time average of P-F bond directions. In addition, studies of the hydrogen-bonding interaction between $NH_3$ and CO adsorbates on Ni(110) have shown a degree of azimuthal freezing of N-H bond directions in $NH_3$ as a result of the attractive interaction between the dissimiliar adsorbates.

## 2. Experimental Methods

The important features of the digital ESDIAD system [4] employed here are shown in Fig. 1. A resistive anode detector behind the dual microchannel

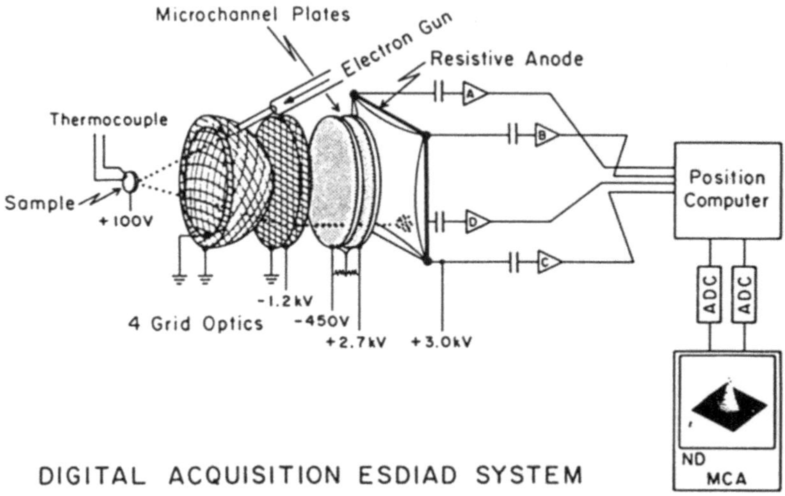


Fig 1. Digital acquisition ESDIAD system [4].

plate amplifier records individual ion collection events as a data pair (x,y) corresponding to the arrival position of positive ions in the ESDIAD pattern. By reversing grid bias, the apparatus may be used for digital LEED. Corrections for a broad x-ray background effect are routinely made [4a], and full data smoothing programs have been developed in this laboratory [4a]. Typical data collection times at $10^4$ counts/s are about 100s.

The vacuum system, employing AES, mass spectrometry, and our molecular beam dosing method has been described previously [5,6]. Details about the experimental methods employed here may be found in references [4-9].

## 3. Results and Discussion: $PF_3$ Chemisorbed on Ni(111)

We have previously shown that the $PF_3$ molecule adsorbs as an undissociated surface species on Ni(111) [7], in agreement with other studies [10-13]. Adsorption occurs on atop Ni sites by means of a σ-donor - π-acceptor mechanism, producing a p(2 x 2) LEED pattern at full coverage. Figure 2 is a schematic summary of the azimuthal orientation of chemisorbed $PF_3$ on Ni(111) which has been deduced by comparison of $F^+$ ESDIAD patterns and the LEED pattern of the Ni(111) substrate. A six-beam $F^+$ ESDIAD pattern, originating from two domains of $PF_3$ molecules is observed. The $F^+$ azimuths correspond to P-F bond directions oriented azimuthally in the nearest Ni-Ni neighbor directions from the Ni site which is bonded to the P atom of each $PF_3$ molecule. Thus the $PF_3$ molecule senses the Ni(111) lattice symmetry in its azimuthal orientation at 85K.

We have shown that electron beam damage of $PF_3$ efficiently produces both $PF_2$ and PF adsorbed species. $PF_2$ species bond to the three types of two-fold Ni-Ni sites, giving again six azimuthal P-F bonding directions WHICH

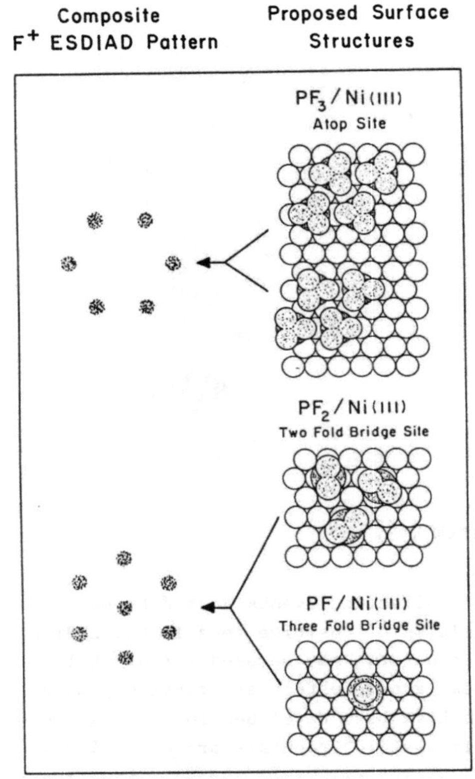


Fig. 2. The bonding of $PF_3$, $PF_2$ and PF species to Ni(111), and the resultant $F^+$ patterns [7].

ARE ORIENTED IN DIRECTIONS ROTATED 30° FROM THAT OF THE ORIGINAL SIX-BEAM $PF_3$ ESDIAD PATTERN. In addition, a normal $F^+$ ESDIAD beam is due to the formation of PF adsorbed species which are thought to occupy three-fold Ni hole sites as shown at the bottom of Fig. 2.

The thermal behavior of oriented $PF_3$ species on Ni(111) is shown below in Fig. 3. The two $PF_3$ coverages shown correspond approximately to 0.04 $PF_3$/Ni and 0.25 $PF_3$/Ni. Upon heating the low coverage layer to 275 K, azimuthal smoothing is quite apparent, and this effect is seen to be reversible on cooling back to 85K. We believe that this effect is due to the thermal population of hindered rotor states, where the molecules rotate about $C_{3v}$ axes which are normal to the surface. A simiilar heating experiment on the fully-covered layer shows some evidence of thermal broadening of the six beams, but under these coverage conditions more hinderance to rotation is obviously present. In both experiments, a tiny normal $F^+$ beam becomes evident in the center of each pattern due to a very slight production of PF species by electron beam damage [7].

Detailed measurements of the $F^+$ intensity ratio in the 6-fold peak and valley regions have been made as a function of temperature with the objective of determining the barrier to rotation of the $PF_3$ molecule [8]. An

Temperature and Coverage Dependence of $PF_3$ Hindered Rotation on Ni(111): $F^+$ ESDIAD

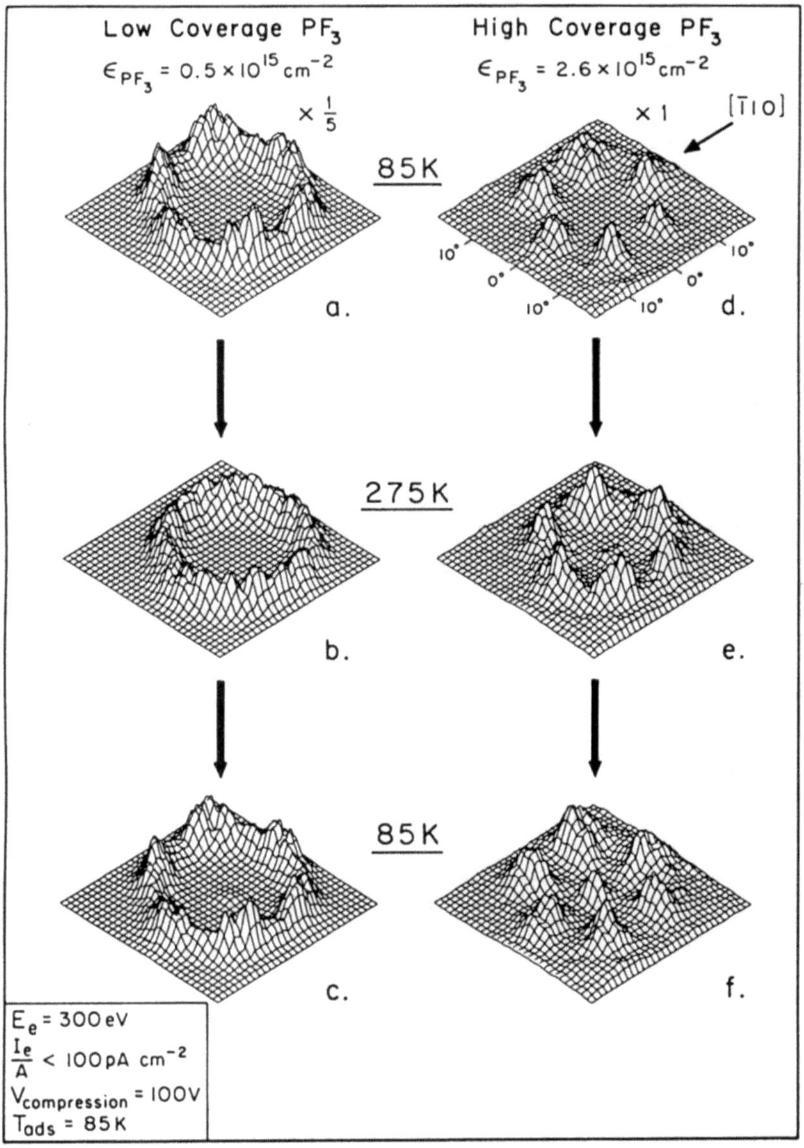


Fig. 3. Temperature and coverage dependence of $F^+$ ESDIAD patterns for $PF_3$ on Ni(111). Low coverage = 0.04 $PF_3$/Ni; high coverage = 0.25 $PF_3$/Ni.

analysis method similiar to that of KOEHLER and DENNISON [14] was used in which the wavefunctions and eigenvalues for 99 states in and above the potential well have been determined. Figure 4 shows the 99 state Boltzmann-weighted averaged ($\psi^*\psi$) for an 80 $cm^{-1}$ high sinusoidial barrier at three temperatures. Here it may be seen that the ratio of valley-to-peak intensity increases with increasing temperature for the six-fold barrier problem.

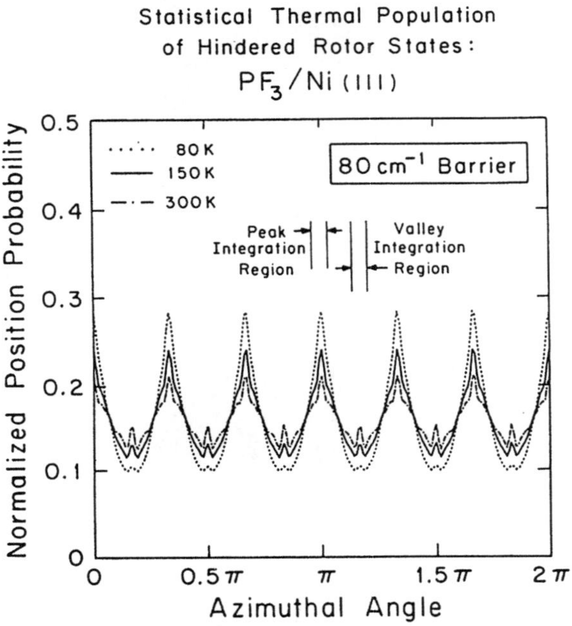


Fig. 4. Statistical thermal population of hindered rotor azimuths – $PF_3$.

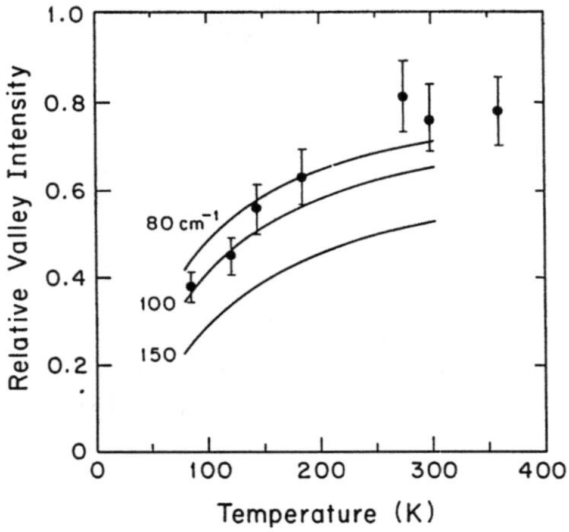


Fig. 5. Comparison of calculated and observed thermal effects in $PF_3$ on Ni(111).

A comparison of the experimental data measuring the relative valley intensity as temperature increases with the calculated relative valley intensity for three barrier heights is shown in Fig. 5. On the basis of this comparison between experiment and theory, we estimate that the barrier height is 80 ± 20 $cm^{-1}$ for $PF_3$ rotation from the preferred F-atop-Ni azimuthal bonding configuration on Ni(111).

The barrier to the aximuthal rotation of $PF_3$ is very low (~0.01eV) and represents a small effect energetically compared to the measured $PF_3$ chemisorption bond strength at the atop Ni sites (~1.1eV) [7]. On this basis, we believe that the azimuthal orientational effect is due to secondary bonding effects influencing the $PF_3$ energy. The enhancement of the barrier height at full coverage may be due to the requirement of cooperative rotational motion in the closely packed layer [8].

## 4. Results and Discussion: $NH_3$ + CO Interaction on Ni(111) and Ni(110) Surfaces

The chemisorption of CO on the Ni(111) and the Ni(110) surfaces has been well studied [15]. These studies have determined the location of the CO adsorption sites as well as the CO bond orientation. $NH_3$ adsorption has also been well studied, and site locations are known [9]. In ESDIAD studies of $NH_3$ on Ni(111), MADEY et al. [16-17] determined that coadsorbed O or Na could freeze the halo $H^+$ pattern of $NH_3$ into a spot pattern indicating a restriction of the dynamical motion of the $NH_3$ by coadsorption. For Ni(111), the formation of an $H_2NH...O$ hydrogen bond was postulated; however, on Ni(110) the effect was absent [18]. In the work to be reported below, we see evidence for the formation of a hydrogen bond between the electronegative O moiety of chemisorbed CO and the H moiety of chemisorbed $NH_3$, leading to modification of the rotor behavior of $NH_3$.

The angular distributions of ESDIAD ions from $NH_3$ interacting with various coverages of CO on the two Ni planes are shown in Fig. 6. In both cases for pure $NH_3$ layers, a halo $H^+$ pattern is evident from the rotor species. As the CO coverage is increased (at constant $\theta_{NH_3}$), Figs. 6B to 6D show systematic changes in the $H^+$ ESDIAD pattern. The broad central region of $H^+$ emission becomes stronger and more peaked [9]. This finding suggests that the $NH_3$ molecule is reoriented with one of its N-H bonds nearly normal to the surface in the case of Ni(111).

The data obtained for a similiar experiment on Ni(110) are strikingly different as shown in Figs. 6E to 6H. As the CO coverage is increased, the $NH_3$ halo is partially attenuated along the <1$\bar{1}$0> directions, breaking the $H^+$ halo into a symmetrical twin-arc pattern. In the later stages of CO adsorption, a sharp normal central $O^+$ beam is observed at the center of the twin-arc $H^+$ pattern [9].

Temperature programmed desorption experiments indicate that we are dealing on Ni(111) with an attractive $NH_3$...CO interaction of about 5 kcal/mol [9]. In this paper [9], we have also discussed $NH_3$ coverage calibrations. In the analysis below, we assume that the CO and $NH_3$ species

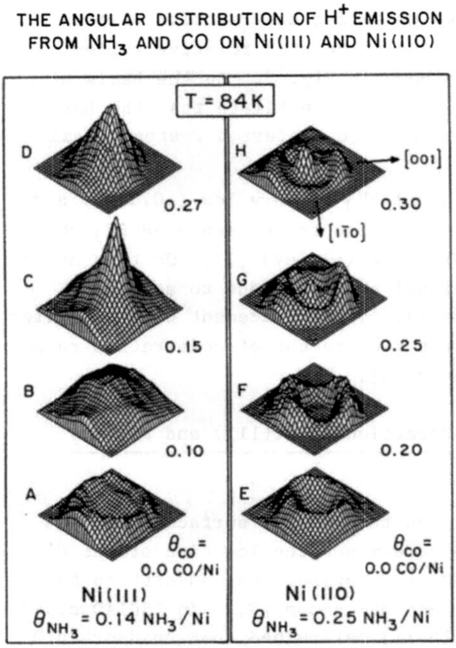


Fig. 6. ESDIAD patterns for the $NH_3$...CO interaction on Ni(111) and Ni(110) for increasing CO coverages.

occupy sites in the mixed layer which are identical to those occupied in the pure layers of each adsorbate [9]. In particular, $NH_3$ occupies three-fold hollow sites on Ni(111) [19], and atop sites on Ni(110) [5,20]. Two-fold short bridge sites are occupied by CO on both Ni planes.

It is clear by comparing results on the two Ni planes that we are dealing with very different types of $NH_3$ configurations following interaction with coadsorbed CO. As will be shown, the more closely packed Ni(111) plane supplies a substrate for short $NH_3$...CO interaction distances, whereas the more open Ni(110) plane involves longer interaction distances. Figure 7 schematically illustrates the effect of the strong interaction expected for the Ni(111) template and the weaker interaction for Ni(110).

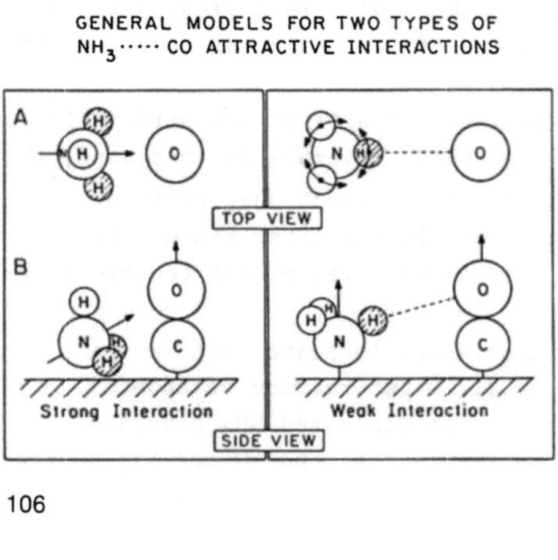


Fig. 7. A general model to explain $NH_3$...CO interactions on Ni(111) (left) and Ni(110) (right).

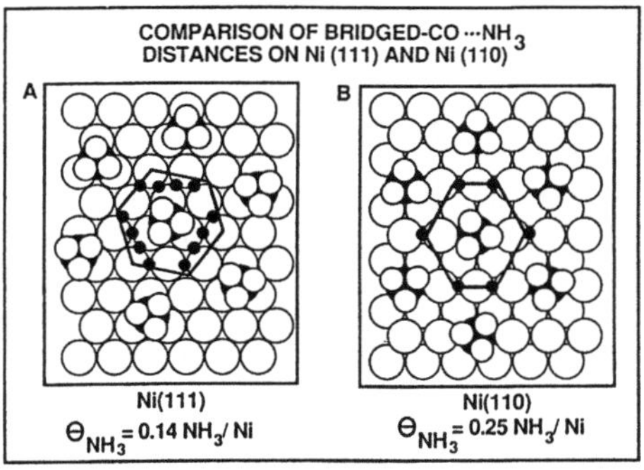


Fig. 8. Wigner-Seitz cells for $NH_3$ adsorbed on Ni(111) and Ni(110) surfaces. Dark circles = possible CO sites [9].

For strong interaction, $NH_3$ molecules are tipped over producing a nearly normal N-H bond direction; in the weaker interaction at longer distances, the N-H bond is oriented toward neighboring CO species resulting in partial attenuation of the rotor motion into hindered torsional $NH_3$ vibrations. The twin-arc $H^+$ pattern is produced because the interaction is along the $\langle 1\bar{1}0 \rangle$ directions with a superposition of two obverse structures like that shown in the right-hand panel of Fig. 7.

For the general understanding of site interaction problems, the Wigner-Seitz formalism is useful, as shown in Fig. 8. Here the two surfaces, containing $NH_3$ at the coverages of our experiments are divided into translationally equivalent hexagonal cells. Thus the sites within the cells typify all those on the surface. For Ni(111) there are 12 possible CO bridge sites shown as dark circles (6 at 259 pm; 6 at 313 pm). For Ni(110) however, all possible CO bridge sites lie near the boundary of the hexagonal cell, placing them approximately 374 pm from the $NH_3$ center.

Thus we can see that a characteristic difference in $NH_3$...CO interaction distance exists for the two Ni planes leading to strong and weak interactions on Ni(111) and Ni(110) respectively. For the Ni(110), a "remnant" of the $H^+$ halo persists as the $NH_3$ species undergo hindered rotations.

## 5. Summary

This paper has shown that ESDIAD is rather uniquely capable of defining the directional character of interactions between adsorbed molecules and their substrates as well as the intermolecular forces between adsorbate molecules themselves. In addition to the geometrical insight provided by ESDIAD, the barrier to the hindered rotation of an adsorbate has been measured using ESDIAD measurements at various temperatures. Molecules such as $PF_3$ and $NH_3$, which can exhibit hindered rotational behavior, make excellent test molecules for the study of forces operating in the chemisorbed layer.

## 6. Acknowledgement

We gratefully acknowledge support of this work by the Air Force Office of Scientific Research under Contract No. 86-0107.

## 7. References


1. S. Andersson and J. Harris: Phys. Rev. Lett., 48, 545 (1982); S. Andersson: J. Vac. Sci. Technol., A1, 1242 (1983).
2. T.P. Beebe, P. Gelin, and J.T. Yates, Jr.: Surf. Sci., 148, 526 (1984).
3. T.P. Beebe, J.E. Crowell, and J.T. Yates, Jr.: J. Chem. Phys., submitted.
4. (a). M.J. Dresser, M.D. Alvey, and J.T. Yates, Jr.: Surf. Sci., 169, 91 (1986); (b). M.J. Dresser, M.D. Alvey, and J.T. Yates, Jr.: J. Vac. Sci. Technol., A4, 1446 (1986).
5. C. Klauber, M.D. Alvey, and J.T. Yates, Jr.: Surf. Sci., 154, 139 (1985).
6. M.J. Bozack, L. Muehlhoff, J.N. Russell, Jr., W.J. Choyke, and J.T. Yates, Jr.: J. Vac. Sci. Technol., A5, 1 (1987).
7. M.D. Alvey and J.T. Yates, Jr.: submitted, J. Amer. Chem. Soc.
8. M.D. Alvey, J.T. Yates, Jr. and K.J. Uram: submitted, J. Chem. Phys.
9. M.J. Dresser, A-M. Lanzillotto, M.D. Alvey, and J.T. Yates, Jr.: Surf. Sci., in press.
10. G. Ertl, J. Küppers, F. Nitschke, and M. Weiss: Chem. Phys. Lett., 52, 309 (1977).
11. F. Nitschke, G. Ertl, and J. Küppers: J. Chem. Phys, 74, 5911 (1981).
12. G. Doyen: Surf. Sci., 122, 505 (1982).
13. H. Itoh and G. Ertl: Z. Naturforsch, 37, 346 (1982).
14. J.S. Koehler and D.M. Dennison: Phys. Rev., 57, 1006 (1940).
15. See reference [9] for listing of recent references.
16. F.P. Netzer and T.E. Madey: Phys. Rev. Lett., 47, 928 (1981).
17. F.P. Netzer and T.E. Madey: Surf. Sci., 119, 422 (1982).
18. T.E. Madey and C. Benndorf: Surf. Sci., 152/153, 587 (1985).
19. T.E. Madey, J.E. Houston, C.W. Seabury, and T.N. Rhodin: J. Vac. Sci. and Technol., 18, 476 (1981).
20. C. Klauber, M.D. Alvey, and J.T. Yates, Jr.: Chem. Phys. Lett., 106, 477 (1984).

# The Influence of Surface Structure on Mechanisms of Stimulated Desorption

*R.L. Kurtz, R. Stockbauer, and T.E. Madey*

Surface Science Division, National Bureau of Standards,
Gaithersburg, MD 20899, USA

## 1. Introduction

There have been many studies of the mechanisms of stimulated desorption and of the influence of surface structure, however few investigations have considered their simultaneous interplay. Our recent studies of stimulated desorption from $TiO_2$ surfaces have shown that there is a significant interaction between both of these aspects.

The work presented here involves the application of an array of surface spectroscopies to characterize the electronic structure and atomic geometries of two $TiO_2$ surfaces: $TiO_2$(110) and (001). Photon-stimulated desorption measurements show that there is a strong correlation between the ion desorption yield and the excitation of specific electronic levels, and that the valence state of the surface cation also influences this yield [1]. Electron-stimulated desorption (ESD) measurements, however, imply that the absolute yield of ions is mediated by the surface geometric structure. When the ESD ion angular distribution (ESDIAD) patterns are compared with models of surface structure, it is seen that ion emission is associated with specific ligand coordinations. The origin of the enhancement in ion emission from these sites is discussed in terms of the core-hole decay processes, excited state lifetimes, reneutralization probabilities and lattice dynamical response [2].

## 2. Experimental

The measurements presented here were made in two separate experimental chambers that have been discussed previously [3,4]. The chamber at the NBS SURF-II synchrotron storage ring contained provisions for electron and ion spectroscopies using a double-pass cylindrical-mirror analyzer (CMA) for energy analysis, electron-stimulated desorption (ESD) time-of-flight (TOF) for mass analysis (also using the CMA), and a small (25 mm) optics for studying electron-stimulated desorption ion angular distributions (ESDIAD). The second chamber contained a higher-quality (40 mm) ESDIAD optics which were used both for LEED and ESDIAD as well as for measurements of total angle-integrated ion yields.

The $TiO_2$ crystal used in these measurements was aligned, cut, and polished to within 0.5° of the specified surfaces. The sample was cut in the shape of a rectangular rod that exposed both the (110) and the (001) surfaces. Annealing was accomplished by electron bombardment of the rear of the crystal and temperatures were monitored with a W-5%Re / W-26%Re thermocouple pressed between the Ta sample mount and the crystal.

## 3. Electronic Structure

As a first step toward understanding the properties of ion desorption from $TiO_2$ surfaces, we must understand the behavior of their electronic structures for different surface preparations.

$TiO_2$ is one of the most studied of the transition-metal oxides, both experimentally and theoretically. It has long been known that the surfaces become O deficient when they are cleaned using Ar-ion bombardment (500 eV $Ar^+$ at 5 x $10^{-5}$ Torr) [5]. These sputtered

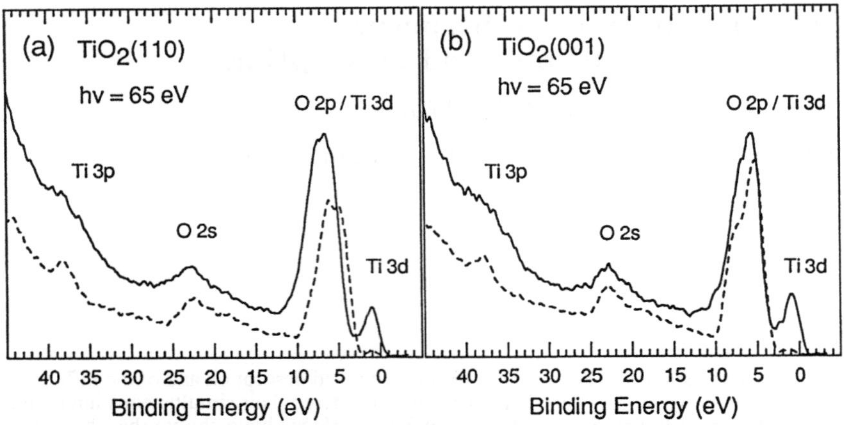


Fig. 1. Ultraviolet photoelectron spectra obtained with $h\nu$=65 eV from (a) $TiO_2$(110) and (b) $TiO_2$(001). The spectra from the sputtered surfaces are given by the solid curves and the spectra from the 1000K annealed surfaces are given by the dashed curves.

surfaces are disordered on a long-range scale and give no LEED patterns. Provided that the bulk is not too O deficient [6], annealing the sputtered surface diffuses O from the bulk to the surface and re-oxidizes the lower Ti valence states. This change in stoichiometry during sputtering and annealing is evident in Auger spectra, in x-ray photoelectron spectroscopy (XPS) data, and in ultraviolet photoelectron spectra (UPS); the signature of lower oxidation states in the latter spectra is the presence of a Ti 3*d*-electron peak ~0.9 eV below $E_F$. Studying the intensity of this feature is a sensitive method for measurement of the departure of the surface from stoichiometry. This is particularly true when the photon energy is adjusted to cause the 3*d* electron emission to resonate with the onset in excitation of the Ti 3*p* level: 3*d*-electron peak intensities are enhanced ~4-5 times compared to off-resonance.

In Fig. 1 the UPS spectra obtained from sputtered surfaces and from 1000K annealed surfaces are shown for the $TiO_2$(110) and $TiO_2$(001) orientations. Figure 1 shows these data for $h\nu$=65 eV; these off-resonance spectra display all of the electronic levels that can be excited by the energies of the photon and electron beams used in this work. It is clear from these data that the annealed $TiO_2$ surfaces show substantial reduction in the intensity of the Ti 3d electron state upon annealing. Both the (110) and the (001) surfaces approach stoichiometry at similar annealing temperatures as evidenced by the intensity of this feature as a function of the annealing temperature. Additional reduction in the intensity of the 3*d*-electron feature is obtained with subsequent oxygen exposure.

## 4. Ion Desorption

The behavior of the ion desorption from these two surfaces as a function of the incident photon energy has been discussed in detail previously [3,7]. The shape of the ion yield curve versus incident photon energy is similar for the different orientations and surface preparations studied in this work. The ion yield displays a sharp onset at the excitation onset of the Ti 3*p* level (~38 eV from Fig.1). The ion emission intensity rises rapidly, peaking at a photon energy of 47 eV and then decays, following the decay in the excitation profile of the Ti 3*p* level. By $h\nu$=65 eV, the PSD ion yield has decayed essentially to the baseline level. From these data, it appears that the creation of a Ti 3*p* core hole followed by an interatomic Auger decay (as described by the Knotek-Feibelman, KF, mechanism, [1]) is associated with the enhanced ion yield.

The magnitude of the ion yield, however, does not behave as one would expect in the light of this mechanism. As the Ti 3*d* electron population increases with increasing surface O

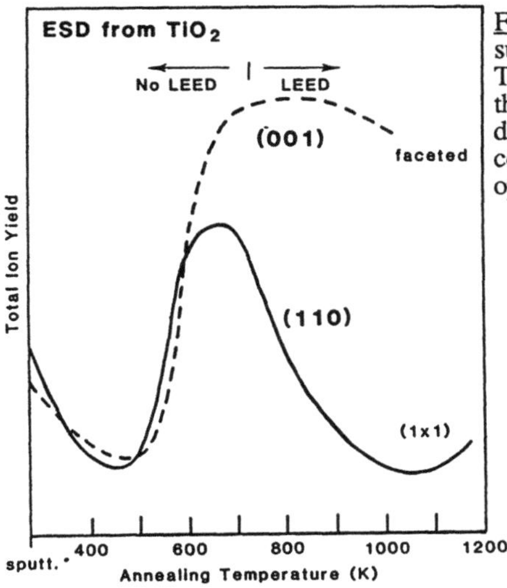


Fig. 2. Ion yield versus annealing temperature for surfaces that have been sputtered. The yield from the $TiO_2(110)$ surface is given by the solid curve while that from the $TiO_2(001)$ surface is given by the dashed curve. The data were obtained by pulse-counting from the phosphor screen of an ESDIAD optics with a 180 eV electron beam at $2 \times 10^{-9}$ A.

deficiency, the interatomic Auger decay is predicted to be quenched and the ion yield is expected to drop. In Fig. 2, the ion yield as a function of the surface annealing temperature is displayed for the (110) and the (001) surfaces; it is clear that the behavior of the ion yield is more complicated than the simple increase in intensity for higher annealing temperatures that would be predicted by the KF mechanism on a stoichiometry argument alone.

For both surfaces, the ion yield drops as the surface is first warmed from the freshly sputtered condition. At an annealing temperature of 600K, O begins to diffuse from the bulk and re-oxidize the surface. LEED patterns appear for both surfaces for annealing temperatures near 750K. It is near this temperature where the ion emission intensity from the (110) surface reaches a maximum. For temperatures greater than this, the yield from the (110) surface decays, giving a minimum near 1000K. The yield from the (001) surface, on the other hand, appears to saturate for temperatures above 750K and does not exhibit the substantial decay that the (110) surface shows.

Although angle-integrated UPS measurements suggest that the overall electronic structures of the two different surfaces are similar, the ion yield behavior from the two surfaces is quite different as a function of their annealing temperatures. This difference suggests that the geometry of the surface may play a role in determining the probability that an excitation will decay in a manner that results in a desorbing ion.

## 5. Geometric structure

When the $TiO_2(110)$ surface is sputtered and subsequently annealed, the surface becomes ordered and gives a (1x1) LEED pattern; the best patterns are obtained for annealing temperatures near 1000K. The $TiO_2(001)$ surface, on the other hand, gives a LEED pattern that implies that the surface is faceted and that the {011} facet planes that form are themselves reconstructed in a (2x1) fashion. These two surfaces therefore give different local site configurations for the desorbing ions and the influence of these sites should be apparent in measurements of their ion angular distributions.

In Fig. 3, the ESDIAD patterns obtained from the two surfaces are presented for three selected annealing temperatures, each. Fig. 3(a)-(c) gives those obtained from $TiO_2(001)$ for the room-temperature sputtered surface, 600K and 1200K annealed while 3(d)-(f) give those from the (110) surface for annealing temperatures of 600K, 800K, and 1000K, respectively. The patterns show reproducible off-normal emission features that are related to the local

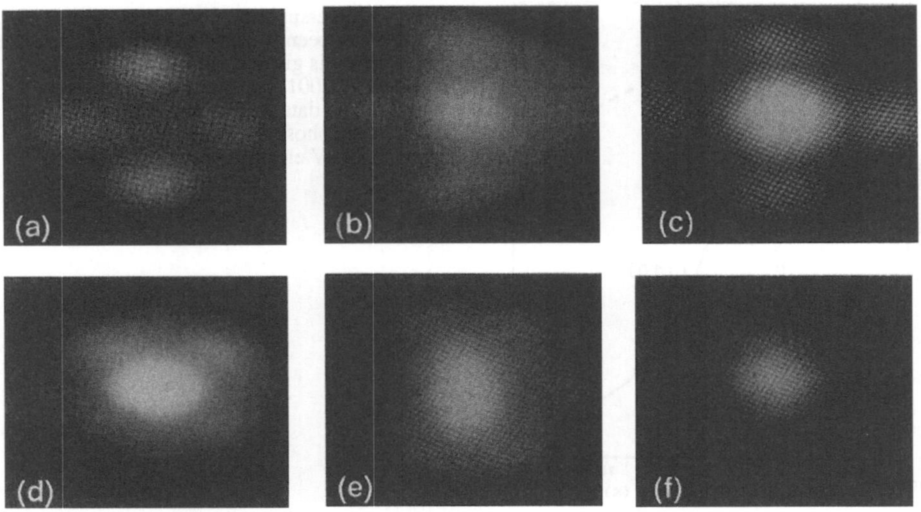


Fig. 3. ESDIAD patterns obtained from $TiO_2$ surfaces. $TiO_2$(001) (a) 300K sputtered; (b) 600K; and (c) 1200K annealed. $TiO_2$(110) (d) 600K; (e) 800K; and (f) 1000K annealed. The ion desorption directions are determined by the local atomic order.

geometric structures that form on the surfaces when they are annealed. It is interesting to note that we observe well-ordered ESDIAD patterns from surfaces that have been sputtered and do not exhibit LEED patterns. This is a consequence of the sensitivity of ESDIAD to the local geometric order that contrasts with the sensitivity of LEED to long-range geometric order.

The features of the ESDIAD patterns obtained from the (001) surface originate, most likely in the facets that have been reported to form [8]. Even after sputtering, the gross nature of the rearrangement due to the faceting is not removed and the ion desorption is perturbed by these remnants. When the annealing temperature is increased, the facets become well-defined and the off-normal emission spots in the ESDIAD patterns become well resolved. The ion desorption patterns from the (110) surface, on the other hand, are quite sensitive to the annealing temperature and show a progression of different off-normal emission spots. At 1000K, however, all of the off-normal emission features disappear and only a single normal-emission feature is observed with a low intensity. This is also the temperature where the LEED patterns indicate that the best surface order is achieved.

## 6. Discussion

The progression of the ESDIAD patterns that are observed as the $TiO_2$ surfaces are annealed is related to the different geometric structures that form. These structures are related to the stability of local atomic coordinations and, although a first-principles assessment of the factors leading to this stability is beyond the scope of this investigation, the ESDIAD patterns can give us important information that is not available in other data. In particular, by relating the observed patterns to structural models, we may be able to draw conclusions about the defect structures that form and about the mechanisms that are involved in producing desorbing ions.

The structure of the faceted surface is most straightforward to model and a drawing of this surface is reproduced in Fig. 4. This model is based on the LEED observation of facets composed of {011} planes [8], however, the planes are drawn assuming a bulk truncation and no reconstruction is considered. In a comparison of the local surface-Ti coordination of a bulk-truncated $TiO_2$(001) surface with the facet plane that is reported to form upon annealing, we find that the driving force for faceting produces an increased O-coordination of the surface

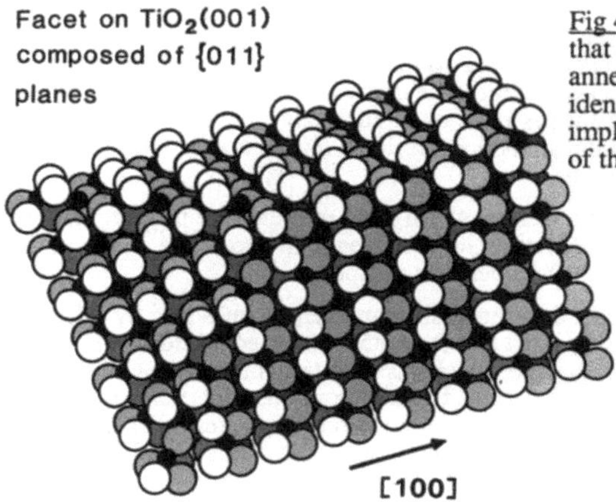


Fig 4. Ball model diagram of the facets that form on the $TiO_2(001)$ surface upon annealing. The three facets shown are all identical (011) planes. The ESDIAD patterns imply that desorption occurs from the edges of the facets. From [3].

Ti cations. The coordination increases from 4-fold (4 near-neighbor O ligands) to 5-fold after faceting. The Ti cation with enhanced coordination is thermodynamically more stable. This is also consistent with the stability of the $Ti_2O_3(10\bar{1}2)$ where the Ti coordination is also 5-fold and the atomic coordinations are similar.

When the observed ion emission directions are compared with surface models, however, it can be seen that the ion emission directions are not consistent with desorption from the (011) faces of the facets. Rather, the ions are emitted along the azimuths defined by the normal to the facet *edges*. This means that the enhanced coordination, and the closer approximation to a truely maximal-valent electronic configuration, is not associated with high ion yield. The edges of the facet, from the idealized ball-model drawings, contain reduced coordination: along the top edges the Ti are all in 4-fold coordinated sites while along the side edges of the facets, the Ti are in alternating 4- and 5-fold coordinated sites. This implies that high ion yield is associated with these reduced coordinations and perhaps even with 4-fold coordinated Ti cations.

If similar notions are applied toward understanding ion desorption from the $TiO_2(110)$ surface, the ESDIAD patterns can be understood in terms of ion desorption originating from the edges of steps. This can give rise to the off-normal emission spots in Fig. 3(d)-(f). As the roughened, sputtered surface is annealed and the structure approaches the flat, well-ordered (1x1) surface observed by LEED, a distribution of steps is expected, whose orientations depend on particulars of their thermodynamic stabilities at the annealing temperature. These off-normal emission ESDIAD features can therefore be related to the orientation of these steps. Figure 5

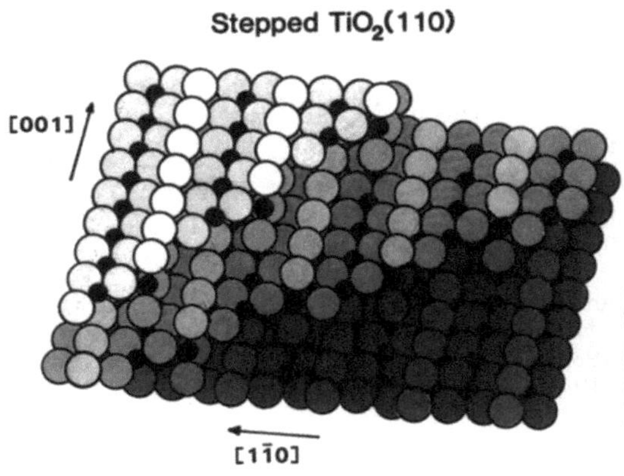


Fig. 5. Ball model diagram of a stepped $TiO_2(110)$ surface. The steps shown are implied by the desorption directions that are observed in the ESDIAD patterns of Fig. 3. From [3].

shows a ball model of step edges that can give rise to the ion emission directions that are observed in Fig 3(e) and (f). These step edges are drawn in a stoichiometric fashion, and it is clear that there are reduced-coordination sites present. The normal distribution of 5- and 6-fold coordinated Ti sites on the flat (110) surface is reduced at such edge sites to 4- and 5-fold coordinated sites. This is consistent with the prediction, from the (001) surface that these sites are more efficient in producing desorbing ions.

The origin of the enhancement in ion yield at these sites may be due to a number of factors; we can separate these factors by delineating their influence in two of the stages of the desorption event: the decay process and the subsequent ion escape.

Once the core hole has been produced, the atomic site can exert its influence by affecting the likelihood that the decay results in a long-lived repulsive state. At an edge site, the Ti cation has a reduced O-coordination. This may increase the chance that only one of the O ligands is involved in the interatomic Auger decay process, since fewer are present to participate. In turn, this increases the probability that all of the charge that is transferred originates from one O site. A similar site-specificity has been seen in ESD of $O^+$ from stepped crystalline W(110) surfaces [9]. As well, the magnitude of the charge at these edge-site O ligands may be reduced.Since three electrons must be removed from a lattice $O^{2-}$ to give an $O^+$, these two effects may be significant. Also, a surface site with a reduced O-coordination will be associated with a smaller overlap of the O valence electronic wavefunctions. If this is the case, the bands would be expected to be narrower. With narrower bands, the final states of the decay (which are O 2*p* holes) will be longer-lived and result in enhanced desorption [10].

In the stage involving the ion escape, the atomic site geometry will clearly influence the lattice dynamics. The details of the lattice response can be examined through calculations such as those of Walkup and Avouris [2] and Walkup and Kurtz [11]. These calculations are useful in examining both the possibility of lattice trapping of a departing ion as well as studying the final kinetic energies of the departing ions. Since the probability for reneutralization is related to the time that it takes for the ion to escape, an understanding of the lattice dynamics is essential. The sites that are proposed to give rise to desorption may be assessed in this manner; by comparison of calculated ion trajectories and relative kinetic energies with the experimentally observed ESDIAD patterns and kinetic energy distributions, we should continue to progress toward a more complete understanding of the process of stimulated desorption.

## 7. Summary

This work has shown that the electronic structure of a surface provides an incomplete description of the ion production probability. The absolute yield of ions is mediated by the surface geometric structure and that specific sites can be significantly more active in the stimulated desorption process.


## Acknowledgement

This work has been supported, in part, by the U. S. Office of Naval Research.


## References


1. M. L. Knotek and P. J. Feibelman, Phys. Rev. Lett. 40, 964 (1978).
2. R. E. Walkup and Ph. Avouris, Phys. Rev. Lett. 56, 524 (1986)
3. R. L. Kurtz, Surf. Sci. 177, 526 (1986).
4. D. M. Hanson, R. Stockbauer, and T. E. Madey, Phys. Rev. B 24, 5513 (1981).
5. V. E. Henrich, Prog. Surf. Sci. 9, 143 (1979).
6. R. L. Kurtz, H. H. Chen, R. Stockbauer, and T. E. Madey, unpublished.
7. R. L. Kurtz, R. Stockbauer, and T. E. Madey, in: Desorption Induced by Electronic Transitions, DIET II, Eds. W. Brenig and D. Menzel (Springer, Berlin, 1985) p. 89; R. L. Kurtz, R. Stockbauer, and T. E. Madey, Nucl. Instr. and Methods B13, 518 (1986).
8. L. E. Firment, Surf. Sci. 116, 205 (1982).
9. T. E. Madey, Surf. Sci. 94, 483 (1980).
10. P. J. Feibelman, Surf. Sci. 102, L51 (1981).
11. R. E. Walkup and R. L. Kurtz, this volume.

# DIET in the Second Layer: An ESDIAD Study of $NH_3$ on a CO Layer on Ni(111) and Ni(110)

*M.J. Dresser*[1], *A.-M. Lanzillotto*[2], *M.D. Alvey*[2], *and J.T. Yates, Jr.*[2]

[1]Department of Physics, Washington State University, Pullman, WA 99164-2814, USA

[2]Surface Science Center, Department of Chemistry, University of Pittsburgh, Pittsburgh, PA 15260, USA

## 1. Introduction

We have recently reported on the interactions of CO and $NH_3$ on nickel surfaces [1,2], as determined by the electron stimulated desorption ion angular distribution (ESDIAD) method. In this paper we review that work and present the results of calculations that show the role of the CO layer in DIET phenomena. Tochihara et al. [3] have also studied this system using MQS spectroscopy, and their results are of importance in our conclusions.

Our interest here concerns ESDIAD from ammonia adsorbed on top of the saturated layer of CO on both the (111) or (110) surfaces of nickel. The work of Tochihara et al. [3] was the first to demonstrate that ammonia adsorbs on the surface of a CO monolayer. This observation was also confirmed by this laboratory [2] where it was found that strong $H^+$ signals arose from the ESD of adsorbed $NH_3$ above the CO monolayer. A work function decrease measured by Tochihara et al. suggests that the ammonia may be adsorbed with a net positive outward dipole moment, and they conclude that the nitrogen end of $NH_3$ then bonds to the O moiety of CO. This seems to be an unlikely bonding configuration and is not known to occur in any other chemical situation. An alternate model that we propose suggests bonding through one or two of the H ammonia moieties to the O moiety of one or two surface CO. If the $C_{3v}$ axis of the ammonia molecule lies close to parallel to the surface the remaining hydrogen moieties provide the source of $H^+$ for ESD. The apparatus used in this experiment was described earlier in this volume by Yates et al. [4], as well as in Reference 5.

## 2. Results

In Fig. 1 we present a survey of ESDIAD and LEED information for $NH_3$ and CO separately adsorbed on the Ni (111) and (110) surfaces. These data were presented previously in References 1,2,5,6. The upper three frames (A-C) of Fig. 1 are for the (111) surface and the lower three frames (D-F) correspond to the (110) surface. The two left frames (A & D) show the digital ESDIAD $H^+$ patterns for ammonia from clean nickel surfaces. Note that both patterns are azimuthally symmetric although the polar angular distributions are obviously different. These patterns were identified as "halos" in photographic ESDIAD and we will continue that

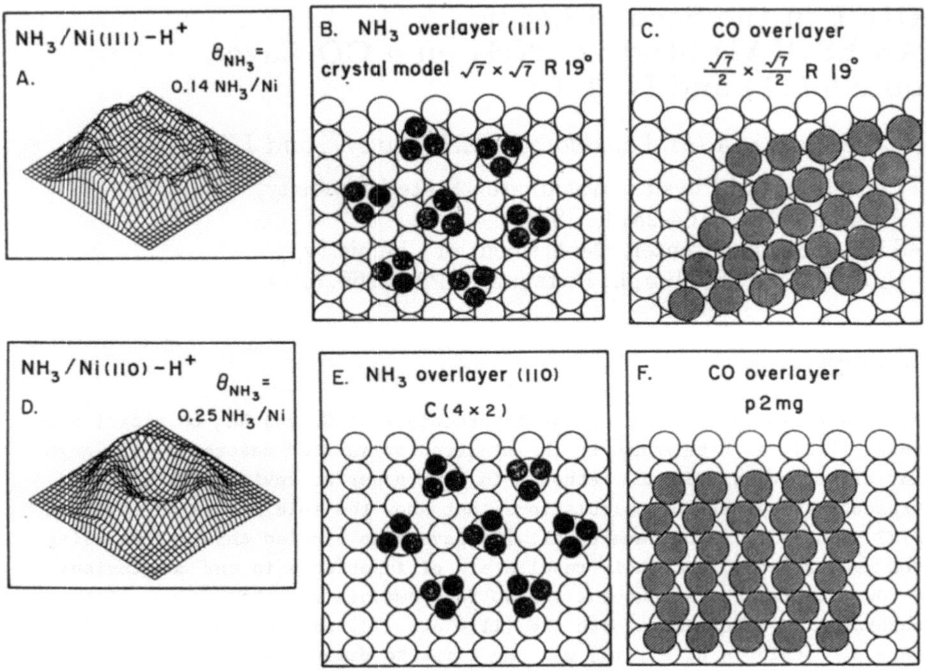


Fig. 1. A review of $NH_3$ and CO ESDIAD behavior on Ni(111) and (110).

designation. The two frames in the center (B & E) indicate the structure and orientation of rotating $NH_3$ molecules responsible for the $H^+$ halo pattern. The two frames on the right (C & F) show the full coverage structure of the chemisorbed CO layer used as a substrate in these experiments. In the case of CO on Ni(111) the CO molecules are oriented normal to the surface, but on Ni(110) the molecules are alternately tipped 19° from the surface normal to either side of the (110) surface ridges.

The digital ESDIAD patterns for $NH_3$ above the monolayer of CO are shown in Fig. 2. The left side of Fig. 2 corresponds to $NH_3$ from the ($\sqrt{7}/2$ x $\sqrt{7}/2$)R19° layer of CO on Ni(111). The upper figure, (A), is the pattern just after exposure to ammonia at T=84K, and the lower figure, (B), was taken after mild annealing, (T=160K). In each case the patterns show an azimuthal symmetry which is reminiscent of the halos seen on bare nickel; however, the polar angle of this circular ridge is much smaller than the halos found for $NH_3$ chemisorbed on clean Ni. We will subsequently refer to this pattern as the "volcano with crater". The ESDIAD patterns on the right side of Fig. 2 are the result of $H^+$ emission from $NH_3$ on the p2mg configuration of CO on Ni (110). Again the A pattern is for ESDIAD just after exposure to the ammonia at T=84K, and B is the pattern after mild annealing (T=200K). On this surface we also have a volcano with crater but there is an azimuthal twofold symmetry manifest as an attenuation of the

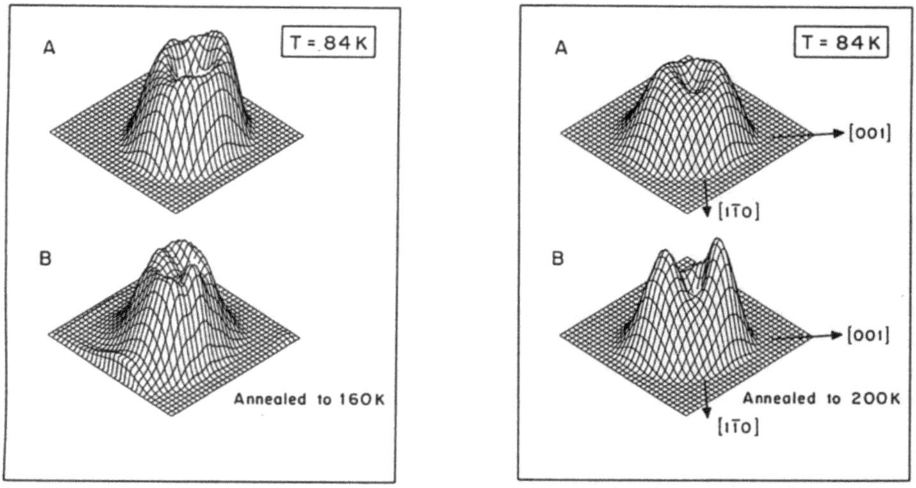


Fig. 2 The $H^+$ ESDIAD patterns for $NH_3$ on a monolayer of CO on Ni (111) and (110) at 84 K.

crater rim in the directions parallel to the Ni (110) surface ridges. Thus the twofold Ni(110) symmetry is transmitted through the CO layer and is manifest in the outer layer ammonia ESDIAD patterns. Rough estimates of binding energies have been made with temperature-programed desorption techniques. Table 1 provides a summary of those estimated energies. Note that $NH_3$ bonds to CO more strongly than it does to itself in multilayers, but not as strongly as is found in its chemisorption bond to the nickel surface.

Table 1: $NH_3$ Desorption Activation Energies

| Desorption From: | Energy (kcal/mol) |
|---|---|
| $NH_3$ multilayer | ~7 |
| Ni(111) | ~17 |
| CO layer on Ni(111) | ~12 |

## 3. Discussion

The important observations that a satisfactory model of this system must explain are:

A. A decrease in work function with ammonia adsorption.
B. Hydrogen moieties oriented to yield strong $H^+$ ESD.
C. Polar angles of $H^+$ release in the range of $20^o$–$40^o$.

The Tochihara model [1], with the ammonia oriented with the nitrogen down satisfies criteria A and B. Criterion C is not met because, if the $C_{3v}$ axis of the molecule is normal to the surface, then the hydrogen polar bond angles would be the same as on bare nickel. It is tempting to argue that the CO layer serves as a spacer layer to reduce the image potential effects on the released $H^+$ and thus place the detected ESDIAD peak at smaller polar angles, closer to the true polar angle of release. Using the formula of Miscovick et al. [7] for the shift in ESDIAD peak polar angle due to image effects, we find that there is only a 4° change in predicted ESDIAD peak polar angle for reasonable atomic dimensions above a CO layer. In fact the maximum positive shift predicted by Miscovick et al., for their example, is only 12°, a change which is much smaller than we observe experimentally. The other weakness in this model is that there are no other known examples of $H_3N\cdots OC$ type bonding.

Our proposal, as shown in Fig. 3, is that the $NH_3$ molecule is oriented with its $C_{3v}$ axis roughly parallel to the surface and with at least one of its hydrogen moieties bonded to the CO surface layer. One hydrogen moiety will then provide the source of $H^+$ ions for ESD processes, and its polar angle of release can then be closer to ~30° which is consistent with the observed crater rim radius. The left frame of Fig. 3 shows two possible orientations that satisfy criteria B and C. The important distinction between these two is the double coordination of the left orientation as opposed to the single coordination of the right. These two orientations could be shifted 10° in any direction and still satisfy criteria B and C, but within that range these are the only possible orientations that can satisfy the criteria. We favor the left, or double coordinated, orientation because the desorption activation energy of 12 kcal/mol is too large for a singly coordinated hydrogen bond, but is quite reasonable for a doubly coordinated bond [8]. Both structures involve an outward pointing $N\text{-}H^{\delta+}$ bond which could contribute to an outward directed positive dipole moment related to the work function shift observed by Tochihara et al., and our criterion A. However, additional charge transfer may occur between CO and Ni as a result of $NH_3$ adsorption thus providing an additional contribution to the work function shift.

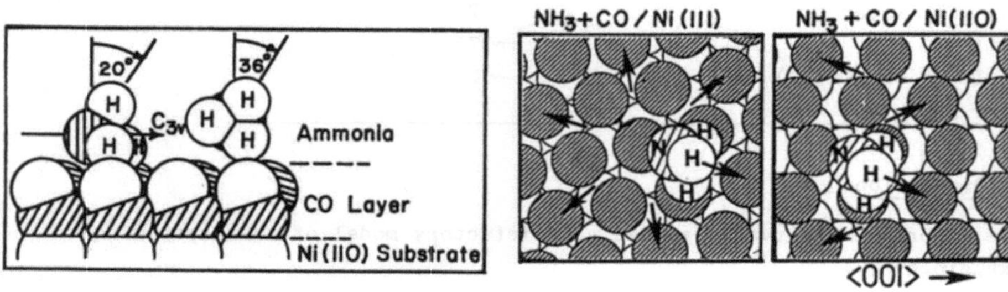


Fig. 3 Possible models for the orientation of $NH_3$ on CO on nickel substrates that are consistent with the experimental results. The arrows indicate the azimuthal $H^+$ emission directions for other equivalent bonding sites.

A full explanation of the observed ESDIAD patterns requires that the location and azimuthal orientation of the $NH_3$ molecule on the substrate must also be specified. The right side of Fig. 3 shows possible positions for the doubly coordinated adsorption sites on the two CO overlayers of Ni(111) and Ni(110). The arrows show the azimuthal direction of $H^+$ release from the outer hydrogen moiety. For the CO/Ni(111) surface there are six unique possible directions of release if the hydrogen moieties are held to strict positions over the CO molecules. Considering the weak nature of hydrogen bonding and its inherently electrostatic nature we postulate that the six directions are smoothed into a circular symmetry by the electrostatic potentials at hydrogen bond lengths from the surface and/or thermal disorder in both the $NH_3$ and CO layers. There are four unique directions on the CO/Ni(110) and these are plus or minus ~40° from the <001> surface direction which is, in turn, perpendicular to the (110) surface ridges. The ridge of oxygen moieties may be considered as a ridge of negative charge zigzagging along the <1$\bar{1}$0> direction. The $NH_3$ hydrogen bonding can occur anywhere along this zigzag ridge so that the remaining hydrogen of $NH_3$ can be emitted azimuthally anywhere within the range of ± ~40° from the <001> surface direction. This explains the enhancement of the ESDIAD pattern's crater rims along the <001> surface direction on the (110) surface. When the ammonia is initially adsorbed at 84K some molecules may span across CO molecules on adjacent Ni ridges thus emitting $H^+$ ions along the ridges. The mild annealing of this system causes the molecules that span the ridges to find the lower energy sites along the ridge and thus give a more pronounced attenuation of $H^+$ along the ridges as seen in the lower right pattern of Fig. 2.

## 4. Acknowledgement

We gratefully acknowledge support of this work by the Air Force Office of Scientific Research under Contract No. 86-0107.

## 5. References

1. M.J. Dresser, A-M. Lanzillotto, M.D. Alvey, and J.T. Yates, Jr.: Surf. Sci., to be published.
2. A-M. Lanzillotto, M.J. Dresser, M.D. Alvey, and J.T. Yates, Jr.: Surf. Sci., to be published.
3. H. Tochihara, G. Rocker, J.D. Redding, J.T. Yates, Jr., R.M. Martin, and H. Metiu: Surf. Sci. 176, 1 (1986).
4. J.T. Yates, Jr., M.D. Alvey, M.J. Dresser, and A-M. Lanzillotto: In Desorption Induced by Electronic Transitions DIET III, ed. by M. Knotek and R Stulen, Springer Series in Surf. Sci. to be published.
5. M.J. Dresser, M.D. Alvey, and J.T. Yates, Jr.: Surf. Sci. 169, 91 (1986).
6. M.D. Alvey, M.J. Dresser, and J.T. Yates, Jr.: Surf. Sci. 165, 447 (1986).
7. Z. Miscovick, J. Vukanic, and T.E. Madey: Surf. Sci. 141, 285 (1984).
8. G.C. Pimentel, and A.L. McClellan, The Hydrogen Bond, Freeman and Co., San Francisco, (1960) p. 169.

# ESDIAD of Small Molecules on Surfaces: A Few Caveats

*T.E. Madey, M. Polak*, A.L. Johnson**, and M.M. Walczak****

Surface Science Division, National Bureau of Standards,
Gaithersburg, MD 20899, USA

## I. Introduction

ESDIAD (Electron Stimulated Desorption Ion Angular Distributions) has proven to be an extremely useful method for characterizing the local bonding structure of molecules at surfaces and a number of examples are indicated below. In order to optimize ESDIAD measurements, however, care must be taken to avoid such pitfalls as beam damage, "defective" regions of the sample surface, and impurity adsorbates. Several recent illustrations of these problems are discussed.

## II. Uses and Limitations of ESDIAD: A Summary

There are many cases in which ESDIAD has been used to identify the structures of adsorbed molecules; a number of these structures have been confirmed using independent surface-sensitive methods [1-3]. Examples include standing-up CO and NO on metals [4], "inclined" OH and CO [1,3,5,6], $NH_3$ on transition metals [1,7], ordered and disordered $H_2O$ adsorbed on many surfaces (monomers, dimers and 2-d clusters) [8,9], small fluorinated molecules [10], and adsorbed hydrocarbons [11]. Of particular interest are the many observations of impurity-induced ordering of $NH_3$, $H_2O$, and CO observed using ESDIAD [1,2,12]. Oxide surfaces of W and Ti have been characterized [13,14], and the bonding of small protic molecules to Si surfaces has been studied [15].

Although ESDIAD is a useful surface structural tool, a great deal of care is necessary in both the measurements and the interpretation of results. Pitfalls exist to trap the unwary experimenter, including sensitivity to "minority states," coadsorbed impurities which influence ion yields, sensitivity to steps and defects (often correlated with minority states), beam damage, and final state effects (influence of image potential and reneutralization on ion yields and trajectories [16]).

The problem of minority states in ESD (i.e., cases in which the ion yield arises from a small fraction of the adsorbed surface species) and the resultant non-linear ion yield as a function of coverage are well known, and have been discussed recently by MENZEL [17]. In general, minority states seem to be more of a difficulty with atomic adsorbates (e.g., oxygen) than for molecular adsorbates (e.g., CO). Moreover, surface microstructure is of

---

*Permanent Address: Dept. of Materials Engineering, Ben-Gurion University, Beer-Sheva 84105, Israel.
**NBS/NRC Postdoctoral Research Associate.
***Permanent Address: Dept. of Chemistry and Ames Laboratory, Iowa State University, Ames, Iowa 50011.

importance in minority states. The extreme sensitivity of $O^+$ ESD to atomic steps and defects was shown clearly for oxygen on stepped W(110) [18] and for $TiO_2(001)$ [14]. In contrast, molecular states of CO on tungsten (virgin CO, $\alpha$-CO) give high ESD $O^+$ signals from <u>both</u> majority terraces and minority step sites [13]. Evidence that ESD is sensitive to the majority species for molecular adsorbates is the observation that the ESD ion yield is proportional to coverage, at least for low coverages. The ability of ESDIAD to characterize the influence of both terraces and steps on molecular bonding (CO on W [1,3], $NH_3$ on Fe(100) [19], $C_5H_8$ on Ag (221) [20]) is a positive advantage, although care must be taken [21] in the interpretation. Coadsorbed impurity atoms can also influence ion yields strongly, by introducing electronic and/or geometric changes in bonding [22].

In the following paragraphs we discuss in detail several examples of possible pitfalls in ESD as revealed by ESDIAD studies: (a) Extreme sensitivity to beam-induced surface chemistry in $H_2O$ on Cu(110); electron-stimulated dissociation with a high cross section gives rise to new ion desorption patterns, (b) Decoration of steps and defects in dissociative adsorption of $NF_3$ on Ru(001); the ESDIAD patterns from highly defective regions of the surface (near spot-welds, edges) are far more complex and "interesting" than from the majority terraces. Both of these examples illustrate the desirability of angle-resolved measurements in ESD, using a wide-aperture detector to detect ion desorption at unanticipated angles.

## III. Experimental Procedures

The data are obtained using ultrahigh vacuum display-type ESDIAD systems which have been discussed previously [1-3]. The ESDIAD data are analyzed by means of the digital video data acquisition/analysis system described elsewhere in this volume [23]. The Cu(110) and Ru(001) crystals are cleaned using standard techniques, are cooled cryogenically (to ~100K), and heated resistively. A dedicated gas doser system is used for introducing fluorinated species into the uhv chamber; molecules are beamed directly onto the sample surface.

## IV. Results and Discussion

### A. Electron beam damage in $H_2O$ on Cu(110).

The interaction of $H_2O$ with surfaces is a subject of intensive investigation [8,9,24]. Of particular interest is the existence of attractive lateral interactions between neighboring adsorbed $H_2O$ molecules, due to hydrogen-bonding. This results in the formation of two-dimensional clusters on surfaces. In studies of the structure and chemistry of $H_2O$ on the structurally-similar Ni(110) and Cu(110) surfaces [9,25], we find evidence for the smallest possible adsorbed cluster, $(H_2O)_2$, the water dimer. The dimer has an interesting local structure, in registry with the (110) lattice plane.

As shown in Fig. 1a, the water dimer structures formed at ~100K on Cu(110) yield a characteristic rectangular ESDIAD pattern containing four emission maxima ("four beams"). In the contour plot of the pattern, the four maxima are parallel to [001] and [011] azimuths. There is also a central maximum due to ion emission normal to the surface, from larger $H_2O$ bilayer clusters. This pattern is highly sensitive to electron beam damage, and rapid changes are observed during the measurements. After only 23 seconds at a current of $\sim 5\times10^{-8}$ A (current density $\sim 5\times10^{-6}$ A/cm$^2$, energy= 280eV), pattern 1b is observed. This pattern contains new emission along [001] azimuths. After 45 seconds, the new "two beam" pattern along [001]

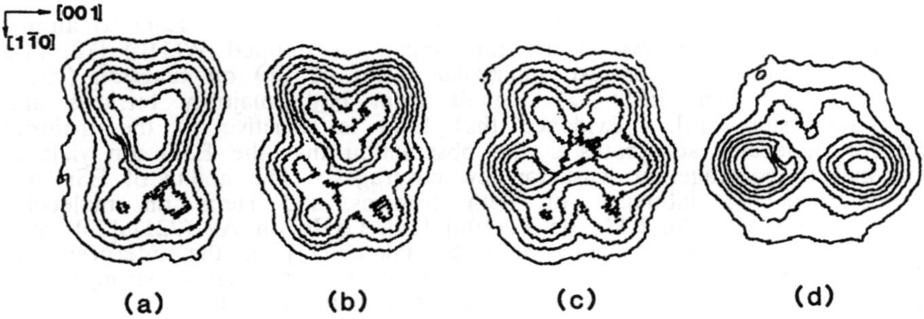


Figure 1. Contour plots of $H^+$ ESDIAD patterns observed during electron beam induced dissociation of $H_2O$ on Cu(110). Water coverage ~0.15 monolayers at 100K. Beam energy = 280eV, beam current density ≈$5x10^{-6}$ A $cm^{-2}$. (a) 4 sec., (b) 23 sec., (c) 45 sec., (d) 160 sec.

azimuths is quite intense [Fig. 1c]. After 160 seconds, the "four beam" dimer pattern has been completely replaced by the intense "two beam" pattern [Fig. 1d].

Figure 2 contains plots of the intensities of the ESDIAD features in Fig. 1, as a function of electron bombardment time under the above conditions. Both the central beam and the four-beams decrease rapidly with bombardment time. The delayed onset for intensity decrease of the four-beam pattern results from a competition between the destruction of dimers, and the formation of dimers from beam-induced dissociation of bilayer clusters. The two-beam pattern increases with bombardment time and eventually dominates the ESDIAD pattern [26].

Similar data are seen for $H_2O$ adsorbed on Ni(110) [9], and are interpreted similarly: the four-beam pattern is identified with dimers and/or $H_2O$ stabilized by OH(ad), and the two-beam pattern is due to inclined OH(ad) species which arise from electron-beam-induced dissociation of adsorbed water. The OH(ad) species have been identified and characterized using other methods. A schematic illustration of one aspect of the dissociation process is shown in Fig. 3.

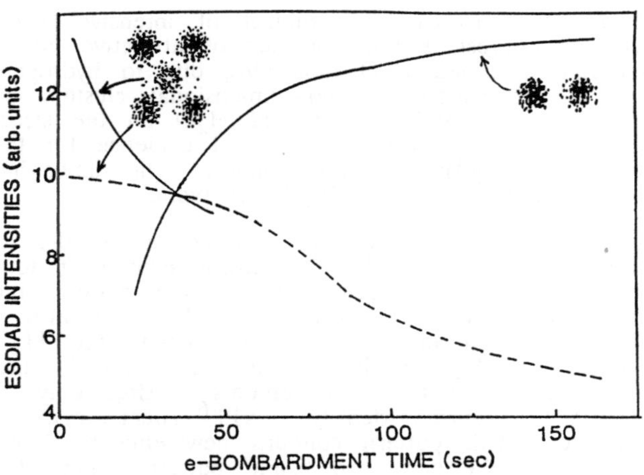


Figure 2. Intensities of $H^+$ ESDIAD beams vs. electron bombardment time for $H_2O$ on Cu(110). Experimental conditions as in Fig. 1.

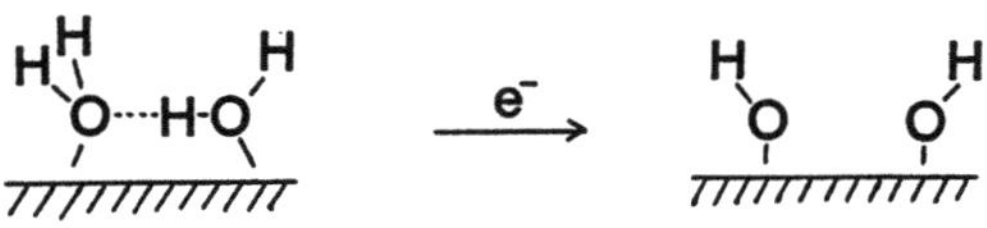


Figure 3. Schematic illustration of electron-beam-induced dissociation of water dimers $(H_2O)_2$ leading to the formation of inclined OH(ad) on Cu(110).

The data of Figs. 1 and 2 provide a warning to experimenters: minimize electron beam exposure in surface studies of "fragile" adsorbates. The cross section for dissociation of $H_2O$ on Cu(110) is $\sim 10^{-16}$ cm$^2$; the critical dose for substantial damage in ESDIAD (Fig. 1c) is $\sim 2\times10^{-4}$ C/cm$^2$. For comparison, in a typical LEED experiment (1 $\mu$A beam, 1mm diameter, 100eV), this dose would be exceeded in 2 seconds.

B. The interaction of $NF_3$ with Ru(001): Decoration of steps and defects.

As part of a larger study of fluorinated molecules on semiconductor and metal surfaces, we are characterizing the adsorption of $NF_3$ on Ru(001) using ESDIAD, LEED, thermal desorption spectroscopy (TDS) and Auger electron spectroscopy [27]. $NF_3$ adsorbs dissociatively with a high sticking probability. Some molecular $NF_3$ desorbs upon heating, with a maximum desorption rate at $\sim$200K. All molecular desorption is complete by 400K, but AES shows that N and F remain on the surface up to 1100K. The LEED pattern remains (1x1) at all coverages and annealing condition. $F^+$ ions are detected in ESD.

As shown in Fig. 4, the observed $F^+$ ESDIAD patterns are different, depending on the region of the surface struck by the electron beam. As the beam is scanned from edge-to-edge and top-to-bottom of the 5mm x 8mm sample, the patterns characteristic of the majority planar terraces of the Ru(001) sample are shown in the left column (Fig. 4a,b) for 100 and 300K. In these cases, the $F^+$ emission is centered about the surface normal, with no noticeable azimuthal variation.

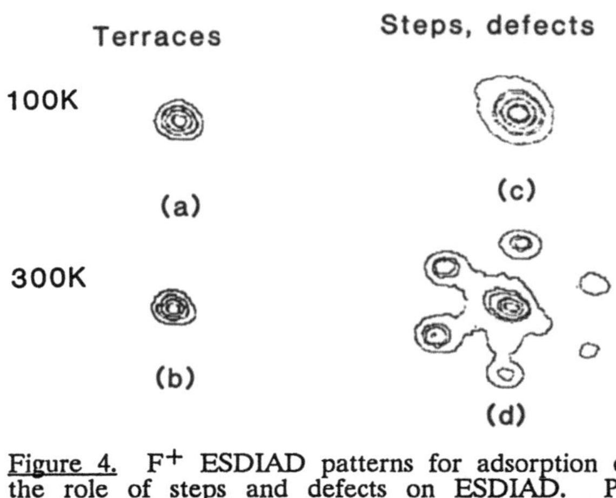


Figure 4. $F^+$ ESDIAD patterns for adsorption of $NF_3$ on Ru(001), illustrating the role of steps and defects on ESDIAD. Patterns (a), (b) are from the majority terraces; patterns (c), (d) are characteristic of defective regions of the surface. All measurements are at 100K, following annealing.

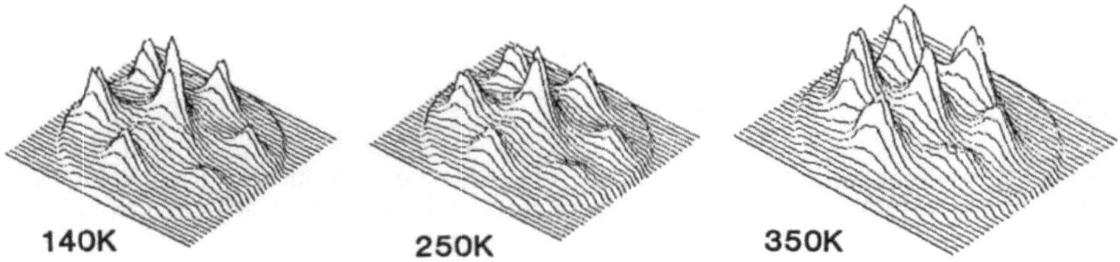


Figure 5. $F^+$ ESDIAD perspective plots illustrating reversible temperature effects for $NF_3$/Ru(001).

In striking contrast, when the beam strikes the edges of the sample, or defective regions near spot-welds, the patterns in the right column (Fig. 4c,d) are seen. At 100K, the pattern of Fig. 4c is centered on the surface normal, and is broader than ESDIAD from the terraces. After annealing to >200K, the sharp, intense hexagonal pattern of Fig. 4d is observed. The angle between the surface normal and the outer beams is ~67°. The hexagonal pattern arises from $NF_x$ fragments or Ru-F species at steps and defect sites which, on average, retain the surface symmetry. Moreover, the pattern exhibits reversible temperature dependence, both in beam widths and relative intensities, as shown in Fig. 5; these data are measured at the indicated temperature. This phenomenon is a consequence of increased vibrational amplitudes of the surface species.

A detailed discussion of these results is beyond the scope of this short paper. The main point of this example is to show the importance of insuring that the signal being measured in ESD (and PSD) is typical of the majority region of the surface under investigation. This is sometimes hard to remember when the patterns from defective regions are as intense and "interesting" as those shown in Fig. 4,5 [See also Ref. 4b,19]. FEULNER, MENZEL and colleagues [21] have presented another example recently of the sensitivity of PSD to crystal surface defects, and why it is important to take care in moving the sample to maximize the DIET signal: one can be studying defects unknowingly!

The data of Figs. 1 to 5 also illustrate the advantage of detecting and resolving ion emission over a wide solid angle in DIET experiments. If ion intensities are measured at only one angle, e.g. normal to the surface, one could miss completely a number of important phenomena.

## V. Summary

ESDIAD is a very useful tool for characterizing the structures of adsorbed species, but care is necessary in carrying out the experiments and interpreting the results. The sensitivity of ESDIAD (and PSD) to minority states, atomic steps and defects on the sample surface, beam damage and other perturbations must be considered before valid conclusions can be made concerning structure.

## VI. Acknowledgement

M.M. Walczak acknowledges support from Ames Laboratory, which is operated for the U.S. Department of Energy by Iowa State University under contract number W-7405-Eng-82. This research has been supported, in part, by the U.S.-Spain Joint Committee for Technical Cooperation.

## References


1. T.E. Madey, F.P. Netzer, J.E. Houston, D.M. Hanson, and R. Stockbauer in N.H. Tolk, M.M. Traum, J.C. Tully and T.E. Madey (eds.), Desorption Induced by Electronic Transitions DIET-I, Springer Series in Chemical Physics 24, Springer, Berlin (1983), p. 120.
2. T.E. Madey, C. Benndorf, N.D. Shinn, Z. Miskovic and J. Vukanic, in W. Brenig and D. Menzel (eds), Desorption Induced by Electronic Transitions, DIET-II, Springer Series in Surface Sciences 4, Springer, Berlin (1985); p. 80.
3. T.E. Madey, in Inelastic Particle-Surface Collisions (Edited by W. Heiland and E. Taglauer), Springer Series in Chemical Physics 17, Springer, Berlin (1981); p. 80.
4. a. F.P. Netzer and T.E. Madey, J. Chem. Phys. 76, 710 (1982);
   b. F.P. Netzer and T.E. Madey, Surface Sci. 110, 251 (1981).
5. W. Riedl and D. Menzel, Surface Sci. 163, 39 (1985); M.D. Alvey, M.J. Dresser and J.T. Yates, Jr., Surface Sci. 165, 447 (1986).
6. C. Benndorf, C. Nöbl and T.E. Madey, Surface Sci. 138, 292 (1984).
7. C. Benndorf and T.E. Madey, Surface Sci. 135, (1983).
8. D.L. Doering and T.E. Madey, Surface Sci. 123, 305 (1982).
9. C. Nöbl, C. Benndorf and T.E. Madey, Surface Sci. 157, 29 (1985).
10. M.D. Alvey and J.T. Yates, Jr., in press; J.T. Yates, Jr., M.D. Alvey, M.J. Dresser and A.-M. Lanzillotto, this volume.
11. T.E. Madey and J.T. Yates, Jr., Surface Sci. 76, 392 (1978).
12. F.P. Netzer and T.E. Madey, Phys. Rev. Lett. 47, 928 (1981); Surface Sci. 119, 422 (1982).
13. H. Niehus, Surface Sci. 80, 245 (1979).
14. R. Kurtz, R. Stockbauer and T.E. Madey, Nucl. Instr. Meth. B13, 518 (1986); R.L. Kurtz, Surface Sci 177, 526 (1986).
15. A.L. Johnson, M.M. Walczak and T.E. Madey, Langmuir, in press; C.U.S. Larsson, A.L. Johnson, P.A. Thiel and T.E. Madey, J. Vac. Sci. Technol. A5. 842 (1987).
16. Z. Miskovic, J. Vukanic and T.E. Madey, Surface Sci. 169, 405 (1986).
17. D. Menzel, Springer Series in Surface Sciences (Proceedings of ICSOS-II, Amsterdam, 1987) in press.
18. T.E. Madey, Surface Sci. 94, 483 (1980).
19. C. Benndorf, A.L. Johnson and T.E. Madey, Surface Sci. (1987), in press.
20. M.D. Alvey, K.W. Kolasinski, J.T. Yates, Jr. and M. Head-Gordon, J. Chem. Phys. 85, 6093 (1986).
21. P. Feulner, S. Auer, W. Riedl, A. Cassuto and D. Menzel, Surface Sci. 182, L221 (1987).
22. T.E. Madey and C. Benndorf, Surface Sci. 164, 602 (1985).
23. A.L. Johnson, D. Barak, R. Stockbauer and T.E. Madey, this volume.
24. P.A. Thiel and T.E. Madey, Surface Sci. Reports (1987), in press.
25. M. Polak and T.E. Madey, Surface Science, submitted.
26. M. Polak and T.E. Madey, in preparation.
27. M.M. Walczak, A.L. Johnson, P.A. Thiel and T.E. Madey, J. Vac. Sci. Technol. (1988), in press.

# An Ellipsoidal Mirror Analyzer for the Study of Photon Stimulated Desorption

*R. Stockbauer, R.L. Kurtz, and T.E. Madey*

National Bureau of Standards, Gaithersburg, MD 20899, USA

## 1. Introduction

Recent efforts in our laboratory to understand the mechanisms involved in stimulated desorption have forced a re-evaluation of the factors that influence surface stability under ionizing radiation. Our experimental results in desorption from different surfaces of $TiO_2$ [1] have shown that the local surface atomic arrangements can be as important in determining ion desorption probability as the electronic excitations. Likewise, our studies of desorption from MgO [2] have shown that in this highly ionic compound with a large Madelung potential at an $O^{2-}$ site, desorption of $O^+$ does not readily occur when the Mg 1s core level is excited. The Knotek-Feibelman model [3] predicts a large ion signal in this system. It appears that the geometry, lattice dynamics, and excited state lifetime have large effects on the ion desorption probability [2, 4]. To expand our efforts in futher elucidating the role of geometry and lattice dynamics in ion desorption, it is important for us to be able to measure the angular distribution of ions leaving the surface under photon bombardment, and at the same time determine their kinetic energy and mass. In addition, since ion signals often are weak, we also require an analyzer that collects over a large solid angle.

Since there are no commercial charged particle analyzers that presently meet these requirements [5], we embarked on a program to design and construct one that would. We chose a design developed by EASTMAN et al. [6] which is an ellipsoidal mirror analyzer (EMA). Previous experience [7] with this analyzer showed that it is capable of measuring simultaneously kinetic energy, angular distribution and mass of ions emitted into a large solid angle in front of the sample.

## 2. Ellipsoidal Mirror Analyzer

The principle of operation has been described fully in Ref. 6. Briefly, the EMA energy analyzes the ions using a combination of low and high pass filters adjusted to pass ions with only a narrow range of kinetic energy. Both filters preserve the angular distributions of ions leaving the surface. The low pass filter consists of an ellipsoidal surface of revolution generated about an axis containing its two focal points. The sample is placed at one focal point and illuminated with monochromatized synchrotron light (see Fig. 1). Ions leaving the surface travel in straight line trajectories toward the ellipsoidal mirror. A field established between grid $G_3$ and the mirror surface reflects ions with kinetic energy below the electrostatic potential placed on the mirror. The reflecting properties of the mirror directs these ions to the second focal point which is also the entrance of a 3-grid retarding analyzer. Only the ions with kinetic energy above the electrostatic potential on the center grid, $G_5$, are transmitted through the retarding analyzer and arrive at $G_6$ maintaining straight line trajectories. These energy analyzed ions are then accelerated into the microchannel plate particle multiplier and detected by a resistive anode area detector. The voltages on the mirror and $G_5$ determine not only the mean kinetic energy transmitted by the analyzer but also the energy resolution since they can be adjusted relative to each other.

The pair of grids, $G_1$ and $G_2$, allow the ions to be accelerated or decelerated before they enter the filters so they transverse the instrument with a pass energy independent of initial kinetic energy and can be analyzed with constant energy resolution. Mass analysis is achieved

ELLIPSOIDAL MIRROR ANALYZER

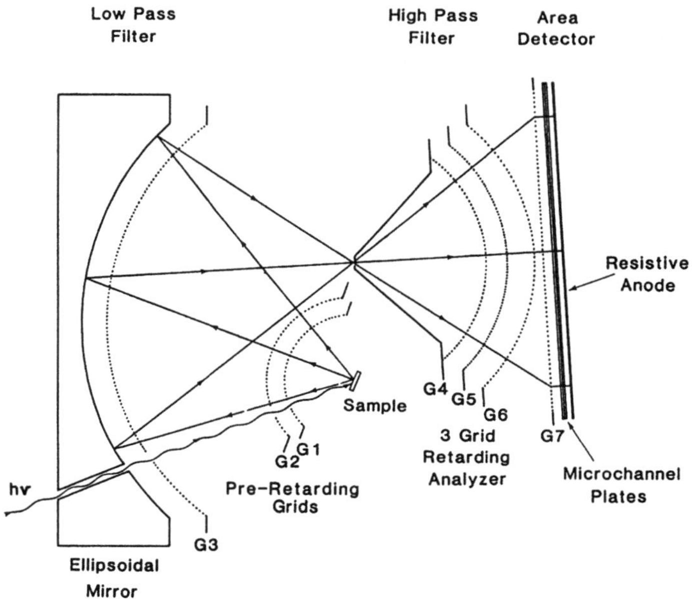


Fig. 1. Ellipsoidal mirror analyzer showing the low pass (ellipsoidal mirror) and high pass (3-grid retarding analyzer) filters, the pre-retarding stage ($G_1$ and $G_2$) and the area detector (microchannel plates and resistive anode). Three trajectories of charged particles emitted from a sample are also shown.

by gating the ions, i.e., by pulsing one of the elements in the analyzer, and measuring the time-of-flight to the area detector.

## 3. Construction

Several unique features have been incorporated into the mechanical and electron optical design of the EMA. These have been described in detail elsewhere [8] and will be discussed here only briefly. Special care has been taken to avoid all magnetic material within the three layers of magnetic shielding and we consider any stainless steel to be a magnetic material. Titanium is used as an alternative to stainless steel for the principal structural components, i.e., grid supports, central mounting assembly, fasteners, and drive shafts. Titanium serves well as a stainless substitute, since it is ultra-high vacuum compatible, strong, machineable, readily available, and spot weldable. This latter property was dictated by our design which required the grids to be spot welded to their supports. This requirement eliminated aluminum as a possible choice since it is difficult to spot weld.

The grids themselves are constructed of chromium plated tungsten mesh instead of the usual stainless steel mesh, again, to avoid magnetic materials. The tungsten mesh is first etched to about 70% of its 25 µm original diameter, plated with a copper strike, then a thick layer of copper and finally a thin chromium layer. The copper serves an intermediate material since it is not possible to plate chromium directly to tungsten.

The mirror is machined from a solid OFHC copper blank, 6.75 inch diameter by 1.70 inch thick. Instead of the ellipsoidal surface used in previous analyzers of this type, a slightly different surface was machined to correct for an aberration. This aberration arises from the fact that charged particles, unlike photons or billiard balls, do not reflect specularly from the surface, but rather follow parabolic trajectories in the space between $G_3$ and the mirror [7]. As a result the focal points for the charged particles are slightly displaced from those for specular reflection. This aberration can be corrected by machining the mirror to a shape which deviates slightly from an ellipsoid [8].

The last major change incorporated into the design of the EMA is a resistive anode area detector. This is a commercial device manufactured by Surface Science Laboratories [9] and consists of a specially terminated resistive sheet. The charge packet produced by a single ion striking the microchannel plate multiplier spreads radially over the sheet and is detected at 4 points. The pulse height at each of these points depends on its distance from the point of impact. Digital electronics calculate the position of the charge packet and produce both analog and digital signals proportional to the x and y coordinates.

## 4. Status

The EMA is currently undergoing its initial testing. We expect to install it shortly at the NBS synchrotron light source, SURF-II. Some of the more interesting questions concerning the mechanisms of ion desorption that we expect to address with the help of the EMA are: (1) how does the ion desorption vary as a function of excitation energy, in particular, what are the energy and angular distributions of ions near the desorption thresholds; (2) how do the angular profiles of individual ion beams vary with adsorption coverage and the presence of coadsorbates; (3) how do ion energy distributions vary as a function of desorption angle; (4) what are the angular profiles for individual masses; (5) how do measured ion trajectories compare with theory?

## 5. Summary

The unique property of the EMA to measure simultaneously kinetic energy, angular distribution and mass makes it a formidable component in our quest for answers to the questions about ion desorption mechanisms. It should provide us with a rich supply of data for many years. In addition to the ion analysis, it will also be extremely useful for angle resolved and angle integrated photoemission studies.

## 6. Acknowledgements

The authors wish to thank the members of the National Bureau of Standards Fabrication Technology Division whose skill in machining and problem solving brought the EMA to fruition. Special thanks are extended to Anthony Pararas, Ernst Neitzel, and Patrick Connelly for their help in constructing various parts of the analyzer. This work was supported in part by the United States Office of Naval Research.

## References

1. R.L. Kurtz, Surf. Sci. 177, 526 (1986); R.L. Kurtz, R. Stockbauer, and T.E. Madey, this volume.
2. R.L. Kurtz, R. Stockbauer, R. Nyholm, S.A. Flodström, and F. Senf, Phys Rev. B 35, 7794 (1987), and this volume.
3. M.L. Knotek and P.J. Feibelman, Surf. Sci. 90, 78 (1979); Phys. Rev. Lett. 40, 964 (1978).
4. R.L. Kurtz and R. Walkup, this volume.
5. R. Stockbauer, Nucl. Instr. and Meth. 222, 284 (1984).

6. D.E. Eastman, J.J. Donelon, N.C. Hien, and F.J. Himpsel, Nucl. Instr. and Meth. 172, 327 (1980).
7. T.E. Madey, R. Stockbauer, J.F. van der Veen, and D.E. Eastman, Phys. Rev. Lett. 45, 187 (1980).
8. R. Stockbauer and A. Pararas, Nucl. Instr. and Meth. (To be published).
9. In order to describe this equipment adequately, it is necessary to identify it by manufacturer's name. In no instance does such identification imply endorsement by the National Bureau of Standards nor does it imply that the particular item is necessarily the best available for this application.

# Digital Video Data Acquisition/Analysis for Existing ESDIAD Apparatus

*A.L. Johnson*, R. Stockbauer, D. Barak**, and T.E. Madey*

Surface Science Division, National Bureau of Standards, Gaithersburg, MD 20899, USA

*NBS-NRC Postdoctoral Research Associate

**Nuclear Research Center, Negev, Israel

## 1. INTRODUCTION

In most ESDIAD (Electron Stimulated Desorption, Ion Angular Distributions) instruments [1] the ion angular distributions are displayed on a phosphor screen, and data acquisition is accomplished using photographic techniques. In this paper, we discuss the advantages of digital imaging, and describe a digital video system we have recently assembled for this purpose. The advantages of digital data acquisition and analysis include greater sensitivity, convenient data storage, and improved data analysis. Digital data analysis includes the generation of line, contour, and perspective plots of the ESDIAD data (Fig. 1), which make identification and quantification of ESDIAD features practical. The system uses readily available video and computer equipment, requires no vacuum modifications, can be used at more than one experimental station, and can be adapted to do two-dimensional photometric measurements (photographic negatives [2], etc.).

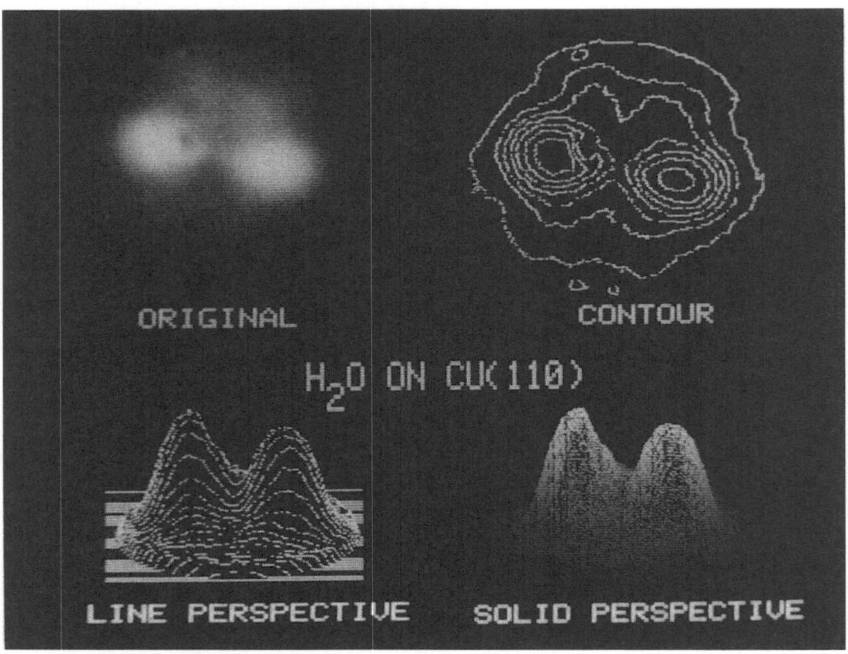


Fig. 1 Example of digital ESDIAD data analysis and presentation

There are two main types of digital ESDIAD: video and position sensitive anodes (PSAs). Position resolved measurements, such as are required for ESDIAD, are used in HREELS [3], electron microscopy [4], and LEED [5-7]. Moving detectors [8] and PSAs [9,10] have been used for ESDIAD previously. We will show the merits of video systems and indicate the choices of equipment for the data system.

PSAs have high sensitivity, good intensity linearity and fair position linearity, limited signal intensity (typically ~100,000 counts/second), and potential for giving ion flight-time information. Video data acquisition has good sensitivity, good intensity linearity, and high position linearity. Response at high intensity is limited only by the recovery time of the phosphor and microchannel plates, not the camera. This is important if the signal is confined to a small time interval (i.e. time-of-flight or multiphoton ionization experiments). Data storage (our images take 512x480 bytes) can be handled by video tape recorders. Using video tape to record an entire experiment with timing and temperature on the audio channel allows later analysis of such things as beam damage kinetics and temperature dependent phenomena. The video camera/recorder can be moved between experimental stations and one digitizer can serve many video cameras/recorders. Video data acquisition requires 1/60th of a second to get a field; thus it is difficult to obtain time-of-flight and position data simultaneously, compared to PSAs.

## 2. HARDWARE

The data acquisition system is composed of a video camera, video recorder (and time base corrector), image processor (digitizer and computer), and graphics display (Fig. 2).

The video camera is a RCA TC2811 [11], a monochrome CCD camera with a resolution of 403 pixels horizontal by 512 pixels vertical, a sensitivity of 0.045 foot-Lamberts scene brightness for full video, and a S/N >59 dB. The camera has a motorized 11-75 mm zoom and focus, f1.0-1.4, autoiris lens. The autoiris feature provides protection from accidental exposure to bright light but introduces ambiguity in comparisons between bright and dim scenes. The camera has sufficient sensitivity for single ion detection, giving a S/N of 5/1 when the potential across the microchannel plates in the ESDIAD optics is 980V each.

Our video cassette recorder is a Panasonic NV-9240XD [11], in 3/4" format and modified for high resolution black and white use, i.e., the color burst signal used for color information has been suppressed since it degrades resolution [12]. The recorder has >500 lines horizontal resolution and 48 dB S/N. Two audio channels for single frame editing signals or audio (time, sample conditions, etc.) are

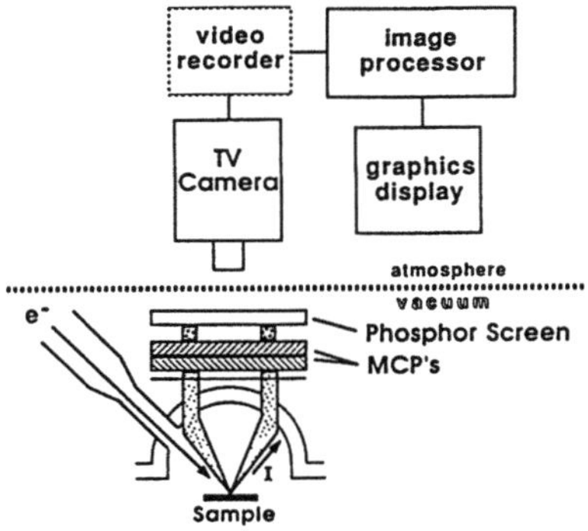


Fig. 2 ESDIAD apparatus and video data acquisition and analysis system

available. 60 minute cassettes typically have enough recording time for one or two experiments. The recorder decouples the experiment from the computer, so the experiment becomes independent of the computer. The sensitivity of our digitizer (see below) to signal timing instabilities of the recorder require the use of a time base corrector which is a dedicated image processor with timing error detection and correction. Newer digitizers do not need time base correction.

Our image processor is an IP-512 system, manufactured by Imaging Technology, Inc., mounted in an DEC LSI 11/73 computer [11]. The image processor converts the analog video signal to digital data to be subsequently processed. The IP-512 consists of an analog processor, an arithmetic unit, and three frame buffers (Fig. 3). Two of the frame buffers are paired to operate either independently or as a 16 bit frame accumulator. A 16 bit frame accumulator does not overflow as frequently as an 8 bit buffer during image operations such as convolutions. Figure 3 is a block diagram of the IP 512 system, but discussion of it will indicate features desirable in any image processor for ESDIAD.

First, the video signal has timing information extracted and is digitized with an 8 bit analog to digital (A/D) converter. We find for most images the least significant two bits have no information, so 6 bit A/Ds would be adequate. 8 bit A/Ds, while more expensive, give greater linearity and sensitivity.

The digital data is sent to the input lookup table (ILUT)/arithmetic processor. The ILUT converts each input value into a new value (for linearization of the video source or discrimination against noise). The arithmetic processor operates on image data to form arithmetic and logical products of the video and operator data. The arithmetic processor can be incorporated into the ILUT by presenting both new and stored image data to the ILUT and programing an operation table into the ILUT. Therefore, it is possible to obtain ESDIAD patterns where the signal appears visually as scattered ion arrivals with long accumulation times and background removal. Image convolutions are used for smoothing and contour plots (see below).

The output lookup table (OLUT) allows the production of false-colored plots. False color renditions make intensity modulations obvious. The color graphics monitor is a Sony PVM-1271Q [11] analog RGB monitor with 640 x 550 resolution. The monitor has an underscan option, allowing the entire video frame to be viewed. Good quality slides are obtained using a 35mm camera to photograph the monitor directly. A dot matrix printer is used for black and white hard copy of the line plots using standard screen dump techniques.

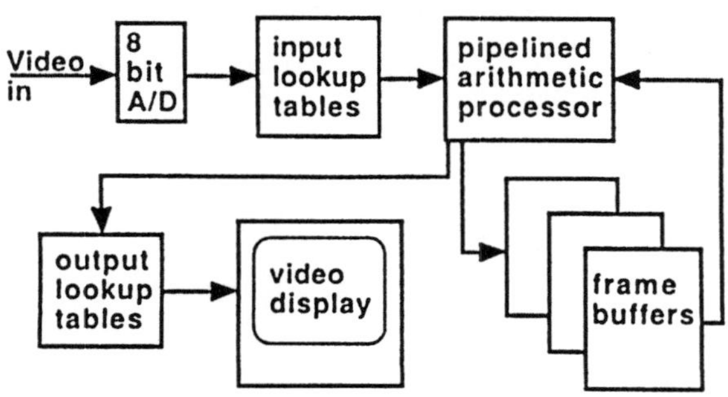


Fig. 3 Video digitizer schematic

## 3. SOFTWARE

The software is based on a set of drivers determined by the capabilities of the hardware. Table 1 is a list of video software functions that have proven useful for ESDIAD data acquisition and analysis. The first four categories are self-explanatory. We will elaborate on contour and perspective plots and the use of command files.

Perspective plots present the two-dimensional intensity data as a surface in three dimensions. These are produced by assigning each intensity an elevation, scaled for visually appealing results. The surface is rotated using the 3-D rotation matrix (in our plots, generally 45° about Z and Y is used), and then is projected onto the X-Y plane and lines parallel to the X (and/or Y) axis in the original data are plotted. Hidden line segments are removed by plotting the figure from "front" to "back", i.e., choose the lines to plot such that the lines of larger Z are drawn first. A "horizon" matrix is made during the plot: if the Y value of the data point to be drawn to is lower than the matrix value at the same value of X, then the line is hidden and is not drawn. If the Y value is above the horizon, the matrix is updated with the current Y value and the line is drawn. Striking perspective figures can be made by drawing 480 lines with each line segment the intensity of the data it represents. The figure appears solid giving a "mountainous" representation of the ESDIAD data. Using a false color map in the OLUT enhances the effect.

**TABLE 1**

| Category | Functions |
|---|---|
| hardware | initialize, acquire and accumulate image, set input and output look-up tables, hardware status |
| image movement | move between frame buffers, file input and output |
| image arithmetic and logic operations | (image, scalar) or (image, image) operations [+, -, etc.], bit operations on images (shifts, clear) |
| image calculations | rescale image intensity or size, find minimum and maximum intensity, convolutions, set intensity range to value |
| analysis and graphics | line, contour, and perspective (intensity verses position) plots, window image, insert text |
| procedural | get help, exit program, execute command file, calculations on passed parameters |

Contour plots on the monitor are made by masking out the $2^5$ bit of the data to obtain rings of intensity corresponding to changes of intensity of 32, and then convolving with the Lawrencian kernel [13] to enhance edges. Methods exist for the generation of contour plots [13] which are appropriate for plotters.

Complex and repetitive tasks can be automated with the use of command files. Command files (macros) are files that hold lists of commands to be executed. Procedures can be made from primitive operations using command files without reprograming the video software. An enhancement is to allow scalar results (minima and maxima, etc.) to be passed between operations. The calculator command performs arithmetic on passed scalars.

## 4. SUMMARY

We have applied digital video techniques to the acquisition and analysis of ESDIAD data. The technique uses readily available hardware to get quantitative ESDIAD patterns from display type analyzers without modifications in vacuum. The apparatus is capable of single ion sensitivity, time dependent investigations, can be used at many different stations, and can be made for $10,000-$40,000.

## 5. ACKNOWLEDGEMENTS

This work was supported in part by the Office of Basic Energy Sciences, U. S. Department of Energy.

## REFERENCES

1. T. E. Madey, R. Stockbauer, in Methods in Experimental Physics, eds. R. L. Park and M. G. Lagally (Academic, NY, 1985), 22 465
2. T. N. Tommet, G. B. Olszewski, P. A. Chadwick, S. L. Bernasek, Rev. Sci. Instrum., 50 147 (1979)
3. L. J. Richter and W. Ho, Rev. Sci. Instrum., 57 1469 (1986)
4. J. T. Brenna, M. G. Moran, and G. H. Morrison, Anal. Chem., 58 428 (1986)
5. P. C. Stair, Rev. Sci. Instrum., 51 132 (1980)
6. M. G. Lagally and J. A. Martin, Rev. Sci. Instrum., 54 1273 (1983)
7. J. W. Anderegg and P. A. Thiel, J. Vac. Sci. Technol., 4 1367 (1986)
8. H. Niehus, Surf. Sci., 78 667 (1978); 80 240 (1979); 87 567 (1981)
9. R. H. Stulen, in Desorption Induced by Electronic Transitions, DIET-II, W. Brenig and D. Menzel, eds., Springer Series in Surface Science 4 (Springer Verlag, Berlin 1985) p. 130
10. M. J. Dresser, M. D. Alvey, and J. T. Yates, Jr., Surf. Sci., 169 91 (1986); J. Vac. Sci. Technol. A, 4 1446 (1986)
11. Specification of a manufacturer is included to identify experimental conditions and does not imply endorsement for the product by NBS.
12. "VCR idiosyncrasies complicate frame-grabber interfacing", 1987 Applications Handbook, Data Translation Inc., Marlboro, MA USA
13. See for example: B. M. Dawson, Byte, 12,3 169 (March 1987); S. Ciarcia, Byte, 12,5-8 (May, June, July 1987); P. D. Bourke, Byte, 12,6 143 (June 1987)

# Part V

# Theory

# Dynamics in Neutral DIET from Chemisorbed Molecules

*E.B. Stechel, D.R. Jennison, and A.R. Burns*

Sandia National Laboratories,
Albuquerque, NM87185, USA

## 1. Introduction

Experimental advances since the time of DIET II [1] challenge the theorist to analyze desorption events in quantitative detail. We have chosen to begin such an endeavor by studying the dynamics of neutral DIET. There are several theoretical advantages in studying high-probability neutral, as opposed to low-probability ion, stimulated desorption. While in either case the initial state is often well characterized since the ground state potential energy surface is mapped out by thermal desorption and molecular scattering experiments, only the low-lying excitations are well known from photoemission and inverse-photoemission. Currently the experimental capability exists to look in unprecedented detail at these neutral events. This supplies the theorist with experimentally measured translational, vibrational, rotational [2,3] and angular distributions [4] arising from a small number (possibly only one) excited electronic configuration. (Typically one valence excitation is much longer-lived than all others, that is the shallowest excitation of each symmetry providing it is not degenerate with the bulk [5]). In contrast, the low probability events, such as positive ion desorption from metals, may result from a multitude of excitations which are all short-lived, but none of which can be eliminated on the basis of lifetime. Furthermore, the role of minority sites cannot be separated. In this paper we consider in quantitative detail the dynamics of the neutral NO desorption event. Although this is the only system we consider here many of the concepts and methods are general and have already been extended to desorption of physisorbed rare gases [6].

Before proceeding we first review the experimental results [2,3] for neutral NO desorption from Pt(111) with "low" translational energy (~0.05 eV). Then we summarize what we know about the electronic structure of the excited state which leads to this event [5]. The best estimate of the yield (desorption probability given that the adsorbate is initially excited) is ~0.01-0.1. The time-of-flight (TOF) distribution peaks at a low energy (~0.05 eV) relative to that for neutral CO (~0.5 eV). The observed threshold is ~6-10 eV whereas the threshold for CO desorption is much higher ~18-20 eV [2,7]. The internal NO vibrational distribution is very hot with the yield in $\nu=1$ a factor of 0.58 that of $\nu=0$. (It would take a temperature of 5000K (~0.43 eV) to produce this amount of excitation.) The rotational distribution is also hot (the best Boltzmann fit implies ~500-600K) relative to the surface temperature (80K). The low-lying excitations are $2\pi^{-1}$, $1\pi^{-1}$, $5\sigma^{-1}$ and $4\sigma^{-1}$. From threshold measurements [2,3] we infer either the $1\pi^{-1}$ or $5\sigma^{-1}$ excitations dominate. Moreover, theoretical arguments [5] indicate that the $2\pi^{-1}$, $4\sigma^{-1}$ and $1\pi^{-1}$ are all very short lived. Furthermore, calculations show that the relaxed $5\sigma$ hole is longer lived than the relaxed $1\pi$ hole by a factor of 20 (a result of symmetry). Hence we assigned [3,5] the $5\sigma$ hole as the dominant desorption producing channel, all other channels being ruled out on the basis of lifetime. After the creation of the hole, screening relaxation occurs on a time-scale which is fast compared with atom motion [5]. We believe that a screening electron is almost completely transferred into the molecular $2\pi$ orbital. The screened (relaxed) $5\sigma$ hole then decays via "intra-atomic" Auger, $2\pi \rightarrow 5\sigma$, $2\pi \rightarrow \vec{k}$ (where $\vec{k}$ is an electron in the continuum) with a time-scale of ~100 $eV^{-1}$ (~$10^{-13}$ sec.). The $5\sigma$ hole is non-bonding to both N-O and NO-Pt [8] hence it, itself, probably has no major effect on the bond strengths. However, the $2\pi$ screening electron does affect the nature of the excited state, in that it weakens the NO-Pt bond and it lowers the NO stretching frequency. The first effect can be understood in terms of the "equivalent binding" model; i.e., an NO adsorbate with an extra $2\pi$ electron has the same number of $2\pi$ electrons as an $O_2$ adsorbate and it is the $2\pi$ electrons which largely determine the bonding to the surface. An $O_2$ adsorbate is more weakly bound to the surface (~0.33eV) than is NO (~1 eV) and $O_2$ prefers the lying down orientation whereas NO prefers to be upright. Hence, excited NO with its extra $2\pi$ electron should emulate $O_2$-Pt(111) and be weakly bound and a relatively free rotor.

The outline for the remainder of this paper is as follows: (1) first we discuss a one-dimensional model which considers the effect of the softer NO-Pt bond in the excited state. However, this fails to explain the observed high yield. (2) We then consider the role of the rotational degree of freedom (dof) which leads to the mechanism of "desorption by electronically stimulated adsorbate rotation" [3]. (3) We review the six-dimensional nature of the full potential and show that dynamics occurs in two dimensions by a separation of time scales. (4) We discuss the two-dimensional model and what can be calculated in terms of one basic equation. We show results in which vibrational, rotational and kinetic energy distributions have been calculated and compared to experiment [2,3]. (5) Finally, we discuss the observed coverage dependence and offer a model to explain the data.

## 2. One-Dimensional Model

We consider a "Raman" (excitation de-excitation) mechanism for desorption when the excitation threshold is below the ion threshold: (STEP 1) a Franck-Condon (vertical) excitation to an excited state, (STEP 2) evolution on the excited state and finally (STEP 3) incoherent lifetime decay back to the ground state. For this mechanism the desorption probability is

$$P_D = 1 - \sum_n \int_0^\infty dt \ |\langle n|\phi(t)\rangle|^2, \tag{1}$$

where $|n\rangle$ are the bound eigenstates of the ground state (Born-Oppenheimer) hamiltonian, $H_0|n\rangle = E_n|n\rangle$, $|\phi(0)\rangle = |0\rangle$ and

$$|\phi(t)\rangle = \exp^{-i(H_1 - i\Gamma/2)t/\hbar} \ |\phi(0)\rangle, \tag{2}$$

where $H_1$ ($\Gamma$) is the Born-Oppenheimer hamiltonian (lifetime width) for the excited state. We have assumed that the transition moment is independent of the nuclear coordinates across the extent of the wavepacket. Time-dependent wavepacket calculations with only one dof (the distance of NO from the surface, z) and with parameters relevant to NO-Pt and NO$^*$-Pt indicate $P_D < 10^{-3}$ in the isolated molecule limit. This is much less than observed [2,3]. In these calculations the ground state potential was taken as a Morse oscillator with dissociation energy, 1 eV [9] and frequency 307 $cm^{-1}$ (38 meV) [10]. The excited state was also taken as a Morse oscillator with the following parameters: the equilibrium position, 0.8 bohr further from the surface than that of the ground state, the binding energy, ~0.33 eV (that of $O_2$ [11]), and the lifetime width, ~0.01 eV [5]). The excited state frequency is completely unknown and was thus varied from 100 $cm^{-1}$ to 300 $cm^{-1}$ without any significant effect on the results. The low yield produced by this one-dimensional Raman model with a bound-to-bound transition is well understandable in terms of the classical time evolution. The maximum energy after decay, obtained by the classical trajectory (the center of the wavepacket) is ~-0.4 eV (it must be greater than zero to desorb) and hence desorption occurs only from the extreme tails of the wavepacket.

## 3. The Rotational Degree of Freedom

In the last section we discussed a one-dimensional model which might describe the dynamics for some systems, but clearly does not for neutral NO desorption from Pt(111). The most likely reason for this is the role played by the rotational dof. The coordinate for this dof is the angle, $\theta$, between the NO bond axis and the normal to the surface. NO on the surface is a strongly hindered rotor which means that the potential energy is a sensitive function of $\theta$ and the initial state is strongly peaked about $\theta = 0$ (Pt-NO; $\theta = \pi$ corresponds to Pt-ON). It also means that one-dimensional slices through the potential at fixed $\theta$ exhibit differing binding potentials (as a function of z) with the greatest binding potential when $\theta = 0$. This then implies that it is easier to desorb (the desorbate fights less of an uphill climb) if $\theta > 0$ than if $\theta = 0$. The question remains, how might the NO molecule find itself at $\theta \gg 0$ if initially its wavepacket is strongly peaked about $\theta = 0$? The answer lies in the quantum spreading of the wavepacket while evolving in the excited state.

The observed hindered rotor frequency for NO/Pt(111) is $\hbar\omega = 465\ \text{cm}^{-1}$ [10]. An effective hamiltonian which leads to this is $H_0 = BJ^2 + V(\theta)$ where J is the angular momentum conjugate to $\theta$, $B = 1/2I$ and I is the moment of inertia for NO (gas phase value for $B = 1.6725\ \text{cm}^{-1}$). Then with $V(\theta) \approx -V_0[\cos^2(\theta)]^\alpha$ (with $V_0$ = ~1 eV, the observed binding energy), $H_0|\psi_n\rangle = \epsilon_n|\psi_n\rangle$, determines $\alpha$ such that $h\omega = \epsilon_1 - \epsilon_0$. The ground state, $|\psi_0\rangle$, of $H_0$ is the initial rotor wavepacket. It is very localized about $\theta = 0$ with a FWHM $\approx 14^0$. In the excitation process $|\psi_0\rangle$ is vertically excited to $H_1 = BJ^2$ (a free rotor hamiltonian). The excited state is approximately a free rotor (at least in the region near $\theta = 0$) because first it is weakly bound at all angles and second the minimum for the excited state analog, the $O_2$ adsorbate, is at $\theta = \pi/2$. The consequence is that the initial wavepacket spreads while it is in the excited state. This then allows the molecule to access the region of coordinate space where $\theta \gg 0$ and thus regions of the ground state potential (after de-excitation) which are only weakly binding.

A summary of the mechanism, "desorption by electronically stimulated adsorbate rotation" [3] is as follows: (STEP 1) Vertical excitation of a hindered rotor ground state to a free rotor excited state. (STEP 2) Spreading of the hindered rotor wavepacket to large angles. (The rotational energy is available from zero-point motion.) (STEP 3) De-excitation of the wavepacket to a weak binding region of the ground state potential and hence high probability for escape. The requirements for this mechanism to be operable are first a change in rotational geometry in the excited state (this gives the wavepacket the initiative to spread), second, large zero-point energy (this gives the wavepacket the components of large angular momentum which determine how fast it spreads) and lastly a long lifetime in the excited state (this gives the wavepacket time to spread to large angles). This mechanism does not cause CO to desorb because there is no change in the rotational geometry in the excited state (by the "equivalent binding" model CO with a $5\sigma$ hole emulates NO and both prefer the upright orientation on the surface).

## 4. The NO/Pt(111) Ground State Potential

For a rigid surface, the system has six dof. We have discussed two of these above; i.e., the normal translation (z) which has a frequency of 307 $cm^{-1}$ (38 meV) and hindered rotation ($\theta$) which has a frequency of 465 $cm^{-1}$ (58 meV). The remaining four are (1) N-O vibration (r) with frequency 1518-1720 $cm^{-1}$ (0.188-0.213 eV) [10], (2) two hindered lateral translations (x and y) which have frequency 60 $cm^{-1}$ (7 meV) [10] and (3) the projection of the rotation on the normal axis ($J_z$). It can be seen from these frequencies that there are three relatively distinct time scales, the fastest is the N-O vibration whose vibrational period is ~2 X $10^{-14}$ sec. The next is the normal translation and hindered rotation (~7-10 X $10^{-14}$ sec.) and the last is the hindered lateral translations (~6 X $10^{-13}$ sec.) Motion thus occurs in two dimensions with the following assumptions and approximations: (1) energy in motion lateral to the surface is conserved in the desorption event (the freqencies are so different that energy cannot be transferred in or out of these dof), (2) the projection of the rotation onto the normal axis is conserved (uncorrugated surface, hence at low temperatures $J_z$=0) and (3) the NO vibration is adiabatic (vibrational quantum number is conserved, although excitation and de-excitation does change the vibrational quantum number, the atom motion cannot). The ground state potential is parameterized as follows:

$$V_g^{(\nu)}(\theta,z) = D_g(\theta)\left\{\exp^{-2\beta(z-z_g(\theta))} - 2\exp^{-\beta(z-z_g(\theta))}\right\} + (\nu+1/2)\,h\omega_g(\theta,z),$$

where $D_g(\theta)$ $\left(z_g(\theta)\right)$ is the binding energy (equilibrium position) respectively, at fixed $\theta$. The first two terms are a Morse potential at each angle and the last term is the adiabatic NO vibrational energy. Other characteristics of the potential energy surface include a binding energy of ~1 eV (at $\theta$=0) and (from ab initio calculations [8]) a binding energy of 0.33 eV at $\theta$ = 60° with an equilibrium shift of 0.8 bohr between $\theta$ = 0° to 60°. The excited state is assumed to be separable in z, $\theta$ and r with the following functional form:

$$V_e^{(\nu)}(\theta,z) = D_e\left\{\exp^{-2\beta(z-z_e)} - 2\exp^{-\beta(z-z_e)}\right\} + (\nu+1/2)\,h\omega_e \;.$$

Finally, the kinetic energy operator is $T = \frac{1}{2M}\hat{p}_z^2 + B\,J^2$, where M is the mass of the NO molecule.

## 5. The Basic Equation

There is one basic equation from which most observables can be calculated, such as the yield, the translational, rotational and vibrational energy distributions and the angular distribution. The coverage dependence will be discussed in the final section. The basic equation is

$$Y\left[|\Psi_f\rangle\right] = \int_0^\infty dt\; |\langle\Psi_f|\sqrt{\Gamma}|\phi(t)\rangle|^2, \tag{3}$$

which is to say that the yield (Y) in a final state, $|\Psi_f\rangle$ is the probability that the time evolved state ($|\phi(t)\rangle$) is in $|\Psi_f\rangle$, where

$$|\phi(t)\rangle = \sum_\nu M_{O\nu}\exp\left\{-i(T+V_e^{(\nu)}-i\Gamma/2)t/\hbar\right\}\,|\phi_0\rangle.$$

The lifetime weighting (which may be a function of the internal coordinates) is contained in $\Gamma\exp\{-\Gamma t/h\}$. The initial state is $|\phi_0\rangle$. The weighted sum (with cross-section weighting $M_{0\nu}$) is over all possible excited states. For neutral NO desorption the sum on $\nu$ includes only one electronic state, i.e., the $5\sigma$ hole, but it also includes all the possible NO vibrational levels of that electronic state. Then $M_{0\nu}$ is a Franck-Condon overlap integral.

## 6. The Rotational and Translational Distributions

Two steps lead to the rotational distribution (and ultimately to the kinetic energy distribution as well): (1) the freeing of the hindered rotor in which energy (entirely zero-point ~29 meV) is conserved while the wavepacket spreads in angle space. (2) When the system decays back to the ground electronic state, the adsorbate sees a gradient in the potential with angle. Furthermore, an adsorbate cannot escape from all angles, consequently there is a "window" from which it can escape. The quantum result is heating; i.e, the rotational energy increases. The quasi-classical result is those adsorbates with larger angular momentum can get to larger angles first and have a greater probability of escaping. The experimental observation is a mean rotational energy 40-50 meV. Figure 1a shows the experimental and calculated rotational distribution. The calculated distribution assumes that the initial state has projection $J_z = 0$ and that this is conserved in the desorption event. The experiment measures the total intensity for fixed J. The figure plots $\ln(I(J))/(2J+1)$ ($I(J)$ is the measured intensity in rotational state J). This is a typical Boltzmann plot of a non-Boltzmann distribution. The agreement first shows than the distribution has curvature and second that the mechanism completely describes the observed rotational distribution. The TOF distribution shown in Fig. 1b is necessarily peaked at low energy since the adsorbate can never attain large kinetic energy from its evolution in the weakly bound excited state. The CO distribution is also plotted in Fig. 1b for contrast. This level of agreement for both distributions strongly suggests that the model adequately describes the physics of the event.

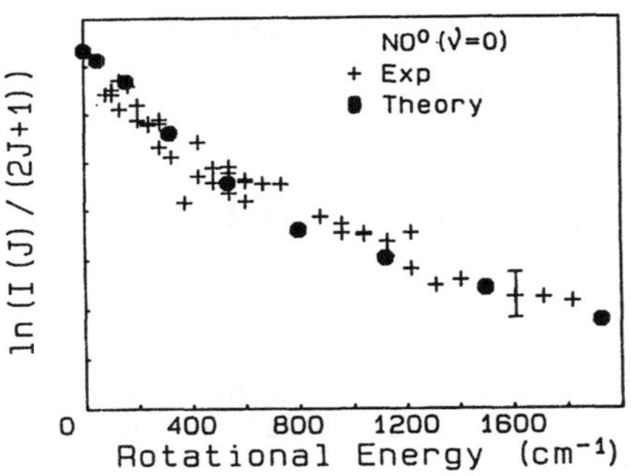


Fig. 1a Boltzmann plot of the theoretical and experimental rotational distribution for neutral NO desorbates in the $\nu$=0 vibrational state

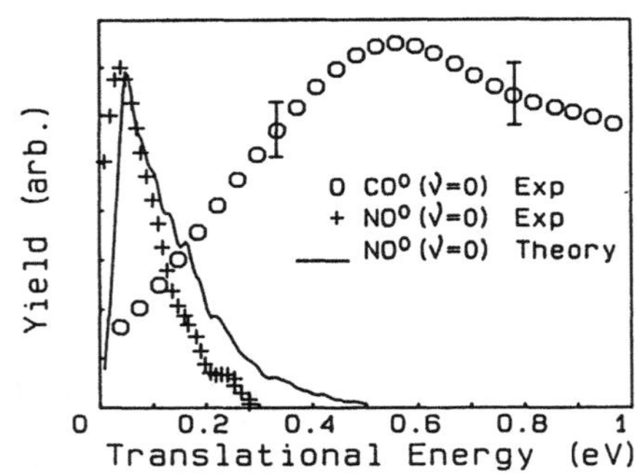


Fig. 1b Theoretical and experimental time-of-flight distribution for neutral NO and experimental time-of-flight distribution for neutral CO

## 7. Total Yield in NO Vibrational State $|\nu\rangle$

The origin of the vibrational excitation is the reduced frequency in the excited state which is a result of the extra $2\pi$ electron from screening of the $5\sigma$ hole. This also causes an enlarged NO equilibrium distance relative to the ground state. Three factors determine the yield as can be seen in Eq. 4,

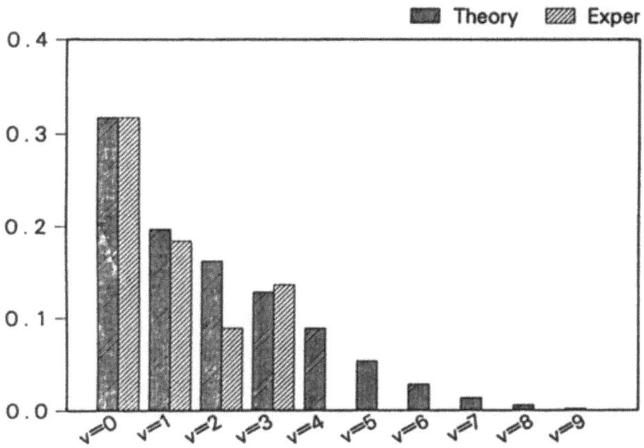


Fig. 2 Calculated and experimental vibrational distribution. $\nu > 3$ was not measured

$$Y[|\nu\rangle] = \sum_{\nu'} |\langle 0|\chi_{\nu'}\rangle|^2 |\langle\chi_{\nu'}|\nu\rangle|^2 \left(1-S_{\nu'}^{(\nu)}\right), \tag{4}$$

where the first and second factors are vibrational Franck-Condon factors up and down, respectively, while the third factor is the escape probability (1 - trapping probability). Figure 2 shows the calculated and the experimentally measured vibrational distribution. The calculations are a best fit with an excited state NO stretching frequency of 1450 $cm^{-1}$ ($\Delta\omega \approx -300$ $cm^{-1}$ shift from the ground state) and $\Delta r \approx 0.13$ bohr. Comparing these calculations to the experiment suggests that another channel is depleting the desorption into $\nu = 2$. This also shows a need for ab-initio calculations which can determine the excited state frequencies and equilbrium positions as a further test of the theory.

## 8. Coverage Dependence

Thus far we have implicitly assumed that the specific yield (the desorption probability given that an adsorbate is excited) is itself the product of two terms, $Y_* = L(\theta)\ D_*$, where $L(\theta)$ is the probability that an excitation will localize [12,13] on an adsorbate at a coverage given by $\theta$ and $D_*$ is the desorption probability given an excitation on an isolated adsorbate. We first point out that there is a major distinction between the concepts of reneutralization, relaxation and delocalization. Reneutralization is the decay of an ion state $M\text{-}A^+$ to $M^*\text{-}A$ after atom motion; i.e., the hole on adsorbate A causes atom motion and subsequently hops into the metal making the excitation $M^*$ and returning adsorbate A to its ground state. Relaxation is the decay of an ion state $M\text{-}A^+$ to a screened state $[M\text{-}A]^*$ which is an excited state of the metal-adsorbate system. Finally delocalization is excitation hopping through the adsorbate overlayer; i.e., $\underline{A^*\ \ A} \dashrightarrow \underline{A\ \ A^*}$. The localization probability ($L(\theta)$) is a function of the probability of resonant transfer at the mean residence time ($P(\Delta\tau)$) [13]. $\Delta\tau$ is the amount of time that an excitation would remain on a site in the absence of atom motion. This is a function of the local coordination (how many nearest adsorbate neighbors) and the transfer integral ($\beta$) which enables the excitation to hop from site to site. At any fixed coverage ($\theta$) we can in principle write

$$L(\theta) = \sum_{n=0,6} F_n(\theta/\theta_{sat})\ L_n,$$

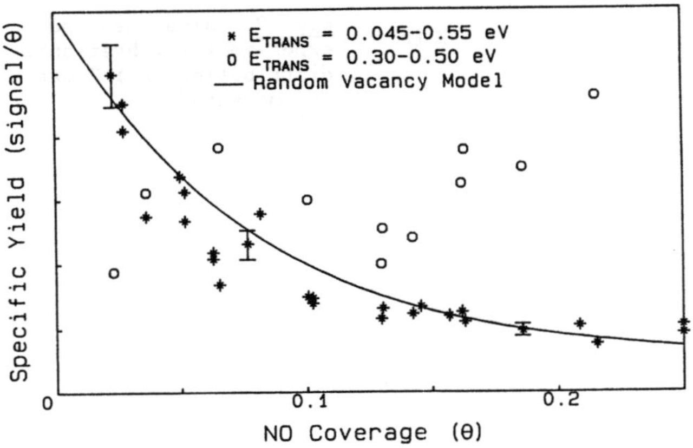


Fig. 3 Coverage dependence for the low (~0.05eV) and high (~0.5eV) energy channels. Also shown is the theoretical fit

where $F_n(x)$ is the fractional number of adsorbates with n nearest adsorbate neighbors at a coverage $\theta/\theta_{sat} = x$. For NO/Pt(111) at saturation, each adsorbate has 6 nearest-neighbors. We next consider $L_n$. The mean residence time is determined by the uncertainty principle, $\Delta\epsilon(n)\Delta\tau(n)/h = 1$ and $(\Delta\epsilon(n))^2 = n\beta^2$ with $\Delta\epsilon(n)$ the variance in the two-dimensional adsorbate overlayer band for an adsorbate with n nearest neighbors. The probability of resonant transfer $P(t) = |\langle\phi|\phi(t)\rangle|^2$ where $|\phi\rangle$ ($|\phi(t)\rangle$) is the initial (time-evolved) adsorbate wavefunction. For short times $P(t) \approx \exp\{-(\Delta Et/h)^2\}$ with $\Delta E$ the variance in the energy of the initial state when projected onto the excited state hamiltonian. Finally, [13]

$$L_n = 1 - P\left[\Delta\tau(n)\right] = 1 - \exp\left\{-\Delta E^2/n\beta^2\right\} = 1 - \exp\left\{-\alpha/n\right\}. \tag{5}$$

For the purposes here, we assume a random vacancy model [13] and take $F_n(x) = \left[\ 6!/n!(6-n)!\ \right]\ x^n\ (1-x)^{6-n}$.

The coverage dependence is well fit with $0.75 < \alpha < 1$. Figure 3 shows the experimental coverage dependence and the fit for $\alpha = 0.75$. The rather noisy coverage dependence for the "high" translational channel is also shown for contrast. This does not exhibit any definite coverage dependence. Furthermore, our first-principle calculations to determine $\alpha = \left[\Delta E/\Delta\epsilon\right]^2$ indicate that $\alpha$ ~1. We should point out that the deviation in the coverage range from θ ~0.2-0.4 is expected since the NO adsorbates undoubtedly do not follow the random vacancy model in this range but more likely form some islanding.

## 9. Conclusion

The new quantum-state resolved experiments [2,3] has provided data of unprecedented detail which puts many constraints on any proposed model. We have combined this with electronic structure calculations [5] which put limits on the lifetimes and the potential energy surfaces of the possible excited states. From this we were able to rule out all but the $5\sigma$ hole as substantially contributing to the desorption yield. We have been able to construct a very detailed picture of the desorption process of neutral NO from Pt. We have tested the model with quantitative multi-dimensional quantum dynamical calculations and have obtained remarkable agreement.

## Acknowledgments

We are indebted to T. A. Green for a critical review of this manuscript. This work was performed at Sandia National Laboratories supported by the U.S. Department of Energy under Contract No. DC-AC04-76DP00789.

## References

1. Desorption Induced by Electronic Transitions, DIET II, edited by W. Brenig and D. Menzel (Springer-Verlag, N.Y., 1985).
2. A. R. Burns, E. B. Stechel and D. R. Jennison, this volume.
3. A. R. Burns, E. B. Stechel and D. R. Jennison, Phys. Rev. Lett. 58, 250 (1987)
4. P. Feulner, in Ref. 1, pp. 142-151.
5. D. R. Jennison, E. B. Stechel and A. R. Burns, this volume.
6. D. R. Jennison and E. B. Stechel, in preparation.
7. P. Feulner, S. Auer, T. Mueller, A. Puschmann and D. Menzel, this volume.
8. C. W. Bauschlicher and P. S. Bagus, J. Chem. Phys. 80, 944 (1984).
9. J. L. Gland, Surf. Sci. 71, 327 (1978).
10. J. C. Tully and M. J. Cardillo, Science 223, 445 (1984); C. W. Muhlhausen, L. R. Williams and J. C. Tully, J. Chem. Phys. 83, 2594 (1985).
11. J. L. Gland, B. A. Sexton and G. B. Fischer, Surf. Sci. 95, 587 (1980).
12. E. B. Stechel and M. L. Knotek, Surf. Sci. 167, 297 (1986).
13. A. R. Burns, D. R. Jennison and E. B. Stechel, J. Vac. Sci. Tech. A 5, 671 (1987).

# The Nature of Repulsive States and the Role of Nuclear Dynamics in Desorption Induced by Electronic Transitions

*Ph. Avouris, R. Walkup, R. Kawai, D.M. Newns, and N.D. Lang*

IBM Thomas J. Watson Research Center,
Yorktown Heights, NY 10598, USA

## I. Introduction

Electrons and photons cause desorption by exciting transitions to electronic states that are repulsive, i.e. by breaking bonds. Thus, understanding desorption means understanding the behavior of excited electronic states at surfaces and the coupling of repulsive potential energy into atomic motion. In the first attempt to provide a microscopic description of the ESD/PSD process, Menzel, Gomer[1] and Redhead[2] (MGR) discussed the role of Franck-Condon transitions to repulsive excited states and the role of the substrate in the quenching of these excitations. The MGR model has provided the conceptual framework to discuss DIET processes in the valence region of the spectrum. Unfortunately, however, there has been little progress towards identifying the specific nature of these repulsive states (particularly in the valence region), and understanding the subsequent electron and nuclear dynamics leading to desorption. In the case of ionic solids, a conceptually simple positive-ion desorption mechanism has been identified based on the reversal of the Madelung potential, following multiple ionization of a negative ion in the crystal[3,4]. Despite the fact that the repulsive character of such a state is beyond any doubt, both theoretical [5] and experimental [6] evidence does not always support a desorption mechanism based on Coulombic repulsion.

Here we present the results of density functional and cluster calculations of the electronic structure of atomic and molecular adsorbates. These results provide substantial insight into the nature of the repulsive excitations involved in DIET processes. We also present results of classical trajectory calculations which demonstrate the crucial role that lattice relaxation can play in DIET processes. Our calculations provide new insight into DIET processes and suggest new desorption mechanisms.

## II. The nature of the repulsive states

### (A) Fluorine on metals

We first discuss the desorption of fluorine from an Al surface. $F^+$ desorption has been observed from a variety of metal[7,8] and semiconductor surfaces[8 – 10]. The ESD cross-sections are high, and in fact $F^+$ desorption is observed without the need to dose the surface with a fluorinating agent; the minute level of F impurity is often sufficient to yield strong $F^+$ ESD signals. We chose Al as a substrate because it is a simple free electron metal which can be successfully represented by a jellium model. In addition, this particular system (F/Al) has been the subject of a recent experimental inquiry into the $F^+$ ESD mechanism [7]. The calculations[12] utilize the Kohn-Sham density functional formalism[13] with exchange and correlation effects treated within the local density approximation as described by Lang and Williams[14]. This approach has been very successful in calculating the potential energy (PE) surfaces of adsorbates on metals.

In Figure 1(A), we show the PE curve for the ground state of the F/Al system. The electronic configuration of fluorine at its equilibrium position on the surface is found to be $2p^6$; i.e. fluorine adsorbs as $F^-$ on Al. This is to be expected on the basis of the large electron affinity of free fluorine and its further increase as a result of the adsorption on the metal. We also show in Fig. 1(A) the behavior of the image interaction. It can be seen that at large distances (>4 a.u.) the energy closely follows the image potential. At shorter distances the attraction due to the effective metal potential and image-like interaction combine with Pauli repulsion to produce a minimum in the potential for $F^-$

We now want to consider what kind of excitations can lead to the desorption of neutral ($2p^5$) and ionic ($2p^4$) fluorine. The simplest routes to $F(2p^5)$ would be either a charge-transfer transition to the metal or direct ionization of $F^-$ . Using techniques similar to those described by Lang, Norskov and Lundqvist[15], we have calculated [12] the PE curve for $F(2p^5)$ interacting with Al, which is shown in Fig. 1B. Interestingly, we find that the $2p^5$ state is repulsive over the entire Franck-Condon region of the ground state. This behavior can be understood on the basis of the "image" attraction plus Pauli (electron kinetic energy) repulsion description of the bonding in the ground state. In the $2p^5$ neutral state, the "image" attraction is lost, leaving a net repulsive interaction. The repulsive character of the $2p^5$ state suggests that this ionization channel may be a source for F-atom desorption.

Since the bonding of $F^-$ to Al is by the "image" interaction, single ionization of <u>any</u> level should remove this interaction and lead to a repulsive state. In Fig. 1B, we show (open circles) the PE curve for the state resulting from the 2s-ionization of the adsorbed $F^-$. Indeed, the resulting core-excited state of neutral fluorine, $F^*(2s^1 2p^6)$, is repulsive. The $F^*(2s^1 2p^6)$ state is expected to have a relatively long lifetime with respect to metal → fluorine charge-transfer because the 2p shell is already full and the 2s level is located below the bottom of the Al sp-band. However, F* still cannot desorb because in this case a very fast Auger process (Coster-Kronig) leading to $F^+(2s^2 2p^4)$ + e is energetically allowed. Since the $F^*(2s^1 2p^6)$ state of free F* has not been observed experimentally, no linewidth measurements are available, but theoretical calculations[16] give an Auger decay rate of $3 \times 10^{15} s^{-1}$ . In the present case, the auto-ionization lifetime of $F^*(2s^1 2p^6)$ is too short to allow desorption, however, this may not be true for other species. If the core-excited state lifetime is sufficiently long, because, for example, it decays by a CVV process where the valence levels (V) show significant delocalization on the substrate or on other atoms of a molecular adsorbate, core-ionization initiated desorption of excited neutrals can take place. In principle, the core-excited neutrals could decay away from the surface to give positive ions. Although the experimental manifestations of this mechanism are similar to those of the "Coulomb explosion" mechanism (i.e. there should be thresholds at core-ionization energies leading to energetic positive ions), the details are fundamentally different. The driving force would be mainly due to Pauli repulsion, and as the particle desorbs in a neutral state, there are no image forces involved, and no need for Coulombic 2h-localization[17].

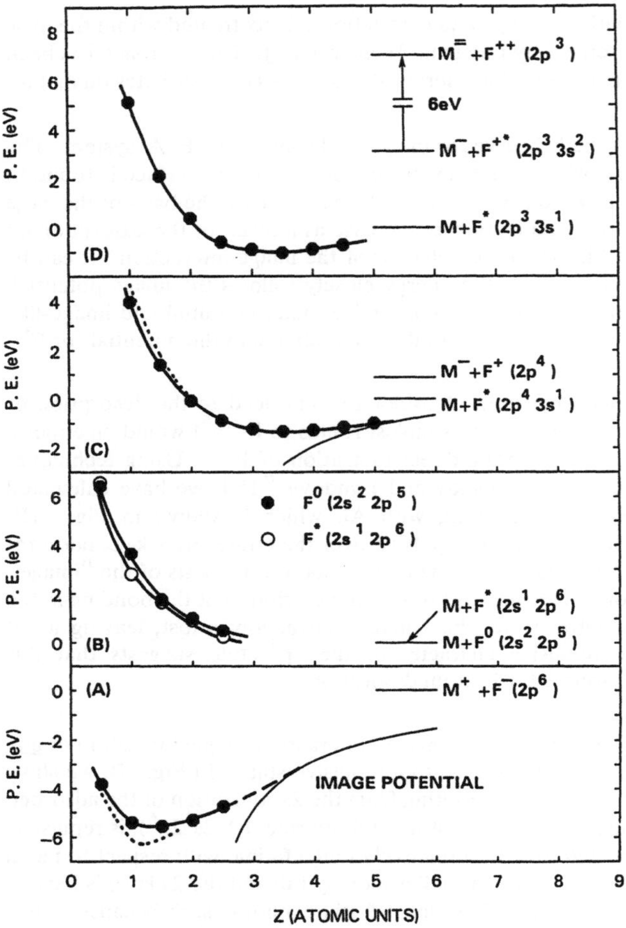


Figure 1. Local density functional calculations of the interaction of fluorine with an aluminum surface. (A) The solid curve shows the potential energy for $F^-(2p^6)$, with Al treated as jellium with $r_s = 2$, and the dotted curve shows the potential for $F^-$ in a top-site position for Al(111). (B) Potential energy for $F^0(2p^5)$ (solid circles) and $F^*(2s^1 2p^6)$ (open circles) interacting with Al (jellium approximation). (C) Potential energy for $F^+(2p^4)$ interacting with Al (jellium) (solid curve), and at a top-site of Al(111) (dotted curve). (D) Potential energy for $F^{++}(2p^3)$ interacting with Al.

Next, we consider the bonding of the $F^+(2p^4)$ state produced by, for example, the Coster-Kroning decay of the $F^*(2s^1 2p^6)$ state. According to current thinking, the $F^+(2p^4)$ state at the surface should be the source of the ejected $F^+$ ions. Figure 1C shows the PE curve for the $F^+(2p^4)$ state. We see that overall this state is bound with respect to $F^+$ desorption. This should not be surprising; from the point of view of image attraction by the substrate, the $F^+$ and $F^-$ states should be similarly bound. However, the calculations show that the bonding of $F^+$ to the metal is significantly different from

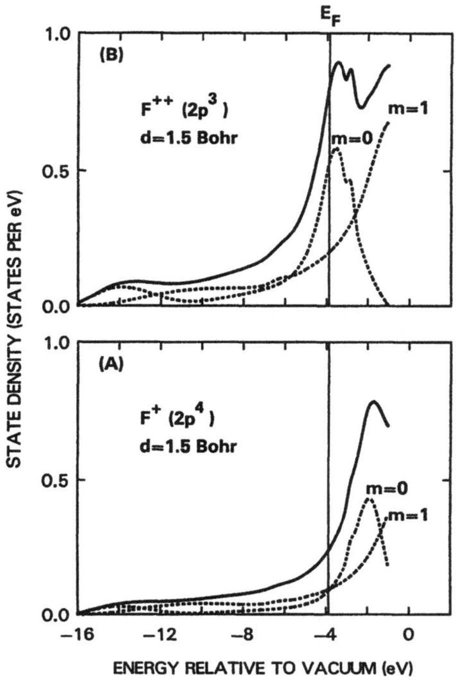


Figure 2. Density of 3s and 3p states of $F^+(2p^4)$ and $F^{++}(2p^3)$ on Al. Both the total density of states and the decomposition into m-subcomponents, i.e. m=0 (3s, $3p_z$) and m = 1 ($3p_{x,y}$), are shown.

that of $F^-$. First note that in the $2p^4$ configuration, the fluorine 2p levels are pulled below the bottom of the Al sp-band, eliminating resonant electron tunneling from the substrate into the 2p levels; thus only the somewhat slower interatomic Auger neutralization can take place. Thus, the $2p^4$ state may be long lived. More importantly, the fluorine 3s and 3p levels, which were empty in the $2p^6$ and $2p^5$ states are now pulled down in energy, they cross the Al fermi level and are partially occupied. This can be seen in Fig. 2 which shows the 3s and 3p density-of-states (DOS) at the equilibrium Al-F distance. This DOS can be further decomposed to its m-components, i.e. m=0 (3s, $3p_z$) and m=1 ($3p_{x,y}$). It is evident from Fig. 2 that significant metal → adsorbate charge-transfer (~0.9 electrons) takes place in the "$2p^4$" configuration.

There are two very important ramifications resulting from the partial 3s and 3p occupation: (a) the net charge on the adsorbate, and thus the image attraction, is reduced, and (b) the Pauli repulsion is increased because of the larger overlap of the 3s- and 3p-orbitals with the metal wavefunction. As a result of the above effects, the minimum of this "screened" 2h-state occurs further away from the surface than the minimum of the ground state.

The sequence of core-ionization followed by fast Auger decay then can be viewed as a single Franck-Condon transition from the ground state which brings the system to the repulsive part of the CT-screened 2h-state. The fluorine can thus acquire kinetic energy and desorb as $F^+$. Our calculations show that the PE curve at larger distances closely follows the image potential curve indicating a correlation with the free $F^+$ final state. Thus, we stress again that at the surface, the adsorbate is in a screened state due to the partial occupation of the affinity level, and that Pauli repulsion due to this occupation provides the main driving force for desorption.

In the jellium model, the ionic cores are smeared out as a homogeneous positive background. We have investigated the possibility that the presence of discrete ion cores may produce qualitative changes in the DIET mechanism compared to expectations based on a jellium model[12]. To do this we used pseudopotentials situated at the sites of a regular semi-infinite lattice. The resulting change in energy was calculated using first order perturbation theory[14]. In Fig. 1A and 1C (dotted lines) we show the PE curves thus computed for F on the on-top site of the Al(111) surface. Top adsorption sites (low coordination sites in general) are known to be particularly active in DIET processes. As Figure 1 shows the effect of the discrete lattice in this case is to increase somewhat the repulsive character of the screened-2h state (Fig. 1C). Thus, our conclusion that $F^+$ desorption should occur from a screened $2p^4$ state remains valid.

Experimental support for our picture of ion DIET comes from consideration of desorption threshold energies. Although the $F^+$ desorption threshold from F/Al has not been measured, the corresponding threshold for $F^+$ PSD from F/W(001) has been determined to be ~ 24eV[8]. This excitation energy is significantly lower than the F 2s ionization threshold of free F which is ~ 31eV[18]. In fact, this discrepancy led the authors of Ref. 8 to suggest that $F^+$ desorption from W(001) is not related to core-ionization. However, our calculations give a threshold of 23 eV for $F^+$ desorption from the CT-screened 2s-ionized state from $F^-$ on Al, in agreement with the experimental threshold for F/W(001). Since the chemisorption bond is largely ionic, we should not expect any significant differences in the thresholds for the two different substrates.

Finally, we consider multi-hole states. In Fig. 1D, we show the PE curve for the interaction of $2p^3$ $F^{++}$ with Al. Again, this state is overall bound, and the corresponding DOS diagrams (Fig. 2) show that the 3s, 3p levels are heavily occupied ( ~ 2 electrons). Considering the possible desorption limits (Fig. 1D) we see that desorption from the bare $2p^3$ $F^{++}$ state is not feasible. Because of the strong "image" attraction, $F^{++}$ is bound by ~ 10 eV. Desorption is only possible from the screened 2h2e or 3h1e states which may lead to $F^+$ desorption. Note, that desorption from bare multi-hole states should lead to multiply-charged ions which however, are not generally observed.

(B) NO and CO on metals

In the above, we saw how repulsive states can arise as a result of a change in the charge state of an atomic adsorbate. Ionic-like bonding is also important in molecular adsorption. To demonstrate this, let us consider the chemisorption of CO and NO on metals. The $2\pi^*$ affinity level of free CO is empty and located at ~ 1.5 eV above the vacuum level, while in the case of NO it contains one electron whose ionization potential is 10 eV. The electron affinity of NO is slightly positive. Given the screening of electron-electron interactions at the metal surface, we would expect that charge-transfer from the metal to the $2\pi^*$ level would be more pronounced in the case of NO bonding to the metal than in the case of CO. Cluster calculations of CO bonding to metals give a $2\pi^*$ occupation of the order of 0.1-0.2 electrons[19]. The bonding of NO however, is quite different[19,20]. This has been demonstrated by MCSCF calculations using the $Cu_5NO$ and $Cu_5ON$ clusters as models for NO linearly bonded on top sites of Cu(100) [20]. Both the high ($^5A_1$) and low ($^1A_1$) spin configurations have been studied. In Table 1 we give some calculated results for the $^5A_1$ state which help establish the nature of the chemisorption bond.

Table I

| | $Cu_5NO$ | $Cu_5ON$ |
|---|---|---|
| Dissociation energy | 0.79eV | 0.74eV |
| Re (Cu-NO) | 3.62 a.u. | 3.54 a.u. |
| $d\mu/dR$ | -1.60e | -1.61e |
| $\sigma$-electrons on NO | 9.97 | 9.97 |
| $\pi$-electrons on NO | 5.99 | 5.98 |

The equilibrium Cu-NO distance, the dipole derivative, and the $\sigma$ and $\pi$ – electron projections on NO indicate an ionic bond in which NO bonds as $NO^-(5\sigma^2 2\pi^{*2})$. This is further supported by the fact that, in sharp contrast with CO, the bonding characteristics are little affected if the NO is bonded by the N or O atom towards the surface. The bonding in the low spin $^1A_1$ state of $Cu_5NO$ has a small covalent contribution but overall is again largely ionic $[n(\sigma) + n(\pi) = 15.88]$. The ionic character of the NO chemisorption bond provides a simple explanation for the NO bonding geometry. At low surface coverages NO tends to exhibit a tilted adsorption configuration because in this case the image attraction is larger. At higher coverages when repulsive lateral interactions become important, NO takes on a perpendicular configuration.

In analogy with the fluorine desorption case discussed above, we would expect that excitation processes that remove negative charge from chemisorbed NO could lead to desorption provided that the resulting state is sufficiently long-lived. The lowest energy threshold for neutral NO desorption could correspond to the onset of $2\pi^*$ ionization. However, the $2\pi^*$ level is in resonance with the metal band and fast charge-transfer to it to reform $NO^-$ is expected. If however, another level like the $5\sigma$ level is ionized, a mechanism for neutral NO desorption may result. In this mechanism, charge-transfer from the metal to form $NO^-(5\sigma^1 2\pi^3)$ may be blocked, thus allowing NO desorption, provided that electron repulsion within the $2\pi^*$ resonance is sufficiently high. Evidence supporting a Coulombic inhibition of a metal $\rightarrow$ NO $(5\sigma^1 2\pi^2)$ charge-transfer is presented below. Recently, it has been argued that the state produced by $5\sigma$ – ionization of chemisorbed NO is long lived, and is in fact responsible for the ESD of neutral NO from NO/Ni[21]. The desorption mechanism which was suggested, however, was based on a "rotationally mediated" process in which neutralization into a repulsive orientation of the ground state was proposed. This mechanism is fundamentally different from the mechanism that we propose, which is based on a change in the bonding from ionic in the ground state to repulsive in the excited state produced by $5\sigma$ ionization.

In adsorbate systems where the chemisorption bond involves a charge transfer from the metal to the affinity level of the adsorbate, intra-adsorbate excitations can also lead to repulsive states[19]. This situation arises when the increase in electron-electron repulsion in the affinity level, which is brought about by placing the excited electron in that level, is stronger than the electron-hole attraction. In this case, the adsorbate affinity level will be pushed above $E_F$, leading to a molecule-to-metal charge-transfer in the excited state. The resulting nearly neutral excited adsorbate does not have an image attraction and thus may be in a net repulsive state. This behavior is illustrated by the bonding characteristics of the core-excited states of chemisorbed CO and NO, modeled here by the clusters $Cu_5CO$ [22] and $Cu_5NO$ [19]. The conclusions, however, should be applicable to other chemisorption systems and to valence excitations which can be sufficiently long-lived to be directly ESD/PSD - active. Figure 4 shows the computed PE curves[19,22]. In the ground state of $Cu_5CO$ the $2\pi^*$ level has a very small occupancy, ~ 0.1 electron, leading to a weak chemisorption bond, De = 0.5 eV. The $C1s \rightarrow 2\pi^*$ excitation places an additional electron in the $2\pi^*$ level and the resulting excited state is also weakly bound to the metal, De=0.26 eV. In the case of $Cu_5NO$, however, the occupancy of the $2\pi^*$ level in the low spin ground state is very high, ~ 2 electrons, leading to a strong chemisorption bond De=1.1 eV. In the $N1s \rightarrow 2\pi^*$ excited state the $2\pi^*$ population is about two electrons, not three as may be expected. This is due to electron-repulsion-induced $2\pi^* \rightarrow$ metal charge transfer in the excited state. Indeed, as Fig. 3 shows, the N1s, $2\pi^*$ excited state of $Cu_5NO$ is repulsive. It is also clear that Coulombic repulsion within the $2\pi^*$ level can prevent formation of $NO^-(5\sigma^1 2\pi^3)$ from $NO(5\sigma^1 2\pi^2)$ , as suggested earlier.

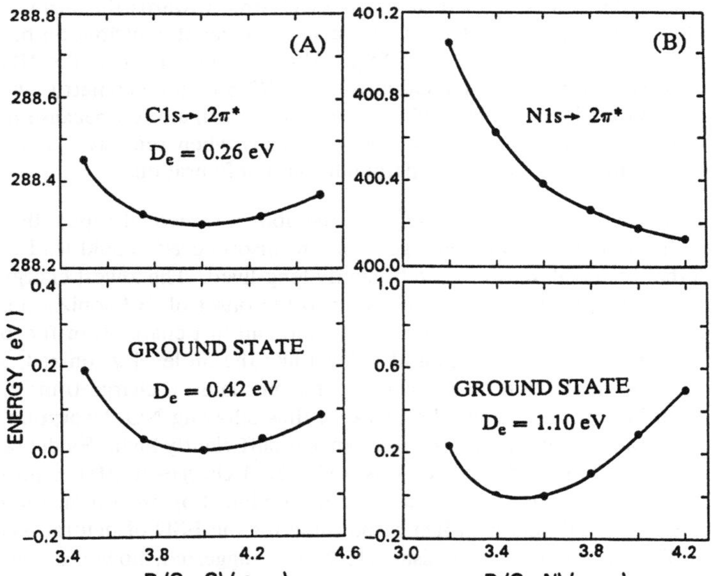


Figure 3. Results of MCSCF cluster calculations using $Cu_5CO$ and $Cu_5NO$ clusters as models for top-site adsorption of CO and NO on Cu(100). (A) PE curves for the ground state and the $C1s \rightarrow 2\pi^*$ excited state of CO. (B) PE curves for the ground state and $N1s \rightarrow 2\pi^*$ excited state of NO.

Electron repulsion-induced destabilization of adsorbate excited states should also cause blue shifts of the energies of the surface excitations with respect to those of the corresponding gas phase excitations. Experiments on the core-excitations of CO and NO on Ni(100) [23] support the above conclusions. Thus, while the C1s→2π* excitation is blue shifted by only 0.2 eV upon adsorption, the N1s→2π* transition of NO, which has a higher 2π* occupation, is blue-shifted by 1.8 eV [23]. Chemisorbed $O_2$ has an even higher occupation of the 2π* level, and indeed the O1s→2π* spectrum of $O_2$/Pt is blue-shifted by ~ 4.5eV [24].

## III. The role of nuclear dynamics

### (A) Desorption from metal halides and oxides

Up to this point we have concentrated on identifying the nature of the repulsive states active in DIET processes. A complete description of a DIET process should also include the time evolution of the electronic and nuclear degrees of freedom, including effects induced by the motion of the desorbing particle. On metal surfaces electronic polarization (screening) and lattice recoil are expected to be important. Lattice recoil should be even more important for ionic surfaces because the forces can be very long-ranged (Coulombic), and there can be very strong polarization effects caused by ion displacement. For this reason and because of the availability of more reliable interaction potentials for ionic systems, we have focused on the role of nuclear dynamics in desorption from ionic solids, such as alkali halides and metal oxides. For these ionic materials, the bonding can be reasonably described by a superposition of charge densities of atomic ions, with electron kinetic energy, exchange and correlation computed using a local-density electron-gas model due to Gordon and Kim[25]. With this method, pairwise interatomic potentials can be calculated which reproduce the lattice constant, cohesive energy, and bulk modulus of many ionic solids to within a few per cent. Since these potentials are determined from properties of atomic ions, rather than having fixed functional forms with parameters fitted to equilibrium data, they should have a relatively large range of validity including substantial distortions away from equilibrium. Using these potentials, we have performed classical trajectory calculations of ion desorption from ionic solids, using a predictor-corrector method[26].

We have previously reported[5] the results of trajectory calculations of positive ion desorption from alkali halides by the "Coulomb explosion" mechanism. In these calculations, a positive halogen ion ($F^+$ or $Cl^+$) is formed at t=0, and the subsequent nuclear motion is followed, assuming that no neutralization occurs during the time of interest. These calculations predict that even in the absence of neutralization, the formation of a positive halogen ion in the first few layers of a perfect crystal would not lead to positive ion desorption from the solid. Instead, the neighboring ions displace within ~ 100 fs to form a bound configuration, and the initial repulsive energy goes into defect formation and heat. Insight into the dynamics of this process can be obtained by considering the forces acting on the ions upon formation of an $F^+$ ion on the surface of NaF, as shown in Figure 4A. The initial force acting on the $F^+$ ion is in the direction of the surface normal, but the forces acting on all of the surrounding ions are much larger. Thus the surrounding ions will displace and reduce the initial repulsive potential energy. This leads to an oscillatory force acting on $F^+$, which becomes trapped above the surface by attractive interactions with $F^-$ ions. The time dependent force and displacement are

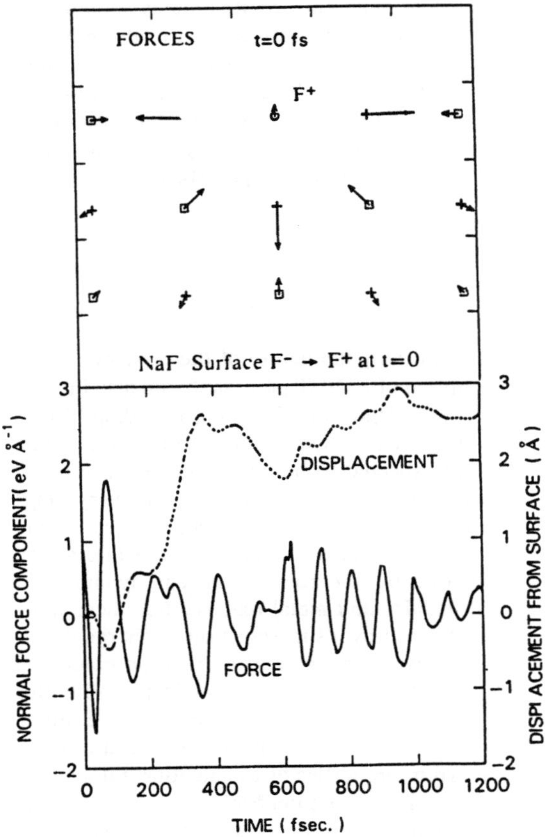


Figure 4. Classical trajectory calculations on the Coulomb explosion mechanism for $F^+$ desorption from a NaF(100) surface. Top: at t=0 a $F^+$ ion is formed at the surface. The magnitudes and directions of the forces are shown. Bottom: the component of force normal to the surface and the $F^+$ distance from the surface are shown as functions of time. The initially repulsive force becomes attractive within 30 fs due to lattice distortions. At long times the $F^+$ ion becomes trapped ~ 2.5Å above the surface.

shown in Figure 4B. It is clear that the displacement of neighboring ions has a dramatic effect on the time development of the system: the displacement quickly changes the initially repulsive configuration into an attractive one. For ionic solids, the notion that desorption occurs along a time independent, one-dimensional repulsive potential energy curve is seen to be unrealistic.

We have recently investigated $O^+$ desorption from metal oxides, including $TiO_2$, the prototype system for the Knotek-Feibelman desorption mechanism[3]. In contrast to alkali halides, direct ion desorption is predicted for certain sites of the $TiO_2$ (110) surface. The different behavior is primarily due to the different geometric structure of the surfaces. The displacement and force in the direction of the surface normal are shown

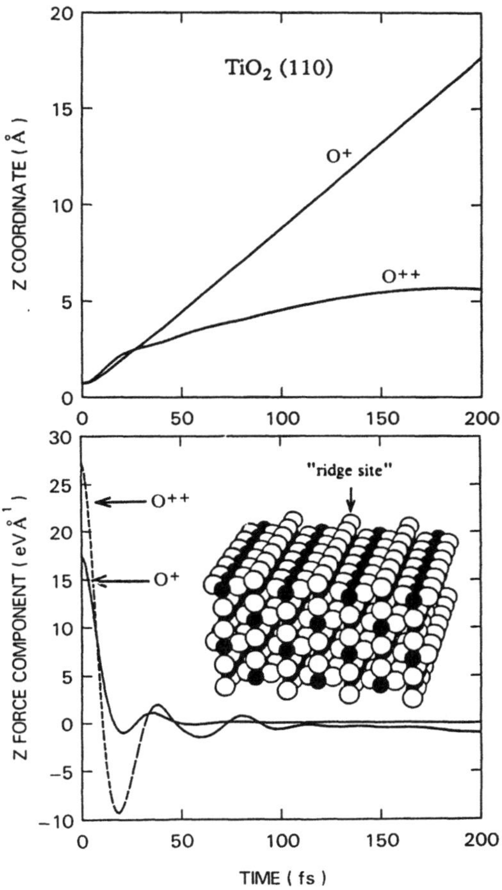


Figure 5. Classical trajectory calculations of $O^+$ and $O^{++}$ desorption from $TiO_2(110)$. Top: the time evolution of the z-coordinate of $O^+$ and $O^{++}$. Bottom: the time evolution of the force component in the direction of the surface normal for $O^+$ and $O^{++}$. Although the initial force on $O^{++}$ is ~ 1.5 times greater for $O^{++}$ than for $O^+$, lattice distortions prevent $O^{++}$ desorption but permit $O^+$ desorption.

in Figure 5 for desorption of $O^+$ from a "ridge" site of $TiO_2$ (110). Lattice recoil has a major effect on the kinetic energies of the ejected ions. If the lattice ions were rigidly held in place, ion kinetic energies of ~ 30 eV would be predicted, based on the Madelung potential. In contrast, the trajectory calculations predict kinetic energies ~ 7 eV. This is much closer to the experimental energies, which are ~ 5eV [27].

The trajectory calculations allow us to compare the desorption of singly and multiply-charged ions. It has recently been suggested that for a desorption mechanism based on Coulomb forces, the kinetic energy of a doubly charged ion should be twice that for a singly charged ion[7], and experimental evidence was reported showing this simple scaling relationship. However, given the crucial role of lattice recoil, such a scaling relationship would not be expected, for reasons outlined below. The force acting on the desorbing positive (or doubly positive) ion depends on the positions of all of the neighboring ions; and the response of these neighboring particles will be different depending on whether the desorbing ion is singly or doubly charged. Denoting the desorbing ion by "a" and surrounding ions by "b", the lattice recoil depends both on a-b interactions and b-b interactions. The Coulombic part of the former would be doubled, but the latter are unchanged, thus there is no basis for a doubling of the final ion kinetic energy. This can also be seen by examining the equations of motion (quantal or classical) for the many-particle system.

Trajectory calculations of $O^+$ and $O^{++}$ desorption from $TiO_2$ (110) were performed for comparison. Identical short-range forces were employed, and only the Coulomb terms were changed. The normal component of the displacement and force for $O^+$ and $O^{++}$ are shown in Figure 4. The force on $O^{++}$ at t=0 is larger than the force on $O^+$ by a factor of $\simeq 1.5$, not 2. The reason for this is that only the Coulomb forces have doubled; the short-range forces have not changed, and these forces are comparable in magnitude to the Coulomb terms. As time develops, the lattice distorts much more severely for the case of $O^{++}$ , and the calculations predict that this particle does not even desorb. In contrast, $O^+$, though initially less repulsive, does desorb with ~7 eV of kinetic energy. The substantial differences in time-development are due to differences in the lattice recoil motion. Although the details of the motion depend on crystal structure, interaction potentials, particle masses, etc., it is clear that a simple doubling of the kinetic energy should not be expected for ion desorption by the Coulomb explosion mechanism. The only case where an energy doubling would be expected is for the motion of an ion in a space-fixed potential. This could occur experimentally if the ion kinetic energy is determined by acceleration in an externally applied electrostatic potential or by the potential developed by macroscopic charging of the surface of the crystal. Very recently, we have investigated the mechanism for positive ion production by electron bombardment of alkali halides[6], and found that the singly and doubly-charged ions are formed by gas-phase ionization of desorbing neutral atoms. In this case, macroscopic charging of the crystal surface determines the final ion kinetic energies, and a simple scaling relation for the kinetic energies of multiply-charged ions would be expected.

(B) Internal energy distributions of desorbed molecules

When the desorbed species are molecules, analysis of the internal (rotational and vibrational) energy distributions can yield insight into the nature of bonding in the chemisorbed state and information on the dynamics of the desorption process. These distributions provide additional information over that obtainable from kinetic energy and angular distributions. For example, vibrational distributions may provide information on changes in bond length between the chemisorbed and free molecules, and rotational distributions may provide information on the orientation of the chemisorbed molecules. In the following discussion, we will consider the factors which determine these distributions and examine in more detail the information that can be obtained.

First consider vibrational excitation of the desorbed molecule. Two excitation mechanisms can be distinguished based on the difference in time-scales for electronic and nuclear motion. The intra-molecular bonding may be different in the ground and excited states, and this difference can lead to vibrational excitation. The electronic transition which initiates the desorption process is fast, thus the molecule in the electronically excited (or ionic) state will initially be in a superposition of vibrational states, with amplitudes determined by Franck-Condon factors between the ground and excited-state vibrational wavefunctions. The vibrational distribution of the desorbed molecule depends on these Franck-Condon factors and on the subsequent time evolution of the nuclear coordinates on the repulsive potential energy surface. If there are further electronic processes, such as neutralization, additional Franck-Condon factors enter in. In general the vibrational distribution is determined by the combined effects of suitable Franck-Condon factors and evolution of the nuclear coordinates on the appropriate potential surfaces. These two processes may be referred to as electronic and mechanical mechanisms for vibrational excitation.

The evolution of the nuclear coordinates after excitation to a repulsive state can be regarded as a "half collision". Thus the mechanical excitation of vibrational modes in ESD/PSD is analogous to vibrational excitation in molecular collisions[28]. The most important factor is the characteristic time that the repulsive force acts compared to the time-scale for molecular vibrations. If the force acts for a time which is small compared to a vibration time, then an impulsive approximation is appropriate and the extent of vibrational excitation is often quite substantial[29]. In practice, the force can be regarded as impulsive only for very low frequency vibrational modes and also for rotational excitation. For diatomic molecules with high vibrational frequencies, such as CO, NO, and OH, the opposite limit is more appropriate. That is, the repulsive force acts for a time larger than a vibration time. In this case, the oscillator can adiabatically adjust to the slowly varying force, thus mechanical excitation is negligible, and the final vibrational state of the molecule is determined entirely by Franck-Condon factors. This situation would be ideal for obtaining information about changes in bond length from the adsorbed state to the free state of the molecule. However, in many cases, the two time scales are comparable, and the effects of mechanical excitation must be considered using techniques of classical or quantal dynamics.

Recently, we have studied the kinetic, rotational, and vibrational energy distributions of $OH^*(A^2\Sigma^+)$ desorbed from surfaces of $TiO_2$, $SrF_2$, and NaCl exposed to $H_2O$ [30]. We have also carried out classical trajectory calculations which simulate the desorption process. In Figure 6 we show the calculated kinetic and rotational energy distributions of OH* desorbed from NaF. In the calculations, a simple exponential repulsive force, directed along bond axes, was utilized. The calculated rotational excitation was low, corresponding to a rotational temperature of ~220 K (for a sample temperature of 300 K). Mechanical excitation of the O-H vibration was found to be entirely negligible, and kinetic energies ~ 0.2 eV were predicted. The low rotational excitation results from the small H/O mass ratio and the assumption that the forces act along bond directions. The experimental rotational distribution was rather close to the sample temperature $T_{Rot}$ ~ 350 K. This is somewhat higher than predicted, which suggests that additional orientation-dependent forces, such as bending forces, are important. In contrast to the low rotational excitation, substantial vibrational excitation was observed, $T_{Vib}$ ~2000 K. Since mechanical excitation of the O-H vibration is negligible, the observed excitation reflects bond length changes upon excitation to the desorbing state. In fact, Franck-Condon factors for the transition from free $OH^-$ to OH* predict the observed vibrational distribution.

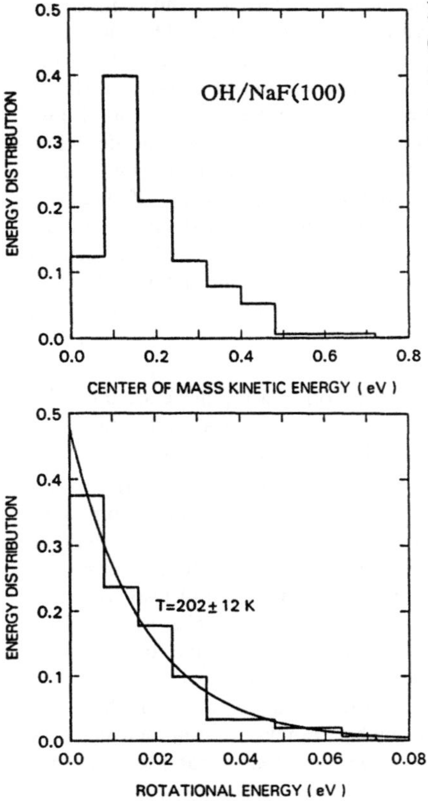


Figure 6. Classical trajectory calculations of $OH^*$ desorption from $OH^-$/NaF(100) at 300 K. Top: predicted kinetic energy distribution of $OH^*$. Bottom: predicted rotational energy distribution.

The interpretation of internal energy distributions is not always straightforward. As a hypothetical example, consider the possibility of NO desorption from Cu as a result of $2\pi^*$ ionization. According to the discussion in section II (B), NO adsorbs on Cu essentially as $NO^-$ However, the bond length and vibrational frequency of chemisorbed NO differ from those of free $NO^-$ . The results of calculations on $Cu_5NO$ indicate that $R_e(N-O)$ and $\omega_e(N-O)$ of adsorbed NO are intermediate between those of free NO and $NO^-$, in agreement with experiment[31]. At the present time, the characteristics of vibrations in the excited state are not known. These characteristics determine the Frank-Condon factors for vibrational excitation. The sensitivity of the vibrational distribution to bond length changes can be seen by comparing the Franck-Condon factors for a transition from free $NO^-$ to free NO with those for a transition from chemisorbed NO to free NO, as shown in Figure 7. As can be seen, substantial vibrational excitation is predicted, but the distribution depends strongly on the change in N-O bonding between the two states. These distributions may be further modified by the subsequent nuclear motion. In general, desorption may involve a complex sequence of electronic transitions together with nuclear motion, and knowledge of the excited-state potentials is often incomplete. Thus it may not be possible to uniquely infer the desorption mechanism from measurements of internal energy distributions.

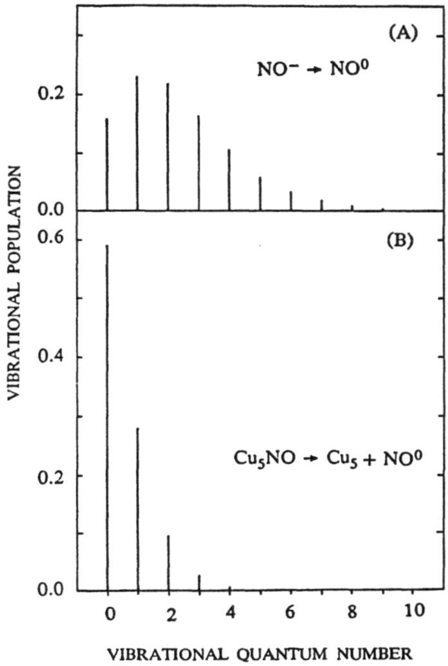


Figure 7. Vibrational energy distributions of NO based on Franck-Condon factors (see text). (A) Distribution obtained using vibrational wavefunctions for a transition from free $NO^-$ to free NO. (B) Distribution obtained using vibrational wavefunctions for the initial state from $Cu_5NO$ cluster calculations (Ref. 20) and free NO for the final state.

## IV. Summary and conclusions

Using density functional and MCSCF cluster calculations, we have explored the role of adsorbate electronic structure, and screening by the substrate free electrons on ion DIET. We find that charge-transfer screening of adsorbate ion states, far from being detrimental to ion desorption as is usually assumed to be the case, in fact provides the driving force for desorption. Although the Coulombic forces are screened, the partial occupation of affinity levels by the charge-transfer screening process increases the Pauli repulsion between adsorbate and substrate and reduces the net charge on the adsorbate. As a result, although overall the screened states are bound, their minima are shifted further away from the surface than the ground state minimum so that Franck-Condon transitions from the ground state populate the repulsive part of the screened-ion state and can lead to desorption. We find that bare multi-hole states (mh-states) are very strongly bound, but screened, mh,ne-states, can be DIET-active. Our calculations also indicate that for electronegative adsorbates such as F, ionization of any adsorbate level, not necessarily a valence level, can lead directly (without the decay of the primary hole) to a repulsive state. The desorption behavior of the particular repulsive state depends entirely on its lifetime. Evidence exists that sufficiently long lifetimes may develop particularly in molecular adsorbates.

Using classical trajectory calculations, we also have investigated the role of nuclear dynamics in DIET processes. As prototypes we have studied systems such as alkali halides and $TiO_2$. We find that dynamic distortions of the lattice can be a crucial factor in determining ion desorption from such solids. In the case of alkali halides (MX) the lattice can rearrange and trap an $X^+$ ion and thus prevent desorption, even when localization of the electronic excitation is favorable. In the case of $TiO_2$ (110), $O^+$ desorption from "ridge"-sites is feasible but the desorbing $O^+$ ion carries only ~ 20% of the available repulsive energy, the remaining energy is dissipated in the solid. We also find that, as a result of lattice relaxation, ion kinetic energies do not scale directly with ion charge, i.e. doubly charged ions are not ejected with twice the kinetic energy of singly charged ions. Finally, we discussed the factors which determine the internal (vibrational, rotational) excitation of desorbing molecules, and the type of information that can be obtained about chemisorption from the analysis of such energy distributions.

## References


1. D. Menzel and R. Gomer, J. Chem. Phys. **41**, (1964) 3311.
2. P. Redhead, Can. J. Phys. **42**, (1964) 886
3. M. L. Knotek and P. J. Feibelman, Phys. Rev. Lett., **40** (1978) 1964.
4. J. H. Varley, Nature **174** (1954) 886.
5. R. E. Walkup and Ph. Avouris, Phys. Rev. Lett., **56** (1986) 524.
6. R. Walkup, Ph. Avouris and A. P. Ghosh, to be published.
7. P. Williams and G. Gillen, Surface Sci., **180** (1986) L109.
8. D. P. Woodruff, M. M. Traum, H. H. Farell, N. V. Smith, P. D. Johnson, D. A. King, R. L. Benbow and Z. Hurych, Phys. Rev. B**21**, 5642 (1980).
9. Ch. Park, M. Kramer and E. Bauer, Surface Sci. (1981) L533.
10. Ph. Avouris, F. Bozso and A. R. Rossi, in **Photon, Beam and Plasma Stimulated Chemical Processes at Surfaces**, MRS Symposium Proceedings, 1987.
11. M. J. Bozock, M. J. Dresser, W. J. Choyke and J. T. Yates, to be published.
12. Ph.Avouris, R. Kawai, D. M. Newns, and N. D. Lang, to be published.
13. W. Kohn and L. J. Sham, Phys. Rev. **140** (1965) 1133; P. Hohenberg and W. Kohn, Phys. Rev. **136** (1964) 864.
14. N. D. Lang and A. R. Williams, Phys. Rev. B **18** (1978) 616.
15. N. D. Lang, J. K. Norskov and B. I. Lundqvist, Phys. Scr. **34** (1986) 77.
16. B. Crasemann, personal communication.
17. M. Cini, Solid State Commun. **24** (1977) 681; G. A. Sawatzky, Phys. Rev. Lett.,**39** (1977) 504.
18. J. A. Bearden and A. P. Burr, Rev. Mod. Phys. **39**, 125 (1967).
19. Ph. Avouris, P. S. Bagus and C. J. Nelin, J. Electr. Spectr. **38**(1986) 269.
20. P. S. Bagus, C. J. Nelin and Ph. Avouris, J. Vac. Sci. Tech. A, 1987, to be published.
21. A. R. Burns, E. B. Stechel and D. R. Jennison, Phys. Rev. Lett. **58** (1987) 250.
22. P. S. Bagus, A. R. Rossi and Ph. Avouris, Phys. Rev. B **31** (1985) 1722;Ph. Avouris, P. S. Bagus and A. R. Rossi, J. Vac. Sci. Technol. B **3** (1985) 1484.
23. J. Stohr and R. Jaeger, Phys. Rev. B**26** (1982) 4111.

24. J. Stohr, J. L. Gland, W. Eberhardt, D. Outka, R. Madix, F. Sette, R. J. Koestner and U. Doebler, Phys. Rev. Lett. **51** (1983) 2414.

25. R. G. Gordon and Y. S. Kim, J. Chem. Phys. **56** (1972) 3122;Y. S. Kim and R. G. Gordon, J. Chem. Phys. **60** (1974) 4332.

26. D. Beeman, J. Comput. Phys. **20** (1976) 130.

27. R. L. Kurtz, Surface Sci. (1987).

28. R. B. Bernstein ed., **Atom-Molecule Collision Theory** (Plenum, New York, 1979) and references therein.

29. K. Snowdon, pg. 116 in **DIET 2**, ed. W. Brenig and D. Menzel, Springer Verlag, Berlin (1985).

30. Ph. Avouris, R. Beigang, F. Bozso, and R. E. Walkup, Chem. Phys. Lett. **129**, 505 (1986).

31. J. F. Wendelken, J. Vac. Sci. Technol., **20**, 884 (1982).

# The Dynamics of $O^+$ Desorption from $TiO_2$

R.E. Walkup[1] and R.L. Kurtz[2]

[1]IBM Thomas J. Watson Research Center, Yorktown Heights, NY 10598, USA

[2]Surface Science Division, National Bureau of Standards, Gaithersburg, MD 20899, USA

## Introduction

Electrons and photons can stimulate positive ion desorption from surfaces provided that: (1) the positive ion is formed by electronic transitions initiated by the incident electron or photon, (2) repulsive forces act to expel the particle, and (3) the positive ion escapes without neutralization or re-capture by the solid. A microscopic mechanism has been proposed by Knotek and Feibelman [1] to explain positive ion desorption from ionic insulators. The prototypical system for this mechanism is $O^+$ desorption from $TiO_2$. Since the Ti cation is maximal valent, i.e. a filled 3p shell and no 3d electrons, a Ti core hole that is produced by electron or photon impact can only be filled by an inter-atomic Auger decay process that involves O valence electrons. There is a finite branching ratio for two Auger electrons to be ejected from a single oxygen center, resulting in the formation of $O^+$. This $O^+$ ion can then be expelled from the surface by repulsive Coulomb forces. Since the repulsive forces are relatively well known, the nuclear motion can be predicted. In this article , we discuss classical trajectory calculations of $O^+$ desorption from $TiO_2$ . These calculations establish the time scale for desorption, and they provide predictions of ion kinetic energy and angular distributions which can be directly compared with experimental results [2,3]. The calculations also provide qualitative insight into the effects of lattice recoil and the origin of site-dependent desorption probabilities.

## Calculations

In the calculations, $TiO_2$ is regarded as a maximal valent ionic solid composed of $Ti^{+4}$ ions (...$3s^2 3p^6$) and $O^{-2}$ ions (...$2s^2 2p^6$). The potential energy surface governing ion motion is taken to be a pair-wise sum of spherically symmetric potentials:

$$V(\vec{R}_1, \vec{R}_2, ...\vec{R}_N) = \sum_{i>j} V_{ij}(R_{ij}) \quad .$$

The pair potentials, $V_{ij}(R_{ij})$ are calculated using a local density functional model due to Gordon and Kim [4]. In this model, the total charge density is assumed to be a superposition of Hartree-Fock charge densities of the individual ions. This method is therefore strictly applicable to the interaction of pairs of closed-shell ions. We also use this method to obtain interatomic potentials for $O^+$ interacting with $Ti^{+4}$ and $O^{-2}$. Since $O^+$ is an open-shell ion, the potentials must be regarded as approximate. However, the potentials do qualitatively account for the "orbital shrinkage" of $O^+$ compared to $O^{-2}$; and the Coulombic terms and interactions among the closed-shell lattice ions are accurately treated. Charge densities for $Ti^{+4}$ and $O^+$ were taken from the tables of Clementi and Roetti [5]. Charge densities for $O^{-2}$ were taken from the calculations of Watson

[6], who solved Hartree-Fock equations for $O^{-2}$ stabilized by a Madelung potential. The Gordon-Kim potentials provide a reasonably good description of the mechanical properties of a variety of ionic metal oxides, including $TiO_2$. For example, a stable rutile structure is predicted for $TiO_2$, with a Ti-O bond length of 1.98Å. This compares reasonably well with the actual structure, in which there are two inequivalent Ti-O bonds with lengths of 1.944Å and 1.988Å [7]. The predicted compressibilities are also reasonably close to the experimental data [8]. However, the use of spherically symmetric pair potentials does not provide an extremely accurate description of the anisotropic elastic properties of $TiO_2$ [9]. Fortunately, the important qualitative aspects of $O^+$ desorption do not depend sensitively on the details of the interatomic potentials - the desorption behavior is dominated by the Coulomb interactions.

We will restrict our attention to $O^+$ desorption from the (110) surface of $TiO_2$. This is the most stable rutile surface. It shows no evidence of faceting upon annealing, and it gives sharp 1 × 1 LEED patterns, which suggests that the surface structure resembles a simple bulk truncation [2,3]. A ball model of the (110) surface is shown in Fig. 1. There are two types of surface $O^{-2}$ ions: (1) ions in bridging sites, and (2) ions in in-plane sites. The bridging ions are bonded to two 6-fold coordinated Ti cations, and they form a ridge (running along the c-axis) which protrudes above the surface. The in-plane ions have three nearest-neighbor Ti ions, all located within the surface plane: one of these is a 6-fold coordinated Ti ion which is under the oxygen ridge, and the other two are 5-fold coordinated Ti ions. The trajectory calculations begin with a fully relaxed solid. First a bulk-truncated charge-neutral solid is formed, consisting of ~ 600 ions. These ions are then allowed to displace to minimize the total potential energy, using the pair-wise potentials mentioned above. The relaxed solid resembles a bulk truncation except that the outer-most oxygen ions displace inward, giving increased Coulomb stabilization. Next, an $O^+$ ion is formed, and the subsequent nuclear motion is followed. In the present calculations, it is assumed that the $O^+$ ion is suddenly produced in a normal $O^{-2}$ lattice site. This is equivalent to assuming that the inter-atomic Auger process forms $O^+$ before any substantial nuclear motion takes place. Since inter-atomic Auger rates are of order ~ $10^{14}\ s^{-1}$ [10,11], this assumption can be questioned. Alternative time sequences may occur, and their consequences can be predicted by appropriate trajectory calculations. For example, after the Ti core hole is formed, the Ti ion can be considered to be in a $Ti^{+5}$ charge state. As a result, the neighboring $O^{-2}$ ions will begin to displace toward the $Ti^{+5}$ ion, due to the increased Coulomb attraction. It is possible that inter-atomic Auger decay occurs after a substantial displacement of the

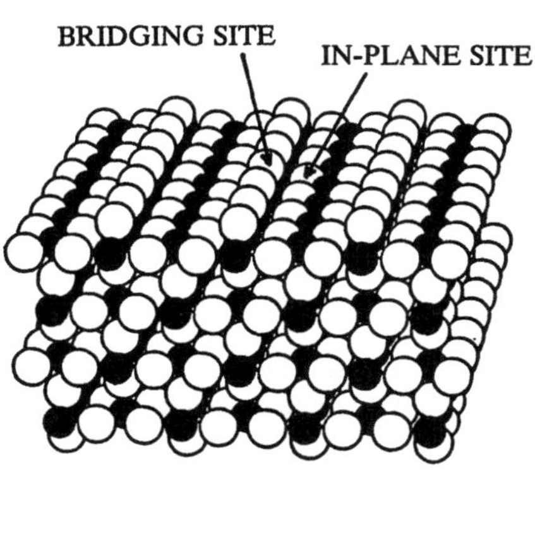


Figure 1. A ball model of the $TiO_2$ (110) surface. Ti ions are black, and O ions are white. Oxygen ions on the surface may be in bridging sites or in-plane sites, as shown.

lattice ions. This possibility will be investigated in future work, however, $O^+$ formation is assumed to be sudden in the present calculations.

If an $O^+$ ion is formed in a bridging site on the (110) surface, there will be a large force (~ 17 eV Å$^{-1}$) in the direction of the surface normal acting on this ion, as shown in Fig. 2. The time dependence of the force and displacement of this $O^+$ ion are shown in Fig. 3 (upper curve). Strong repulsive forces act for ~ 10 fs; and after ~ 20 fs, the ion has displaced ~ 1 Å from the surface. The $O^+$ ion is predicted to promptly desorb, with a final kinetic energy of ~ 7.5 eV, and a final direction along the surface normal. In this trajectory, the initial conditions were chosen such that all ions start in their equilibrium positions with no initial velocity. In order to obtain approximate energy and angular distributions, zero-point and thermal motion were accounted for by starting with random initial displacements and velocities, using suitable Gaussian distributions. For $TiO_2$, the Debye temperature is substantially greater than room temperature [8], thus at room temperature, the zero-point motion is more important than thermally-activated motion. The use of independent (anisotropic) Gaussian distributions amounts to an Einstein description of the lattice vibrations. This treatment is a reasonable approximation, but it under-estimates the importance of low-frequency vibrational modes. Using these Gaussian-distributed initial conditions, the predicted ion kinetic energy distribution has a peak at ≃ 7.5 eV, with a FWHM of ≃ 1.5 eV. The predicted angular distribution is a nearly circular spot (slightly elongated in the direction perpendicular to the oxygen rows) with most of the ions within a 10° cone centered along the surface normal. The calculated spot pattern is shown in Fig. 4 (upper panel).

If the $O^+$ ion is initially formed in a in-plane site, the dynamics of desorption are dramatically different. For the in-plane ions, all of the nearest-neighbor Ti ions are within the surface plane (see Fig. 1), thus the net force in the direction of the surface normal is small (~ 1 eV Å$^{-1}$). In addition, the $O^{-2}$ ions in the bridging sites displace toward the $O^+$ ion in the plane. This reduces the repulsion and leads to an oscillatory force, as shown in Fig. 3 (lower curve). These displacements may have a profound effect on the neutralization probability. The $O^+$ ion formed in an in-plane site is predicted to remain almost stationary for ~ 100 fs, while the surrounding ions move. During this time, the $O^{-2}$ ions from the nearest bridging sites come substantially closer to the $O^+$ ion. Both the time delay and the lattice distortions may enhance the neutralization probability. In our calculations, the $O^+$ ion is assumed to survive without neutralization. This assumption is necessary in order to predict possible kinetic energy and angular distributions. After a substantial delay ( ~ 150 fs), lattice distortions are predicted to push

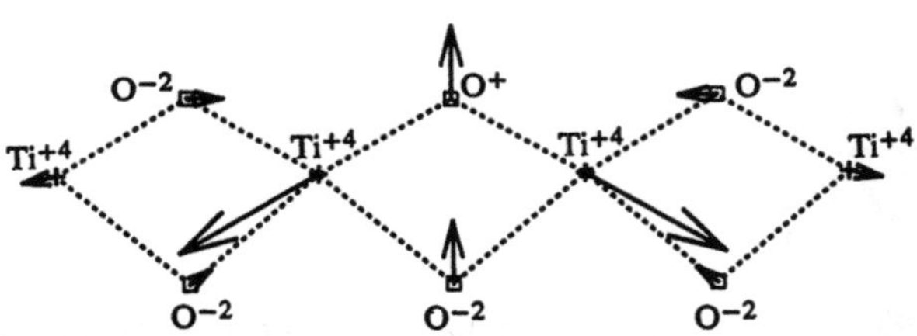


Figure 2. Forces resulting from $O^+$ formation at a bridging site. The surface normal is in the vertical direction. The top row of oxygen ions are in bridging sites, and only a few neighboring ions are shown.

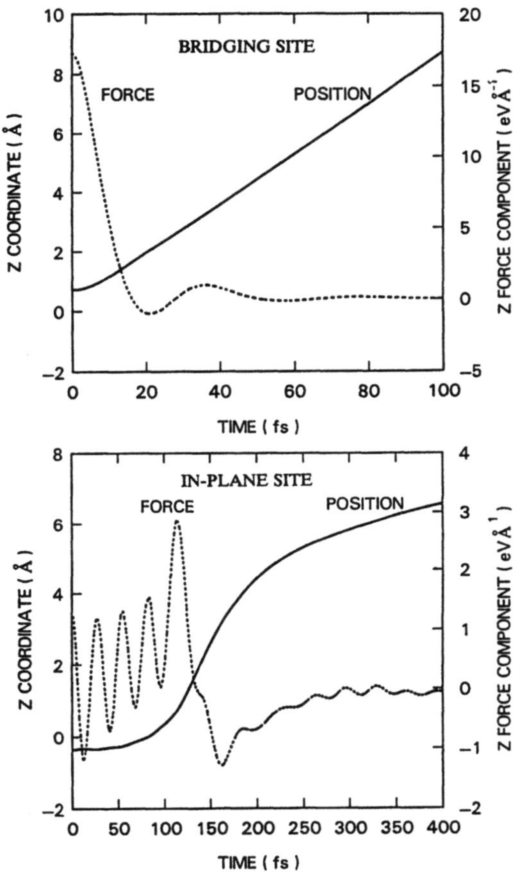


Figure 3. The upper curve shows the time evolution of the normal component of the force (dotted line) and the position (solid curve) for $O^+$ formed in a bridging site. The lower curve shows the corresponding force (dotted curve) and position (solid curve) for $O^+$ ions formed in an in-plane site.

the $O^+$ ion away from the surface. As the $O^+$ ion departs, it is attracted by the nearest row of $O^{-2}$ ions, which bends the $O^+$ trajectories toward this row. As a result, $O^+$ desorption from in-plane sites is predicted to yield an angular distribution with two broad lobes displaced perpendicular to the oxygen rows, as shown in Fig. 4 (lower panel). Moreover, the kinetic energy distribution is predicted to peak at ~ 4 eV, with a ~ 1 eV FWHM.

## Discussion

The results of the trajectory calculations can be directly compared with experimental measurements of the $O^+$ kinetic energy and angular distributions [2,3]. The experimental angular distributions (ESDIAD patterns) show a complex and interesting de-

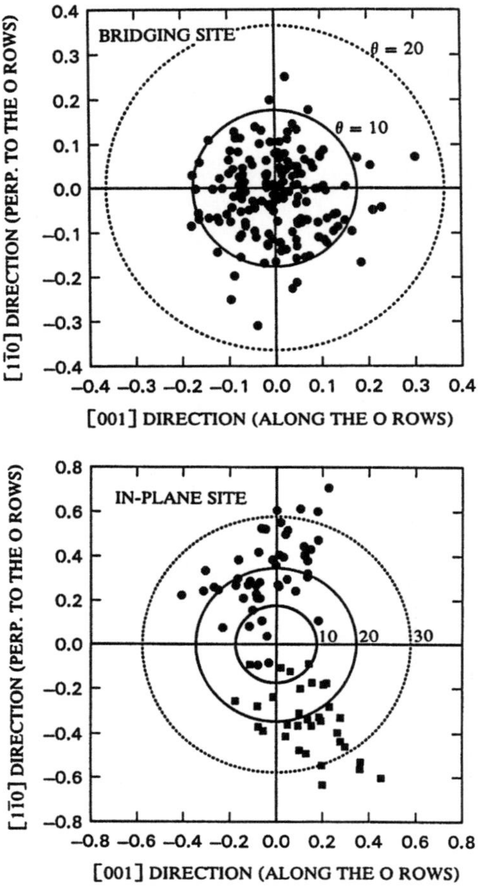


Figure 4. The upper panel shows the predicted angular distribution for $O^+$ ions formed in bridging sites. The lower panel shows the predicted angular distribution for ions formed in in-plane sites. Circles corresponding to polar angles of 10, 20, and 30 degrees are shown.

pendence on surface preparation. In the experiments, $TiO_2$ (110) surfaces are cleaned by sputtering and then annealed. The ESDIAD patterns depend on the annealing temperature. For annealing temperatures above 1000 K, a single ESDIAD spot is observed in the direction of the surface normal, and LEED measurements indicate that the surface is well-ordered and not reconstructed [2,3]. For these conditions, there is no evidence of off-normal emission, and the $O^+$ kinetic energy distributions show a single broad peak at 4-5 eV [2,3].

These experimental observations are qualitatively consistent with $O^+$ desorption from bridging sites of the (110) surface via a Coulomb repulsion mechanism. However, the experimental angular distributions are inconsistent with $O^+$ desorption from the in-plane sites. This suggests that ions formed at the in-plane sites are neutralized, and do not contribute to the desorbed ion flux. The results of the trajectory calculations also suggest a plausible explanation for such site-dependent neutralization probabilities. $O^+$ ions originating from bridging sites can promptly desorb, because the local geometry results in a large force component in the direction of the surface normal. This prompt desorption (~ 10 fs) minimizes the neutralization probability. In contrast, ions originating from in-plane sites experience only a small force in the direction of the surface normal; and lattice distortions reduce the initial repulsion, leading to a long residence time at the surface (~ 100 fs). For these sites, the neutralization probability is enhanced because of the long residence time. The associated displacement of bridging $O^{-2}$ ions toward the $O^+$ ion may also enhance the neutralization probability.

These calculations show that lattice recoil has a dramatic effect on the final ion kinetic energies. When $O^+$ is formed, the repulsive Madelung potential energy is ~ 30 eV, but the final ion kinetic energies are predicted to be ~ 7.5 eV. One reason for the large loss of energy to the lattice is the long-range nature of the repulsive Coulomb forces. When $O^+$ is formed, a large number of ions feel a force imbalance, and the subsequent motion of these ions converts the initial potential energy into kinetic energy of these distant ions. In practice, it is necessary to account for the motion of several hundred ions in order to converge to the behavior of a semi-infinite solid. Much of the recoil energy, however, is lost to the nearest and next-nearest neighbors. The energy lost to the lattice also depends on the anion to cation mass ratio. For example, when $O^+$ is formed at a bridging site, the forces on the neighboring Ti cations are larger than the force acting on $O^+$ (see Fig. 2). However, since the Ti ions are substantially more massive than O, the ionic configuration remains repulsive long enough for $O^+$ to gain substantial kinetic energy. The calculations tend to over-estimate the final ion kinetic energies for a variety of reasons. For example, after the departure of $O^+$, the model solid has a formal charge of +2. This results in substantial Coulomb repulsion at very long times. Also, the interatomic potentials use an incomplete treatment of electronic polarization. Such a polarization partially screens the repulsive forces and contributes to an image-like attraction, which is not explicitly included. Qualitatively, however, it is clear that lattice recoil motion will absorb most of the initial repulsive energy.

The present calculations of $O^+$ desorption from $TiO_2$ contrast strongly with previous calculations of positive ion desorption from alkali halides [12]. In the case of alkali halides, desorption of positive ions from normal sites of (100) surfaces was predicted to be blocked by lattice distortions; i.e. even if neutralization does not occur, the positive ions would be re-captured by the solid. Subsequent experiments suggest that the experimentally observed ions in this case are formed by gas-phase processes [13]. Several factors make positive ion desorption from $TiO_2$ much more favorable. For $TiO_2$, there are oxygen ions (the bridging sites) which protrude above the surface plane. This results in a very large normal force component, and consequently, a prompt desorption process. Also, for $TiO_2$ the lattice distortions are less severe. The $O^{-2}$ ions are held in place relatively tightly by $Ti^{+4}$ ions. This reduces the tendency of $O^{-2}$ ions to displace toward $O^+$ ; and this reduces the likelihood that $O^+$ will be re-captured by the solid. In contrast, for the alkali halides, all of the ion charges are $\pm 1$, thus the surrounding $F^-$ ions (for example) are as strongly attracted to $F^+$ as to $Na^+$. This results in large lattice distortions which can trap the $F^+$ ion.

## Conclusions

We have presented preliminary results of trajectory calculations of $O^+$ desorption from $TiO_2$ (110). A strong site dependence of the desorption dynamics is predicted. The site dependence is closely related to the local bonding geometry, which determines the magnitudes and directions of the repulsive forces. The experimental data appear to be consistent with $O^+$ desorption from bridging sites on the surface, via a Coulomb repulsion mechanism as described by Knotek and Feibelman [1]. Ion desorption from these sites is prompt (~ 20 fs) because of the large initial force in the direction of the surface normal. Prompt desorption helps minimize the neutralization probability. In contrast, desorption from in-plane sites is delayed (~ 100 fs). If these in-plane ions were to desorb, they would have distinct angular and energy distributions which are not observed experimentally. A plausible explanation for this is that ions originating from these sites are effectively neutralized due to their long residence time at the surface. Lattice recoil is seen to have an important effect on the ion kinetic energies. Reversal of the Madelung potential results in ~ 30 eV of repulsive potential energy, but the trajectory calculations predict final ion kinetic energies of ~ 7.5 eV. This is much closer to the experimental values of 4-5 eV.

## References


1. M. L. Knotek and P. J. Feibelman, Phys. Rev. Lett. 40 , 964 (1978); and P. J. Feibelmann and M. L. Knotek, Phys. Rev. B 18 , 6531 (1978)
2. R. L. Kurtz, Surf. Sci. 177 , 526 (1986)
3. R. L. Kurtz, R. Stockbauer, and T. E. Madey, this volume
4. R. G. Gordon and Y. S. Kim, J. Chem. Phys. 56 , 3122 (1972)
5. E. Clementi and C. Roetti, At. Data and Nuc. Data Tables 14 , 177 (1974)
6. R. E. Watson, Phys. Rev. 111 , 1108 (1958)
7. A. von Hippel, J. Kalnajs, and W. B. Westphal, J. Phys. Chem. Solids 23 , 779 (1962)
8. F. A. Grant, Rev. Mod. Phys. 31 , 646 (1959)
9. C. R. A. Catlow and R. James, Proc. Roy. Soc. Lond. A 384 , 157 (1982); and W. C. Mackrodt and R. F. Stewart, J. Phys. C. 12 , 431 (1979)
10. J. A. D. Matthew and Y. Komnios, Surf. Sci. 53 , 716 (1975)
11. T. A. Green and D. R. Jennison, this volume
12. R. E. Walkup and Ph. Avouris, Phys. Rev. Lett. 56 , 524 (1986)
13. R. E. Walkup, Ph. Avouris, and A. P. Ghosh, Phys. Rev. B15, to be published

# DIET and the Electronic Structure of Chemisorbed Molecules and Physisorbed Rare Gases

*D.R. Jennison, E.B. Stechel, and A.R. Burns*

Sandia National Laboratories, Albuquerque, NM87185, USA

## 1. Introduction

Most observables which characterize stimulated desorption, such as the yield (or cross-section) and the translational and internal energy distributions, result from dynamical processes which are now being quantitatively modeled [1,2]. However, such calculations require realistic ground and excited-state potential energy surfaces and excitation lifetimes, which are in turn determined by the electronic structure of the surface-adsorbate system. We also wish to understand the electronic dynamics of excitation creation and decay and the strength of adsorbate-adsorbate interactions which influence the localization (or self-trapping) probability of an excitation [2,3]. Fortunately, some general principles have emerged recently which simplify the interpretation of electronic structure in terms relevant to DIET. We summarize and illustrate these concepts, drawing upon our studies of chemisorbed CO and NO:Pt(111) and physisorbed rare gases on metals.

The presence of the surface (which we assume is metallic) limits the strict application of gas-phase results to the analysis of DIET. In the pioneering work of MENZEL and GOMER [4] and of REDHEAD [5], it was recognized that excitation lifetimes are severely shortened by the presence of the surface. Equally important to the description of the excited state is the fact that an electron can be partially transferred onto an adsorbate due to screening. Thus a species is often created which is without a direct gas phase analogue. We shall see that in the cases which we have studied, screening plays a large role in determining both the excited state forces and lifetimes.

In a given system, a large number of excitations can be relevant to DIET in general. Therefore, we simplify the problem by considering only the highest probability channels but in considerable theoretical detail [1,2]: we study neutral desorption using laser resonance-ionization spectroscopy [1,6] which measures desorbate quantum states. A high neutral cross-section allows us to omit short-lived excitations from consideration; this is a powerful tool for simplifying the analysis.

The longest-lived excitations are those whose energies are sufficiently large that they are not degenerate with simple substrate excitations and therefore cannot decay resonantly. Resonant decay is typically quite fast with lifetime widths $\Gamma$ of the order of 1 eV [7]. Non-resonant excitations decay via a valence Auger process which is usually about a factor of 10 slower ($\Gamma \sim 0.1$ eV) [1]. We have studied this decay mode in some detail and have found that excitations in adsorbates can be even longer lived in certain special circumstances. These arise when a hole must be filled with an electron of different symmetry, thereby reducing the Auger matrix element since angular momentum needs to be transferred between the two electrons involved in the decay. We have found that lifetime widths of $\Gamma \sim 0.01$ eV are possible, which in the NO:Pt(111) system results in the large observed yield [2].

In what follows, we assume that an excitation is localized on a single adsorbate. We discuss the probability for such localization in the companion paper which stresses dynamics [2]. This separation of topics is somewhat artificial but conceptually useful. Rather than discuss the molecular and rare gas examples separately, emphasizing the differences, we instead discuss the excited state forces and lifetimes as two separate topics, thus emphasizing the physics common to both chemisorbed and physisorbed species.

## 2. The Screening Relaxation to Excitations in Adsorbates

It is useful to be aware of the time scales of screening relaxation in response to the creation of an excitation, which we take for simplicity as a single valence hole. A measure of the response time is h/W where W is the bandwidth which in turn is related to the transfer integral, t, between sites. The largest bandwidth is the s-p band of the metal where typically $W \sim 10$ eV. Thus the fastest screening involves the formation of an image charge which produces a (short lived) $\sim e^2/4r^2$ force on the adsorbate. Further relaxation involves the interaction between adsorbate and substrate orbitals. If a normally empty or only partially occupied adsorbate level $\phi_a$ interacts with a metal level $\phi_m$, they produce bonding and anti-bonding ground state hybrid orbitals which are split in energy due to $t(am) = \langle a|H|m\rangle$ ($\ll W$) and $\Delta E = \langle a|H|a\rangle - \langle m|H|m\rangle$. If an excitation is created in the adsorbate, the energy of $\phi_a$ is lowered causing a rehybridization; if the bonding orbital is full and the antibonding orbital empty (as is frequently the case), this rehybridization increases $\phi_a$ occupancy. The degree of occupancy change depends on t(am), $\Delta E$, and $\Delta E'$ where the latter is the energy difference in the presence of the excitation. The timescale for the rehybridation is determined by h/t(am), i.e. the hopping time for an electron from substrate to adsorbate. While oversimplified, this discussion does contain the essential physics of the screening response.

Except for the lightest of adsorbates (e.g. H, $H_2$, and He) the above time scales are not important in that only the fully relaxed state enters the dynamics of atom motion. However, the different relaxation rates do affect the decay rate of excitations in physisorbed systems (see below).

The amount of screening charge added to $\phi_a$ can be large and important. For chemisorbed CO and NO, $\phi_a$ is the $2\pi$ level of the molecule and we believe that essentially an entire electron is transferred for each valence hole produced. There is both theoretical and experimental support for this. An ab initio calculation on a small NiCO cluster reported the electron populations with and without the presence of a carbon core hole [8]. About 0.8 $e^-$ was transferred onto the CO by rehybridization of the occupied $d\pi$-$2\pi$ (bonding) orbital. One may wonder, however, if core hole screening is relevant to that of a valence excitation. Experimental evidence that valence excitations are also well screened comes from several studies of the Auger spectra of chemisorbed CO [9]. By comparing the energy of two-hole valence excitations in the adsorbate (seen in CVV Auger) with the sum of the energies of relevant one-hole excitations (seen in photoemission), it has been determined that the effective hole-hole interaction ($U_{eff}$) is only $\sim$ 0-2 eV in the adsorbate, compared with $\sim$ 15-18 eV in the gas phase. This is only possible if almost one screening electron is transferred onto the CO or NO molecule for each valence hole.

## 3. Excited State Potential Energy Surfaces

Ab initio calculations of the excited states of chemisorbed molecules have not been done in any detail as they are very difficult: valence holes increase the importance of correlation in describing the state and the screening electron must somehow be added to the cluster. (Density functional theory does a better job of describing the ground state than excited states [10].)

Until accurate calculations are available, we must use more approximate methods of finding excited state potential energy surfaces.

In the case of NO, the $2\pi$ orbital which accepts the screening charge is also largely responsible for bonding the molecule to the substrate (the $5\sigma$ and $1\pi$ orbitals play only minor roles). In addition, screening makes the NO $2\pi$ orbital resemble in occupancy that of $O_2$. Thus single-hole excitations in NO:Pt(111) such as the $5\sigma^{-1}$ should cause the molecule to bind to the surface in a manner which resembles the ground state of chemisorbed $O_2$, whose binding strength and orientation on Pt(111) is known: we modeled excited NO as being only ~ 1/3 as strongly bound as the ground state with a preference to lie down (rather than stand upright with the N atom down). We referred to this as the "equivalent binding model", which obviously has limited applicability depending on the nature of the screening [1,2]. However, this model does also suggest that excited chemisorbed CO would resemble the ground state of NO, in that it would be less strongly bound but would still prefer to be upright. This realization was a key factor in understanding the large differences which were observed in neutral NO vs. CO DIET [1,6].

We also performed calculations on $He^*$:Cu and Li:Cu [11] to discover if the "equivalent binding" approximation (more like "equivalent core" in this case) could be used to model excited physisorbed rare-gas atoms as chemisorbed alkali metal atoms. We found that if a screening electron is added to the cluster, He with a 1s hole binds to a Cu atom in an almost identical manner as does Li. Screening results in the production of a polarized $\sigma$-bond between the metal s-p orbitals and the s orbital of the alkali-atom equivalent of the excited rare gas atom. A covalent component in the excited state binding is in agreement with the conclusions of Wimmer et al [12] who studied Cs:W(001).

The presence of the screening electron is essential to have a realistic model of excited chemisorbed and physisorbed systems. Without it (and it has been left out in some recent work by others) one can have neither an image charge nor a realistic description of the rehybridized screening orbital.

Several previous efforts have been made to model the desorption of physisorbed rare gas atoms [13-15]. These workers assumed that the ionized rare gas atom would experience a force produced by the sum of the ground state potential (to have the hard-wall repulsion) and that of a "bare" image charge. Hartree-Fock self-consistent calculations do not support this model. In fact we found that the force in the case of $He^*$ is only about 1/3 what an image charge model would predict, even though with the screening electron added to the cluster, the extra electron was free to localize on the Cu atom and produce an $e^2/r$ potential if that were the lowest energy solution. Although this calculation was a crude approximation to the atop site, we would expect the hollow site to yield even a smaller force as direct movement towards the surface changes the interatomic distances more slowly than in the case of the atop site. Previous work thus assumed a force which was too strong and therefore derived a lifetime for the excitation which varied unrealistically fast with distance from the surface. This compensated for the too-rapid acceleration and was required to fit data such as the temperature dependence of the Kr yield [14]. These workers did recognize that something in their model was creating this problem; we have recently shown that by reducing the force (and using our estimates for the decay rate which we find to vary slowly with distance from the surface), we are able to understand rare gas DIET [11].

There is another simple argument which shows that the image potential model does not describe the excited-state of a physisorbed rare gas atom. Our calculations showed the equilibrium position of $He^*$-Cu to be close to the sum of the covalent radii of Li and Cu. If this observation also holds for Kr, the equilibrium positions of the ground and excited states would nearly coincide since Rb is so large; hence, the Antoniewicz mechanism [17] may not be relevant here. In spite of the above, desorption still occurs as the vibrational wavefunctions in the (physisorbed) ground and (chemisorbed [16]) excited states differ markedly [11].

## 4. The Lifetimes of Adsorbate Excitations

In chemisorbed NO and CO, the lowest energy excitations involve the bonding $d\pi$-$2\pi$ orbital. This orbital is thought to exist only a few eV below the Fermi level and is hence a resonance. Its excitation could involve moving an electron into the metal or to the anti-bonding $d\pi$-$2\pi$ resonance, which has been observed using inverse photoemission to exist ~ 3 eV above $E_F$ in the neutral molecule [18]. However, resonant photoemission experiments [7] show that occupancy of the anti-bonding level in the presence of a core hole is a short-lived excitation, decaying before the core hole decays (i.e. faster than ~ 10 $eV^{-1}$). Therefore, while the presence of the antibonding resonance is certainly important to the primary excitation cross-section near threshold, the DIET cross section for these resonances is small compared with the neutral yield [2] and we consider them no further.

The next higher-energy excitations involve the $5\sigma$ and $1\pi$ orbitals, which are close together in energy (~ 8-9 eV) and are not resonances. For the case of NO, we computed the Auger rate for these excitations. The fastest channels are those in which a $2\pi$ electron fills the $5\sigma$ or $1\pi$ hole and another $2\pi$ electron is ejected. We also considered channels involving metal orbitals such as the $d\sigma$ and found these to be much slower as they are essentially interatomic in nature. A qualitative picture of these decays arises immediately when one considers the symmetry of the orbitals involved. The quantitative picture is more difficult as the outgoing electron is low in energy and therefore such factors as orthogonality to the occupied orbitals, exchange, and the multi-center expansion of the outgoing wave have to be considered. These aspects will be discussed elsewhere [19].

If angular momentum is transferred between the electrons involved in an Auger decay, the expansion of $e^2/r_{12}$ in the matrix element contributes a term $1/(2\ell+1)$ where $\ell$ is the momentum transferred. In the case of a $1\pi$ hole being filled by a $2\pi$ electron, $\ell = 0$. Since the same atomic orbitals are involved in both the $1\pi$ and $2\pi$ MO's, we have a maximal matrix element, reduced only by the fact that in the molecule the $1\pi$ orbital is polarized towards the O atom while the $2\pi$ is towards the N atom. We found that $\Gamma(1\pi)$ is a few tenths of an eV [1]. However, when a $5\sigma$ hole is filled by a $2\pi$ electron, $\ell = 1$ or 2 depending upon whether the (mainly nitrogen) 2s or the 2pz component of the $5\sigma$ is filled by the 2px (or 2py) component of the $2\pi$ orbital. Since the rate goes as the square of the matrix element, we immediately find (ignoring for a moment the different $5\sigma$ and $1\pi$ charge distributions) a reduction in rate of between 1/9 and 1/25 compared with the $1\pi$ decay. The actual reduction turned out to be around 1/20, due in part to the fact that there was more 2pz than 2s character in the $5\sigma$.

We now have a simple rule with which to identify the longest lived excitations in adsorbed species: they will be those for which an electron of different symmetry must fill the hole(s). For example, the $4\sigma^{-1}$ would not be long lived as a $5\sigma$ electron could fill it. We would therefore expect its lifetime to be more like that of the $1\pi^{-1}$ than the $5\sigma^{-1}$. However, the two-hole state $5\sigma^{-2}$ should be equally as long lived as the $5\sigma^{-1}$. Similarly, the three-hole state $4\sigma^{-1}5\sigma^{-2}$ should also be long lived. The recent work reported at this conference by FEULNER et al [20] supported these ideas by identifying the $5\sigma^{-2}$ as a strong PSD threshold for $CO^+$ desorption, as could be predicted based on lifetime arguments alone. (Of course, one still needs to consider dynamical effects [2]).

Until now we have been using a one-electron picture of the electronic structure. In the cases of NO and CO we believe this description to be adequate for our purposes because the splitting between the bonding and antibonding $d\pi$-$2\pi$ orbitals as mentioned above is ~ 5 eV and therefore configuration mixing should not play too large a role for the excitations we have considered (for an example of an Auger decay which has its rate severely reduced by configuration interaction (CI) and for a general discussion of when this can occur, see Ref [21]). In other words, we would expect the orbitals involved in the above decays to be relatively "pure". In the case of the rare gases, the valence hole is also filled from the screening orbital, the polarized $\sigma$, which contains two electrons. However, the $\sigma$-$\sigma^*$ splitting in the case of Cu-He was found by calculation to be only ~ 1 eV, and thus $\sigma$-$\sigma^*$ configuration mixing cannot be neglected out of hand.

The polarization of the $\sigma$-orbital in the rare gas case is less than in the case of alkali halide Auger decay discussed by GREEN and JENNISON [22], where even though the electron-containing orbital is polarized away from the location of the hole, the decay is still "intra-atomic" in that, in the LCAO sense, it is the portion of the wavefunction on the same site as the hole which dominates the contributions to the Auger rate. (The decay is also "interatomic" in the sense that in the final state, the holes end up being mostly on the neighboring atoms.) We can thus conclude that the configuration which will dominate the Auger rate is that in which both $\sigma$-electrons are simultaneously on the rare gas atom, e.g. the $1s^{-1}2s^2$ of He.

At first it seems strange to be thinking about a negative ion resonance in positively-charged excited physisorbed rare-gas atoms; however, the two electron wavefunction for the polarized $\sigma$-bond is still a mixture of three configurations: both on the metal, and on the rare gas, and the singlet coupled configuration in which one electron is on either. In order to determine the Auger rate accurately, we must determine the fraction of time when both electrons are on the rare gas atom.

Remarkably, we cannot use traditional quantum chemical CI methods to correlate the $\sigma$-electrons. The reason is that the response time of the metal substrate, as discussed above, is a factor of ~ 10 faster than the hopping time between the metal and the rare-gas screening orbital, i.e. than the rate of configuration fluctuations. Thus the negative and positive ion resonances have their energy reduced by approximately the image potential. We incorporated these effects in a modified CI model [11] and found that the decay rate was slowly varying (over the distances from the surface of interest) when compared with what is obtained from considering only an interatomic Auger process. In part the reason for the slow variation is the large overlap between the s-orbital of the alkali atom equivalent and the s-p orbitals of the substrate. At the distances at which the excitation is created, the overlap is ~ 0.5. This dampens the sensitivity of the CI result to distance. Further details will be reported elsewhere [11].

## Acknowledgements

We are indebted to T. A. Green for codes, discussions and critical reviews. This work was performed at Sandia National Laboratories, supported by the U.S. Department of Energy under Contract DC-AC04-76DP00789.

## References

1. A. R. Burns, E. B. Stechel, and D. R. Jennison, Phys. Rev. Lett. 58, 250 (1987).
2. E. B. Stechel, D. R. Jennison, and A. R. Burns, this volume.
3. A. R. Burns, D. R. Jennison, and E. B. Stechel, J. Vac. Sci. Technol. A 5, 671 (1987).

4. D. Menzel and R. Gomer, J. Chem. Phys. 41, 3311 (1964).
5. P. A. Redhead, Can. J. Phys. 42, 886 (1964).
6. A. R. Burns, E. B. Stechel, and D. R. Jennison, this volume.
7. C. T. Chen, R. A. DiDio, W. K. Ford, E. W. Plummer, and W. Eberhardt, Phys. Rev. B 32, 8434 (1985); W. Wurth, C. Schneider, R. Treichler, E. Umbach, and D. Menzel, Phys. Rev. B 35, 7741 (1987).
8. D. R. Jennison, G. D. Stucky, R. R. Rye, and J. A. Kelber, Phys. Rev. Lett. 46, 911 (1981).
9. B. E. Koel, J. M. White, and G. M. Loubriel, J. Chem. Phys. 77, 2665 (1982); E. Umbach and Z. Hussain, Phys. Rev. Lett. 52, 257 (1984).
10. R. O. Jones in Many Body Phenomena at Surfaces, D. Langreth and H. Suhl, eds., Academic Press (1984), p. 175.
11. D. R. Jennison and E. B. Stechel, to be published.
12. E. Wimmer, A. J. Freeman, J. R. Hiskes, and A. M. Karo, Phys. Rev. B 28, 3074 (1983); see also P. Soukiassian, R. Riwan, J. Lecante, E. Wimmer, S. R. Chubb, and A. J. Freeman, Phys. Rev. B 31, 4911 (1985).
13. Q.-J. Zhang, R. Gomer, and D. R. Bowman, Surf. Sci. 129, 535 (1983).
14. E. R. Moog, J. Unguris, and M. B. Webb, Surf. Sci. 134, 849 (1983).
15. Z. W. Gortel, H. J. Kreuzer, P. Feulner, and D. Menzel, Phys. Rev. B 35, 8951 (1987).
16. K. F. Wojciechowski, Acta Phys. Polon. 33, 363 (1968).
17. P. Antoniewicz, Phys. Rev. B 21, 3811 (1980).
18. S. Ferrer, K. H. Frank, and B. Reihl, Surf. Sci. 162, 264 (1985).
19. D. R. Jennison, T. A. Green, E. B. Stechel, and A. R. Burns, to be published.
20. P. Feulner, S. Auer, T. Mueller, A. Puschmann, and D. Menzel, this volume.
21. D. R. Jennison, J. A. Kelber, and R. R. Rye, Chem. Phys. Letts. 77, 604 (1981).
22. T. A. Green and D. R. Jennison, this volume; Phys. Rev. B, in press.

# Theory of Desorption of Neutrals by Temporary Ionization of Physisorbed Species at Metal Surfaces

*Z.W. Gortel*[1], *H.J. Kreuzer*[2], *P. Feulner*[3], *and D. Menzel*[3]

[1]Department of Physics, The University of Alberta, Edmonton T6G 2J1, Alberta, Canada
[2]Department of Physics, Dalhousie University, Halifax B3H 3J5, Nova Scotia, Canada
[3]Physik-Department E20, Technische Universität München, D-8046 Garching, Fed. Rep. of Germany

## 1. Introduction

Electron or photon stimulated desorption (ESD or PSD, respectively) is a sensitive probe for studying atomic or molecular systems in interaction with solid surfaces /1,2/. Apart from fragment ions in some cases of physisorbed molecules, only neutral adsorbate species desorb from physisorbed layers /3-7/. This fact is understood in terms of a classical model, due to ANTONIEWICZ /8/ in which the desorption process is a sequence of three steps denoted by 1, 2, and 3 in both panels of Fig. 1. Total electronic energy of the adparticle - solid system depends on the adparticle's distance from the solid surface and on the state of an electronic excitation the solid and the adsorbed particle are in. For physisorption the electrons in the solid and on the adparticle can be considered separately. Thus, the electronic diabatic configuration in which the solid and the adparticle are in their respective electronic ground states corresponds to the energy level $V_o$ and $V_o(z)$ is the surface potential for the physisorbed neutral. In a configuration corresponding to $V_k$ the electrons on the adparticle and those in the solid are in the ground and in an excited state, respectively, with an excitation energy $E_k$ for the latter. Thus, $V_k(z)$'s form a continuum of surface potentials filling all energies above $V_o(z)$. Finally, for the adparticle ionized by removing its valence electron and the solid in the ground state, the corresponding surface potential is $V_+(z)+I_\infty$, where $I_\infty$ - the ionization energy of the isolated particle - was explicitly singled out from the z-dependent part. For physisorption on a metal surface we have

$$V_k(z) = V_o(z) + E_k \tag{1}$$

and

$$V_+(z) = V_o(z) + V_{im}(z) \tag{2}$$

provided polarizabilities of the metal and of the adparticle are not substantially affected neither by an electronic excitation of the former nor the ionization of the latter. $V_{im}(z)$ is the additional image potential the ion experiences in front of the metal surface. In the desorption process considered the only electronic diabatic states of interest are those described above. It is important to notice that in the case of an atom or a molecule bound to the surface by dispersion forces only, ionization will lead to an increase of binding energy and reduction of equilibrium distance from the surface, as seen in the upper panel of Fig. 1.

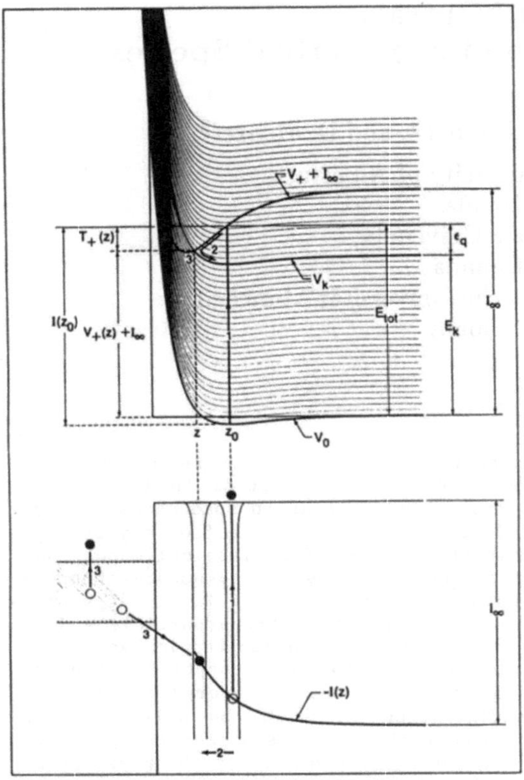


Fig. 1. Schematic diagram of the desorption model. Upper panel: electronic diabatic energy surfaces (surface potentials) for a neutral ($V_0$ and a continuum of $V_k$'s) and an ion ($V_+ + I_\infty$) physisorbed on a metal surface. Lower panel: the same process on a diagram showing the energy levels and excitations of all electrons in the metal-adparticle system

In step 1 the adsorbed particle is ionized by incoming electrons or photons and fast ($\sim 10^{-16}$ sec) rearrangement of electrons in the metal leads to creation of the image force before any substantial displacement of the adparticle takes place. The excited state of the electronic system decays then slowly ($\sim 10^{-14}$ sec) in step 2 by a displacement of the ion towards a new equilibrium position and an associated upward shift of the empty ionization level (Fig. 1, lower panel). The process 2 competes with other decay mechanisms (step 3, $\sim 10^{-16}$ sec) e.g. neutralization of the ion with simultaneous excitation of the electronic system of the metal. One of the possibilities here is the Auger neutralization shown in lower panel of Fig. 1. In the quantum picture, to be considered later on, the energy level $V_+(z)$ is broadened due to its interaction with the continuum of levels $V_k(z)$ it is immersed in. The neutral particle, created in step 3, rebounds and if its kinetic energy is sufficiently large it desorbs.

In this work we provide a quantum-mechanical formulation for the above mechanism of desorption in Section 2 and compare in Section 3 its results with the results of the classical approach for the model surface potentials for which both theories yield numerically manageable results for the kinetic energy distribution of desorbed neutrals and for the overall desorption yield. The physical systems studied are He adsorbed on W and $N_2O$ physisorbed on Ru(001). The latter system is further investigated in Section 4. We will show that both, fitting and the inversion of the experimental data lead to a certain range of possible neutralization parameters compatible with observed kinetic energy distributions. This range contains the physically expected values. However, if as a second constraint the experimental values of the desorption yields are fitted as well, unexpected neutralization parameters result. A short summary is given in Section 5.

## 2. Quantum-mechanical and Classical Models

We start with the quantum theory. Its classical counterpart, known as the Antoniewicz model /8/, was discussed by several workers /3-7/. Our aim is to solve the time-dependent Schrodinger equation for a state vector $|\Psi(t)\rangle$ of a system consisting of a metal (with fixed positions of its ions i.e. phonon degrees of freedom ignored), an adsorbate particle (one only for low coverages), and all electrons. The Hamiltonian contains all interactions between the constituents of the system. In the initial state the adparticle is ionized by an electron or photon impact and the metal is in the electronic ground state. The Hilbert space of states $|\phi)$ of the electronic subsystem is then divided into two mutually orthogonal subspaces corresponding to two classes of diabatic configurations described in Section 1: (i) the "ionic" subspace of state $|\phi_+)$ for a hole present in the valence orbital of the adparticle and the metal electrons in the ground state and (ii) the "neutral" subspace of states $|\phi_k)$ for the filled valence orbital and the metal electrons in any k-th excitation. Projecting the Schrodinger equation onto each subspace and employing the Born-Oppenheimer approximation we effectively decouple the motion of electrons from that of the adparticle and arrive at a set of coupled, time dependent Schrodinger equations

$$(T_N + V_+ + I_\infty)|\chi_+\rangle + \sum_k W_k|\chi_k\rangle = i\hbar \frac{\partial}{\partial t} |\chi_+\rangle \tag{3}$$

$$(T_N + V_k)|\chi_k\rangle + W_k^+|\chi_+\rangle = i\hbar \frac{\partial}{\partial t} |\chi_k\rangle , \tag{4}$$

where $\chi_+(z)=\langle z|\chi_+\rangle$ and $\chi_k(z)=\langle z|\chi_k\rangle$ are the wavefunctions of the ionized and the neutral adparticle, respectively; $T_N$ is their kinetic energy operator; $V_+$ and $V_k+I_\infty$ are expectation values of the electronic Hamiltonian (i.e. the entire Hamiltonian less $T_N$) in the "ionic" $|\phi_+)$ and "neutral" $|\phi_k)$ diabatic states, respectively. They give rise to the surface potentials described earlier. For properly chosen diabatic states the electronic Hamiltonian does not couple two different "neutral" configurations but it couples, via $W_k$, a "neutral" configuration to the "ionic" one.

At t=0 an ion is formed in a state $|\chi_+(t=0)\rangle=|\chi_+^o\rangle$ ($|\chi_k(t=0)\rangle=0$) and for $t\to\infty$ the desorbed neutral with momentum q and kinetic energy $\epsilon_q=\hbar^2q^2/2m$ is detected in an eigenstate of $T_N+V_k$ (i.e. of $T_N+V_o$, c.f.(1)), $|\varphi_q\rangle$. Solving (3) and (4) simultaneously we compute a probability that the particle leaving the surface has, for $t\to\infty$, a momentum $\hbar q$

$$P_q = \sum_k |\langle\varphi_q|W_k^+ G(E_k+\epsilon_q)|\chi_+^o\rangle|^2 ; \tag{5}$$

G(E) is the resolvent operator of $T_N+V_++F$, $E_{tot}=E_k+\epsilon_q$ is the total energy of the system and

$$F(E_{tot}) = \sum_k W_k(E_{tot} - T_N - V_k + i\eta)^{-1} W_k^+ \tag{6}$$

is a nonlocal, energy dependent level shift operator. The forms of (5) and (6) suggest that the ion created at t=0 jumps back and forth between the ionic, $V_+(z)$, and the continuum of neutral, $V_k(z)$, potential energy surfaces executing a motion along $V_+(z)$ between the jumps. Eventually, the last jump (neutralization), desorption and subsequent detection in the state $|\varphi_q\rangle$ follow. A local, energy independent approximation for F, suggested by Fig. 1, is obtained by setting $E_{tot}-T_N \cong V_+ + I_\infty$ in (6). This approximation, employed in several problems of atomic physics /9/ turns $V_+ + F$ into a local complex potential. Its imaginary part, $-\Gamma(z)/2$, describes the neutralization process which acts as a drain on probability amplitude of the ion and broadens the electronic level $V_+$ /10/. Assuming real k-independent $W_k$, transforming summation over k into energy integration with constant average density of states $\langle\rho\rangle$ we get ReF=0, $\Gamma(z)=2\pi\langle\rho\rangle W^2(z)$, and

$$P_q = 2 \sum_{jj'} \Gamma_{qj}^{1/2}\Gamma_{qj'}^{1/2*} \chi_j \chi_{j'}^* /(\Gamma_j + \Gamma_{j'} + 2i(E_j - E_{j'})), \tag{7}$$

where

$$\Gamma_{qj}^{1/2} = \sum_{j'} \langle\varphi_q|\Gamma^{1/2}|f_{j'}\rangle U_{j'j} \tag{8}$$

$$\chi_j = \sum_{j'} U_{jj'}^{-1} \langle f_{j'}|\chi_+^o\rangle \ . \tag{9}$$

$|f_j\rangle$'s are discrete eigenstates of $T_N+V_+$ and $U_{jj'}$ is a complex matrix which transforms $\langle f_j|(T_N+V_+-i\Gamma/2)|f_{j'}\rangle$ into a diagonal form $(E_j-i\Gamma_j/2)\delta_{jj'}$. The kinetic energy distribution of neutrals desorbed in the direction normal to the surface is

$$\rho_{QM}(E) = \text{const}\sqrt{E}\ P_{q=\sqrt{2mE}/\hbar}\ , \tag{10}$$

where the constant factor cannot be determined independently within the one-dimensional theory. Provided the matrix elements in (8) and (9) are known, computation of (10) involves only algebraic operations like matrix diagonalization and multiplication, which can be done numerically.

The classical model /8/ was recently studied in detail in Ref. 7, so only the result is quoted here. Upon ionization at $z=z_o$ (equilibrium position in $V_o(z)$) the ion moves towards the surface with no initial velocity. Neutralized at z it desorbs with kinetic energy (c.f. Fig.1)

$$E = \epsilon_q = V_o(z) + V_+(z_o) - V_+(z) \tag{11}$$

provided it travelled as an ion past the critical distance $z_+$ so the right hand side of (11) is positive. If the probability of neutralization per unit time is $\lambda_R(z)$ then the kinetic energy distribution of desorbed neutrals is /7/

$$\rho_{CL}(E) = \text{const}|E|\left(\frac{\cosh(\theta(z_2)-\theta(z))}{\sinh\theta(z_2)}\left|\frac{dE}{dz}v(z)\right|^{-1}\lambda_R(z)\right)_{z=z(E)}, \tag{12}$$

where

$$\theta(z) = \int_z^{z_o} dz'\lambda_R(z')/|v(z')| \ . \tag{13}$$

$v(z)$ is the velocity the ion acquires at z and $z_2$ is the inner turning point in $V_+(z)$ ($z_o$ is the outer one). $z(E)$ is a solution of (11). The factor in the round bracket in (12) becomes the usual Antoniewicz's result /8/ for $\theta(z_2)>>1$. The factor $|E|$ adopts the result of the one-dimensional theory to the three-dimensional world. The link between the classical and the quantum models is provided by $\Gamma(z)=\hbar\lambda_R(z)$ relating $\lambda_R(z)$ to the imaginary part of the ionic potential. Finally, the estimate of the overall desorption yield is obtained by integrating (12) over E from 0 to $\infty$ and dividing the result by the similar integral from $-V_o(z_o)$ to $\infty$.

To proceed we represent the surface potential $V_o(z)$ and the image potential $V_{im}(z)$ by the Morse

$$V_o(z) = V_o\left(e^{-2\gamma(z-z_o)} - 2\,e^{-\gamma(z-z_o)}\right) \tag{14}$$

and the exponential

$$V_{im}(z) = -\,(e^2/16\pi\epsilon_o\gamma(z_o-z_{im})^2)e^{-\gamma(z-z_o)} \tag{15}$$

potentials respectively. $z_{im}$ is the position of the image plane. $V_{im}(z)$ given above has at $z=z_o$ the same slope as the ideal Coulomb image potential. $V_+(z)$ (c.f. (2)) becomes a Morse potential. It has a shorter equilibrium distance and stronger binding energy than those of $V_o(z)$. The representation (15) is justified as long as the ion does not travel too far away from $z_o$ before neutralization. For $\Gamma(z)$ we choose the form

$$\Gamma(z) = \Gamma_R^o\, e^{-az} = \Gamma_R\, e^{-a(z-z_o)} \tag{16}$$

expected for an electron transfer process between a metal and an adsorbed particle. Thus, each physical system is characterized by the following parameters: $V_o$, $\gamma$, $z_o-z_{im}$ (not $z_o$ and $z_{im}$ separately), $\Gamma_R$ and a. $|\varphi_q\rangle$ and $|f_j\rangle$ in (8) and (9) become now continuum and discrete states in $V_o(z)$ and $V_+(z)$ Morse potentials, respectively. The initial wavepacket $|\chi_+^o\rangle$ is taken as the ground state in $V_o(z)$. The matrix elements in (8) and (9) can be calculated analytically. Similarly v(z) and z(E) in (12) and (13) are given by simple algebraic expressions /7/.

## 3. Numerical Comparison of Quantum-mechanical and Classical Models

We choose two physical systems: for He adsorbed on W one expects quite appreciable quantum effects in contrast to heavier and more strongly bound to the surface $N_2O$ physisorbed on Ru(001). For both systems we choose, rather arbitrarily, $\gamma=1A^{-1}$.

We start with He on W with $V_o=5.23$ meV, $z_o=3.33A$ and $z_{im}=0.5A$ /11/. Kinetic energy distributions (10) and (12) have maxima at the most probable kinetic energy $E_m$. In Fig. 2 we plot $E_m$ as a function of $d_R=\Gamma_R/V_o$ for three values of $\gamma a$. For He/W $d_R=10$ corresponds to $\Gamma_R/\hbar=8\times10^{13}s^{-1}$ - a proper order of magnitude for the Auger neutralization rate /12/. As $d_R$ becomes larger, the ions are neutralized earlier and gain less kinetic energy - $E_m$ decreases. This decrease saturates at higher $d_R$ and the most probable energy remains relatively large - larger than the physisorption energy. The differences between the classical and the quantum approach show up

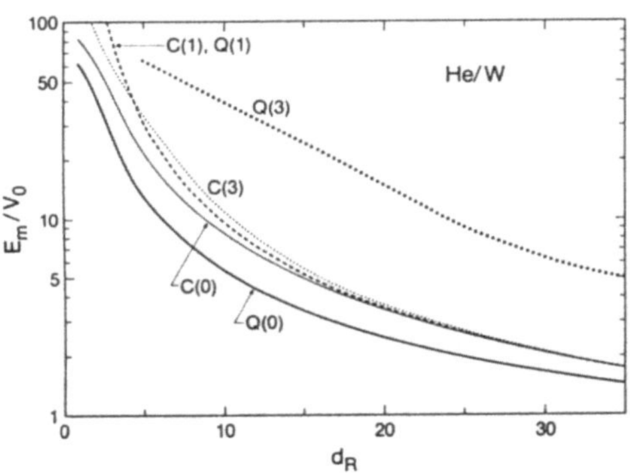


Fig. 2. The most probable kinetic energy of He atoms desorbed from W as a function of the neutralization rate at $z_o$. C - classical, Q - quantum-mechanical theory. The values of $a/\gamma$ are given in brackets

already for the position-independent $\Gamma(z)$ i.e. for a=0. In this case $E_m(QM) < E_m(CL)$ but increase of $E_m$ with increasing $a/\gamma$ is much stronger in the quantum case so for $a/\gamma=3$ $E_m(QM)$ is already three to five times larger than its classical counterpart. Surprisingly, for $a/\gamma=1$ both models predict almost the same $E_m$ over the entire range of $d_R$ considered. The shapes of classical and quantum distributions are also different. In Figs. 3 and 4 a series of distributions normalized to one at $E_m$ are plotted for $d_R=10$ and 30, respectively.

We note first that while the classical distributions are not very sensitive to variations of $a/\gamma$ the quantum ones change quite dramatically with $a/\gamma$, particularly for smaller $d_R$. In general the classical distributions are much narrower but for $a/\gamma=1$ both models yield almost identical results. The total desorption yields, computed in the classical model only, remain high: for $d_R=10$ the yield drops from almost 100% for $a/\gamma=0$ down to ~66% for $a/\gamma=3$. For $d_R=30$ the corresponding values are 72% and 68%. The estimated quantum yields are similar.

For the second physical system, $N_2O$ physisorbed on Ru(001), the kinetic energy distributions and desorption yields were measured (c.f. crosses in all panels of Fig. 5) /5/. Physisorption energy is $V_o=430$ meV, the remaining parameters: $z_o=2$A and $z_{im}=0.5$A, are estimates. A classical approach is commonly believed to be adequate for this system and indeed up to $\gamma a=1$ (Fig. 5a) no differences between the shapes of the distributions calculated in both approaches are noted. The values of $d_R$ are chosen to merge $E_m(CL)$ with the experimental value of 112 meV ($d_R=0.37$ corresponds to $\Gamma_R/\hbar = 2.4\times10^{14}$ $s^{-1}$ for this system). With increasing $a/\gamma$ the deviations between classical and quantum-mechanical results increase as seen in Figs. 5b-d. The quantum (Q) and classical (C) distributions are computed for the same set of parameters. Both become narrower as $a/\gamma$ increases but the quantum-

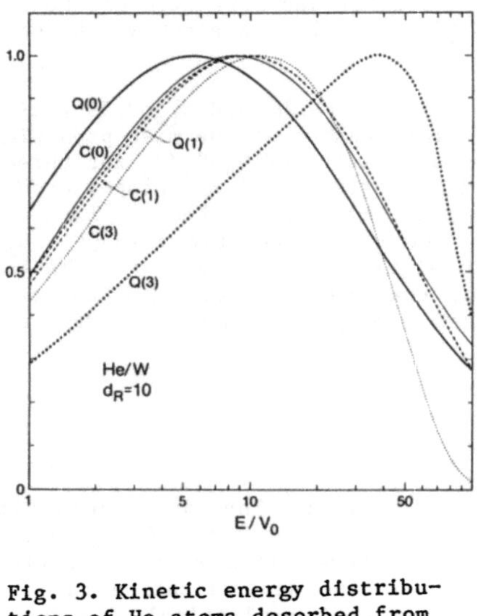


Fig. 3. Kinetic energy distributions of He atoms desorbed from W for $\Gamma_R/\hbar=d_RV_o/\hbar=8\times10^{13}$ $s^{-1}$ in the classical (C) and quantum-mechanical (Q) theory. The values of $a/\gamma$ are given in brackets

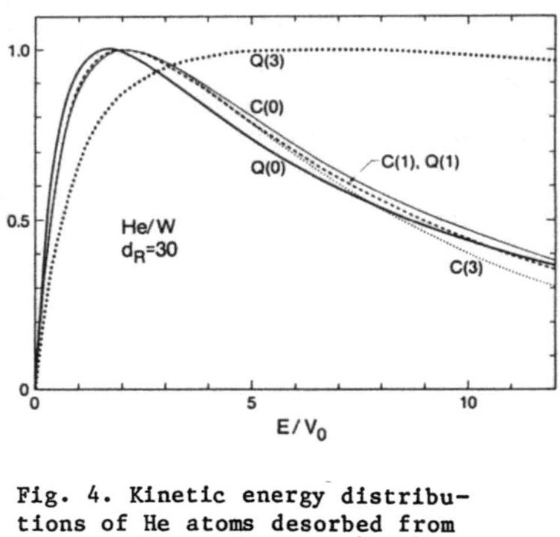


Fig. 4. Kinetic energy distributions of He atoms desorbed from W for $\Gamma_R/\hbar=d_RV_o/\hbar=2.4\times10^{14}s^{-1}$ in the classical (C) and quantum-mechanical (Q) theory. The values of $a/\gamma$ are given in brackets

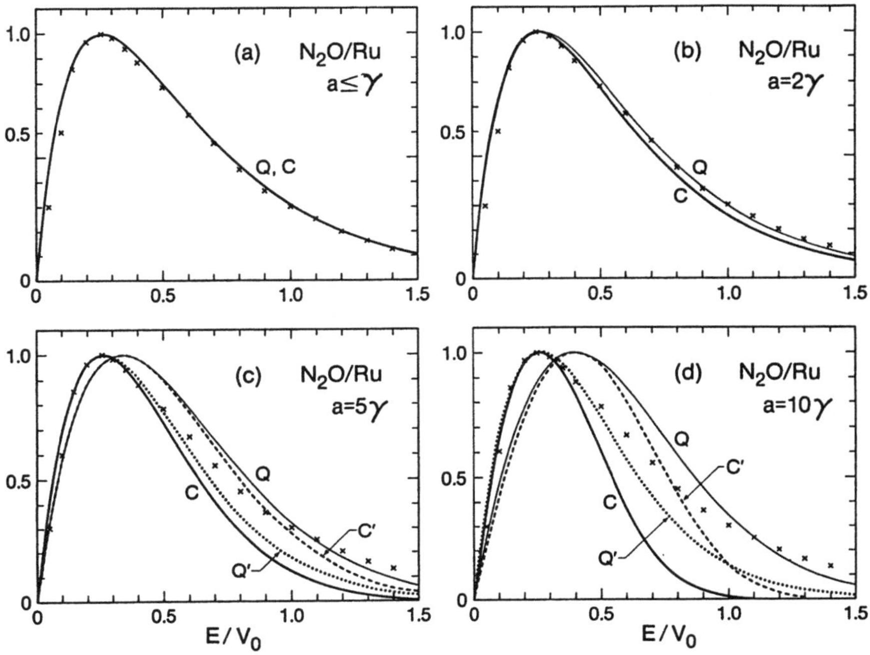


Fig. 5. Kinetic energy distributions of $N_2O$ neutrals desorbed from Ru(001) in the classical (C,C') and quantum (Q,Q') theories. Crosses: experiment. Parameters: (a) $a/\gamma=0$, $d_R=\Gamma_R/V_o=0.370$; $a/\gamma=0.5$, $d_R=0.327$; $a/\gamma=1$, $d_R=0.291$ for C and Q; (b) $a/\gamma=2$, $d_R=0.231$ for C and Q; (c) $a/\gamma=5$, $d_R=0.111$ for C and Q, $d_R=0.150$ for Q', $d_R=0.086$ for C'; (d) $a/\gamma=10$, $d_R=0.032$ for C and Q, $d_R=0.076$ for Q', $d_R=0.019$ for C'

mechanical ones are wider of the two. They remain wider than the classical ones even after $d_R$ is adjusted to bring their maxima together - compare C with Q' and C' with Q in Figs. 5c and d. Obviously, even for the "conventionally classical" system the classical approach becomes inadequate when the neutralization rate (or the imaginary part of the ionic potential) varies significantly across the initial wavepacket of the ion, provided the ion is not allowed to travel too far: the differences between the classical and the quantum distributions disappear even for $a/\gamma$ as large as 10 if $d_R$ is reduced by some four orders of magnitude. In this unphysical case the substantial fraction of ions can make few round trips between the turning points in $V_+(z)$ before being neutralized and $E_m$ can be as high as 10 $V_o$.

## 4. Application to $N_2O$ on Ru(001)

We have already seen in Fig. 5 that the experimentally measured kinetic energy distributions can be almost equally well reproduced theoretically using different pairs of neutralization parameters $a/\gamma$ and $d_R$. The fit is best for $a/\gamma<2$. Obviously an additional experimental information e.g. the overall desorption yield is needed to make the fit unique. The yields resulting from the quantum and the classical theories are almost identical (QM are slightly larger) so the subsequent discussion is limited to the classical approach. The experimentally determined desorption yield for the system considered is of the order of few percent whereas the calculated one increases from ~0.005% for $a/\gamma=0$ (Fig. 5a) up to ~5% for $a/\gamma=10$ (Fig. 5d). Variation of the system parameters, like the range and the position of the minimum of the Morse potential and the position of the image plane within physically justified limits does not change our conclusion: an unusally steep position dependence of the neutralization rate, possibly with $a>10A^{-1}$, is necessary to fit both the experimentally observed kinetic energy distribution with relatively low $E_m$ and quite high overall desorption yield. We note in passing that $\Gamma_R/\hbar$, the neutralization rate at the distance $z_o$ from the metal surface (c.f. (16)), drops from $2.4\times10^{14}s^{-1}$ down to $2.1\times10^{13}s^{-1}$ for curves C in Figs. 5a and d, respectively.

One way out of the low $E_m$ vs. high yield dilemma is a nonexponential dependence of the neutralization yield. The Antoniewicz's expression (12) can be transformed into a nonlinear integral equation allowing to compute $\Gamma(z)=\hbar\lambda_R(z)$ if the kinetic energy distribution is known /12/. This inversion procedure is not unique and for some "initial conditions" yields unphysical solutions. One also has to keep in mind that the kinetic energy distribution is experimentally determined over a finite interval of energies. Fig. 6 presents a few possible solutions. Rejecting those with unphysical behavior within the experimental energy interval we are left with $\Gamma(z)$'s converging to the one which decreases exponentially (dashed line, $a/\gamma=0.83$, $\Gamma_R/\hbar=2.2\times10^{14}s^{-1}$) over almost the entire interval, except for $\gamma z>1.7$, where it decreases much faster. The latter feature is important, as the experimental data in this region are the most reliable. The physical cause of this drop is unclear, it suggests a crossover between the Auger and a direct tunneling mechanism of neutralization. High desorption yield is now achieved because the neutralization is initially very slow so a large fraction of ions reach the critical distance $z_+$ for desorption. Very efficient neutralization at closer distances does not allow the ions to accelerate excessively, so $E_m$ is relatively low.

Some other explanations of the high yield vs. low $E_m$ dilemma can be advanced. (i) If an ion with relatively long lifetime loses part of its energy due to slow dissipative interactions with the solid (i.e. with its electrons and/or phonons) then the mean kinetic energy of desorbed neutrals may be considerably lower than in the case without dissipation. This

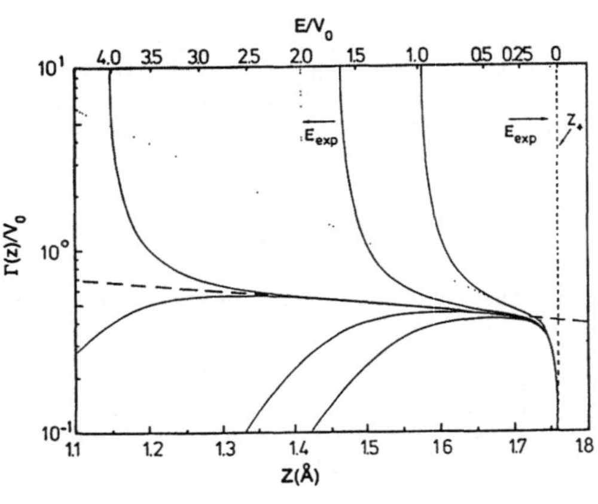


Fig. 6. Some solutions for $\Gamma(z)$ resulting from inversion of the experimental data. The upper abscissa gives the kinetic energy $E/V_o$ of desorbed particle (c.f. (11)) after being neutralized at z (lower abscissa). Experimental data exist only for energies between two vertical lines marked $E_{exp}$. The vertical line marked $z_+$ corresponds to the critical distance the ion must reach in order to desorb. Dashed line: the exponential fit for $\Gamma(z)$ with $\Gamma_R/\hbar = 2.2\times10^{14}s^{-1}$ and $a/\gamma=0.83$. Dotted line: $\Gamma_R/\hbar = 8.6\times10^{13}s^{-1}$, $a/\gamma=4.24$

avenue was tried with rather inconclusive results /7/. (ii) The initial ionization can lead to different ionization states i.e. to different $V_+(z)$ potentials. Double ionization leading to a steeper ionic curve aggravates the problems encountered. One might speculate that an initial excitation leading to a smaller slope of their potential curves might lead to more acceptable results, but no physical model for such diabatic states comes to mind. (iii) Finally, part of the kinetic energy gained by an ion may be transferred to some internal excitations (i.e. rotations or vibrations) of a rebouncing neutral molecule. This cannot be the case for ESD of atomic species so the experimental data on desorption of rare gases, similar to these for $N_2O$, are needed.

## 5. Conclusion

In this work desorption by temporary ionization of atoms or molecules physisorbed on metal surfaces was theoretically investigated. Detailed numerical comparison of the classical and the quantum-mechanical approaches was made for He on W and for $N_2O$ on Ru(001). After potential models are specified the calculations in both versions proceed without further approximations. Mathematical disparities between classical and quantum-mechanical expressions obtained make their comparison, other than numerical, impossible. It is therefore difficult to give specific criteria of validity of the classical model in a form of a mathematical condition imposed on the parameters of the system.

Results of a detailed comparison of the theoretical and experimental ESD results for $N_2O$ on Ru(001) indicate that some important piece of physics is missing in the model considered. Some possible enrichments of the model are proposed and the need for the experimental data on kinetic energy distributions and the overall desorption yield for rare gas atoms electronically desorbed from metal surfaces is stressed.

This work has been supported by Sonderforschungsbereich 128 (Munchen) of the Deutsche Forschungsgemeinschaft (Bonn) and by grants from the Natural Sciences and Engineering Research Council of Canada.

1. Desorption Induced by Electronic Transitions, DIET I, edited by N.H. Tolk, M.M. Traum, J.C. Tully and T.E. Madey (Springer-Verlag, Berlin, 1983).
2. Desorption Induced by Electronic Transitions, DIET II, edited by W. Brenig and D. Menzel (Springer-Verlag, Berlin, 1985).
3. E.R. Moog, J. Unguris and M.B. Webb, Surf. Sci. 134, 849 (1983).
4. Q.J. Zhang, R. Gomer and D.R. Boman, Surf. Sci. 129, 535 (1983).
5. P. Feulner, D. Menzel, H.J. Kreuzer and Z.W. Gortel, Phys. Rev. Lett. 53, 671 (1984).
6. P. Feulner, in Ref. 2, pp. 142-151.
7. Z.W. Gortel, H.J. Kreuzer, P. Feulner and D. Menzel, Phys. Rev. B 35, (in print, June 1987).
8. P.R. Antoniewicz, Phys. Rev. B 21, 3811 (1980).
9. C. Mundel, M. Berman and W. Domcke, Phys. Rev. B 32, 181 (1985).
10. W. Brenig, Z. Phys. B 23, 361 (1976).
11. N. Ernst, W. Drachsel, Y. Li, J.H. Block and H.J. Kreuzer, Phys. Rev. Lett. 57, 394 (1986).
12. H.D. Hagstrum, in: Electron and Ion Spectroscopy of Solids, edited by L. Fiermans, J. Vennick and W. Dekeyser (Plenum, New York, 1978).

# The Magnitude of Secondary Electron Contributions in Photon Stimulated Desorption

*D.E. Ramaker*[1], *T.E. Madey*[1], *R.L. Kurtz*[1], *and H. Sambe*[2]

[1]Surface Science Division, National Bureau of Standards, Gaithersburg, MD 20899, USA
[2]Chemistry Department, George Washington University, Washington, DC 20052, USA

## 1. Introduction

Recent reports on the extent of secondary electron contributions in electron/photon stimulated desorption (ESD/PSD) appear to be contradictory. JAEGER et al. [1] suggested that secondary electrons provide the dominant contribution to the $H^+$ yield from $NH_3$/Ni and called the process x-ray induced ESD (XESD). Others have concluded that the XESD process is the dominant mechanism in the PSD of $N^+$ and $O^+$ ions from mixed condensed gases such as $N_2$ and $O_2$ [2] and in the PSD of $H^+$ ions from OH/YbO-Sm [3]. On the other hand, considerable evidence exists in the literature for the dominance of the direct photon excitation mechanisms. This evidence usually comes from observed differences between the ion yield and the secondary electron yield or from the absence of any clear structure or "turn on" at certain core levels [4].

It seems clear that the magnitude of the XESD effect is generally not known, but this is an important question. Most helpful in sorting out the complex mechanisms in PSD has been the comparison of the photon energy dependence of the ion spectral yield with constant initial state photoelectron spectra [5]. However, if the ion yield is dominated by effects from the secondary and back-scattered electrons, then this comparison provides no information because the ion yield will simply reflect the bulk DOS, regardless of the excitation involved. Moreover, the usefulness of PSD extended x-ray absorption fine structure (EXAFS) measurements for obtaining information on surface structure [6] will be highly limited if XESD contributions dominate. This paper presents a re-examination [4] of the three systems mentioned above where a dominant XESD effect has been postulated to determine the validity of the conclusions previously reached.

## 2. $H^+$ from $NH_3$/Ni

The $H^+$ yield spectrum from $NH_3$/Ni(110), where the $NH_3$ layer was believed by JAEGER et al. [1] to have an appreciable thickness, reveals the Ni L edge [1]. This suggests that the XESD mechanism must be acting, if indeed all Ni atoms are separated by many $NH_3$ layers from the outer surface. Furthermore, at the N K edge, the total electron yield (TEY) and $H^+$ yields are very similar except for the presence of a shoulder in the $H^+$ yield which is not present in the TEY. This shoulder was interpreted by JAEGER et al. [1] to arise from the direct core level excitation of a surface $NH_3$ molecule, the remainder of the yield spectrum arising from the XESD mechanism. Based on this interpretation, the XESD process is estimated to cause 60% of the total yield.

Our thermal desorption studies and other data indicate, however, that the thickness of the $NH_3$ films used by JAEGER et al. [1] were considerably different from those which they reported [4]. We also indicate that the observed shoulder in the $H^+$ yield arises because the $1a_1^{-1}4a_1$ excitation at this energy has a higher branching ratio for the desorption process than for excitation into the Rydberg or continuum orbitals [4]. This is indicated because of the antibonding character of the $4a_1$ orbital. Thus we do not believe that sufficient evidence has been provided to indicate that the XESD contribution is dominating in this case; indeed, we believe that the difference between the TEY and the $H^+$ yield provides evidence that it is not dominant.

## 3. $O^+$ and $N^+$ from $N_2$–$O_2$

Recently PARKS et al [2] reported the TEY and $O^+$ and $N^+$ yields from a condensed mixture of $N_2$ and $O_2$. The presence of large $N^+$ and $O^+$ yields at both the N and O K levels indicates very strongly that the XESD mechanism is playing an important role; indeed, they concluded that the XESD mechanism dominates for $N^+$, but is small for $O^+$ desorption.

PARKS et al [2] analyzed their data on the basis of absolute quantities, namely on their experimental absolute ion yields and photon fluxes, and on previously reported photoionization cross-sections. The latter, in particular, are not known very accurately. We prefer to rely on the TEY which they reported; indeed a simple comparison of the TEY and $O^+$ and $N^+$ spectral yield curves indicates that the $O^+$ yield is very similar to the TEY, while the $N^+$ is very different [4]. This indicates, contrary to the above, that the $O^+$ yield is dominated by the XESD mechanism, while the $N^+$ yield is not. A detailed analysis indicates that the XESD cross sections for $N^+$ and $O^+$ are comparable, but that the direct cross section for $N^+$ is 20 times that for $O^+$ [4].

The primary core level dissociation process in diatomics involves Auger decay of the core hole, resulting in a two-hole (2h) final state, which directly results in dissociation via a "Coulomb explosion". This Auger stimulated desorption (ASD) mechanism is known to be active in the gas phase [7], and it is believed to be active also in the condensed phase, at least for $N_2$ and CO [7]. We believe that a similar mechanism is appropriate in the condensed $N_2$–$O_2$ mixture, only in this case an electron transfer occurs from a resonant $2p^*$ orbital of a neighboring $O_2$ molecule [4]. This transferred electron screens the two valence holes caused by the Auger decay and thus prevents the ASD process. This electron transfer cannot occur in condensed $N_2$ because the comparable $2p^*$ orbital is empty.

Evidence that this charge transfer process can occur in $O_2$ comes from visible absorption spectra. Isolated $O_2$ dimers in a solid neon host have a characteristic absorption spectrum [8] resulting from absorption of a single photon to produce a singlet state in which both halves of the dimer are electronically excited. The electronic energy transfer is believed to result from electron exchange of the $2p^*$ electron [8]. Furthermore, recent absorption and photoelectron data on condensed $N_2$ and $O_2$ reveal that the $2p^*$ band is significantly broadened by normal intermolecular interactions [9]. Finally, direct evidence has been given recently indicating that $2p^*$ electron transfer does occur in condensed NO [10].

## 4. $H^+$ from OH-YbO/Sm

Finally, the $H^+$ yield from an oxidized Yb surface on a bulk Sm metal substrate shows large features which arise from the Sm and Yb 4d–4f giant resonances [3]. Although little of the Sm surface is thought to be exposed, the Sm feature is much larger than the Yb feature. If indeed the Sm surface is totally covered, this strongly indicates that the XESD process is dominant in this system. The $H^+$ which desorbs is most likely bonded to the O atoms in the oxidized Yb. The giant resonance, $4d^{-1}4f^{n+1}$, Auger decays or autoionizes predominantly to $4f^{n-1}$, a one-hole valence state which does not lead to desorption. Thus the direct ASD process is inactive at both the Yb and Sm core levels, and the XESD process is expected to dominate [4].

## 5. Summary

It has been shown for the three systems discussed above that either alternate interpretations of the data are possible, indicating that the XESD effect is not dominant, or that the XESD process dominates only because the direct ASD mechanism is suppressed. This makes the results on these three systems consistent with the significantly larger number of papers, which indicate that the XESD contribution is generally not dominant [4]. We conclude that in cases for which the direct core level ASD process is active, the indirect XESD contribution is generally on the order of 35% or less. The XESD contribution apparently dominates the total yield only in those cases when the direct ASD contribution is suppressed.

## References


1. R. Jaeger and J. Stohr: Phys. Rev. B28, 1145 (1983); R. Jaeger, J. Stohr, and T. Kendelewicz: Surf. Sci. 134, 547 (1983).
2. C.C. Parks, Lawrence Berkeley Laboratory Report LBL-16892, Nov. 1983; V. Rehn and R.A. Rosenberg: Synchrotron Radiation Research: Advances in Surface and Low-Dimensional Science, ed. by R.Z. Bachrach (Plenum Publishing, New York, 1987).
3. J. Schmidt-May, F. Senf, J. Voss, C. Kunz, A. Flodstrom, R. Nyholm, and R. Stockbauer: Desorption Induced by Electronic Transitions, DIET II, ed. by W. Brenig and D. Menzel, Springer Ser. Surf. Sci., Vol. 4 (Springer, Berlin, Heidelberg 1985) p. 94; also Surf. Sci. 163, 383 (1985).
4. D.E. Ramaker, T.E. Madey, R.L. Kurtz, and H. Sambe: to be published.
5. D.E. Ramaker: p.70 in Ref. 3 above
6. R. Jaeger, J. Stohr, J. Feldhaus, S. Brennan, and D. Menzel, Phys. Rev. B23, 2101 (1981).
7. T.A. Carlson and M.O. Krause: J. Chem. Phys. 56, 3206 (1972); B. Brehm and G. de Frenes, Int. J. Mass Spect. Ion Phys. 26, 251 (1978).
8. J. Goodman and L.E. Brus: J. Chem. Phys. 67, 4398 (1977); 67, 4408 (1977).
9. H.J. Lau, J.H. Fock, and E.E. Koch: Chem. Phys. Lett. 89, 281 (1982); J.H. Fock and E.E. Koch: Ann. Isr. Phys. Soc. 6, 285 (1984).
10. B.B. Tonner, C.M. Kao, E.W. Plummer, T.C. Caves, R.P. Messmer, and W.R. Salaneck: Phys. Rev. Lett. 51, 1378 (1983).

# Interatomic Auger Rates for the Sodium Fluoride Crystal

*T.A. Green and D.R. Jennison*

Sandia National Laboratories, Albuquerque, NM87185, USA

Interatomic Auger rates for the NaF crystal are calculated for initial 1s, 2s, and 2p holes in the $Na^+$ ion. The crystal is simulated by an $(NaF_6)^{5-}$ cluster embedded in a large number of point ions, and the one-electron orbitals are obtained from restricted Hartree-Fock calculations of the initial state. The calculated interatomic Auger widths for the transitions Na(1s)Na(2p)F(2p), Na(2s)Na(2p)F(2p), and Na(2p)F(2p)F(2p)' are 0.00173, 0.641, and 0.153 eV respectively. For comparison, the width of the Na(KLL) Auger transition is close to 0.26 eV. Widths for the transitions Na(1s)Na(2p)F(2p) and Na(2p)F(2p)F(2p)' can be inferred from experimental data. Theory and experiment agree to within 50%. This represents a considerable improvement over previous theoretical treatments.

## 1. Introduction

Since their mechanism was proposed by KNOTEK and FEIBELMAN, [1,2] interatomic Auger decay has continued to play an important role in connection with the stimulated desorption of positive ions from ionic solids, particularly oxides. $TiO_2$ is an example of a system of current interest for the electron and photon-stimulated desorption of $O^+$ ions. This occurs via the production of a Ti(3p) core hole, followed by interatomic Auger decay to remove electrons from surface oxygen ions. [3,4] For interatomic Auger to be of experimental significance, its rate must be sufficiently large compared to those of the competing channels for de-excitation, such as intra-atomic Auger and radiative decay. [2] Despite the importance of the interatomic Auger rate, no values for it are available in the literature. In a well-known theoretical paper by MATTHEW and KOMNINOS [5], the rate for the type of interatomic Auger process which leaves both holes on the same ion was estimated to be negligible. In the only other theoretical paper on interatomic Auger decay, the contribution to the Na(2p) hole width from the transition Na(2p)Cl(3p)Cl(3p) was estimated to be of the order of magnitude of 0.02 eV.[6] Here we utilize the notations A(x)A(y)B(z) and A(x)B(y)B(z). The chemical symbols A and B designate neighboring atoms and x,y,z designate one-electron orbitals in which holes occur. A hole occurs in x in the initial state and holes occur in y and z in the final state. The first type of transition is called an intra-inter transition and the second type is called an inter-inter transition. The transitions of interest in the KNOTEK-FEIBELMAN mechanism are of the inter-inter type. The main cause of the discrepancy between Refs. 5 and 6 is not hard to identify. The technique of YAFET and WATSON [6] leads to cluster Cl valence molecular orbitals (mos) which have non-negligible components of the Na atomic orbitals, allowing the valence electron to spend some of its time near the Na ion. As a result, close-in contributions to the interatomic Auger matrix element are included by YAFET and WATSON which are completely missing in the treatmant of MATTHEW and KOMNINOS. For the intra-inter process, the results of MATTHEW and KOMNINOS represent an estimate of the distant contributions to the interatomic Auger matrix element.

The importance of close-in contributions to the interatomic Auger matrix element makes it essential to obtain accurate charge distributions for the valence orbitals in the presence of the initial core hole. For this reason we have chosen to extend the work of YAFET and WATSON by using Hartree-Fock theory to obtain the mos of the $(NaF_6)^{5-}$ cluster which we embed in the potential of a cubic array of point ions to simulate the NaF crystal. This should provide an adequate description of the shift in electronic charge occasioned by the creation of the $Na^+$ hole. The selection of NaF for the initial test of our theory was made because the work of CITRIN, ROWE, and CHRISTMAN [7] provides a comprehensive experimental study of the Auger spectrum from the creation of 1s, 2s, and 2p holes in the $Na^+$ ion in the NaF crystal. In addition, a quantitative value of the ratio of the Na(1s)Na(2p)F(2p) transition rate to that of the $Na(KL_{23}L_{23})$ rate can be obtained from their data. Finally, a second check on the theory can be obtained from the lifetime of the Na(2p) hole state in NaF. [8] In the NaF crystal, energy conservation requirements prevent the decay of a 2p hole in the $Na^+$ cation by an inter-inter process which leaves both holes on the same fluorine ion. [7] Thus this work will not be directly concerned with the KNOTEK-FEIBELMAN mechanism.

The Hartree-Fock calculations and the Auger theory have already been described in detail. [9,10] Therefore we shall present and discuss only the main results in this paper.

## 2. Results and Discussion.

Table I exhibits the results for the interatomic Auger rates in NaF. The pattern of relative intensities is consistent with the lines observed by CITRIN, ROWE, and CHRISTMAN. In Table I, the notation F(2p)F(2p)' indicates that the final-state holes are located on two different neighboring F ions, and not on the same ion. A transition leaving both holes on the same F ion would require about 10 eV more energy, and is energetically forbidden. To distinguish between these final states correctly, a configuration interaction treatment of our independent-particle two-hole states would be necessary. In calculating the Na(2p)F(2p)F(2p)' rate, we summed over all 630 independent-particle two-hole final space and spin states. Thus we included contributions from the 90 states which correspond to two holes on the same Fluorine ion. Therefore, in the crudest approximation of just counting states, our sum is about 17% too high. Experimental rates for two of the transitions have been obtained from the literature and are shown in Table I. The experimental rate for the Na(1s) hole transition was obtained

Table I. Interatomic Auger Rates for the NaF crystal.

| TRANSITION | ENERGY(eV) | Theory<br>Rate(a.u.) | Experiment<br>Rate(a.u.) |
|---|---|---|---|
| Na(1s)Na(2p)F(1s) | 347. | 0.59 (-7) | |
| Na(1s)Na(2p)F(2s) | 1003. | 0.71 (-5) | |
| Na(1s)Na(2p)F(2p) | 1023. | 0.637(-4) | 0.82(-4 ) |
| Na(2s)Na(2p)F(2p) | 16. | 2.36 (-2) | |
| Na(2s)F(2p)F(2p)' | 40. | 0.141(-3) | |
| Na(2p)F(2p)F(2p)' | 7. | 0.557(-2) | 0.37(-2 ) |

from Ref. 7. That for the Na(2p) hole was obtained as an upper bound from Ref. 8. The good agreement should lend credence to theoretical predictions for the more typical systems where no experimental data are available. Our rate for the transition Na(2s)Na(2p)F(2p) is comparable to that of YAFET and WATSON for the analogous transition in NaCl.

In conclusion, we consider the implications of our work for the KNOTEK-FEIBELMAN mechanism. The most important result is that the inter-inter rate for the initial $Na^+$(2p) hole is not extremely small, but is of the size of the upper bound found experimentally. A rate of $0.557\times10^{-2}$ a.u. corresponds to $2.3\times10^{14}$ transitions per sec. This is fast compared to the rates to be expected for competing radiative transitions and to the longitudinal optical frequency, $1.3\times10^{13}$ vibrations per sec. As explained earlier, the production of positive ions by the KNOTEK-FEIBELMAN mechanism is energetically forbidden in NaF. However, the size of our inter-inter rate suggests that in other ionic systems the inter-inter step in the Knotek-Feibelman mechanism can occur with sufficient probability to make the mechanism competitive with alternative channels for energy transfer.

## References


1. M. L. Knotek and Peter J. Feibelman, Phys. Rev. Lett. 40, 964 (1978)
2. M. L. Knotek and Peter J. Feibelman, Surf. Sci. 90, 78 (1979)
3. R. L. Kurtz, Surface Science 177, 526 (1986)
4. R. E. Walkup and R. L. Kurtz, "Dynamics of $O^+$ desorption from $TiO_2$", DIET III (this volume)
5. J. A. D. Matthew and Y. Komninos, Surf. Sci. 53, 716 (1975)
6. Y. Yafet and R. E. Watson, Phys. Rev. B 16, 895 (1977)
7. P. H. Citrin, J. E. Rowe, and S. B. Christman, Phys. Rev. B 14, 2642 (1976)
8. J. A. D. Matthew and M. G. Devey, J. Phys. C 7, L335 (1974)
9. T. A. Green, D. R. Jennison, C. F. Melius, and S. J. Binkley, Phys. Rev. B 36, accepted
10. T. A. Green and D. R. Jennison, Phys. Rev. B, accepted

# Part VI

# Spectroscopy

# Ultraviolet Surface Photochemistry and Charge Transfer

*E.P. Marsh, F.L. Tabares*, M.R. Schneider, and J.P. Cowin*

Chemistry Department, University of California
Santa Barbara, CA 93106, USA
*Present address: Facultad de Quimica, Universidad Completense,
E-28040 Madrid, Spain

## 1. INTRODUCTION

Excimer lasers can initiate surface photochemistry, and are of particular importance in the fabrication of semiconductor electronics [1]. The energy of the photons, 3.5 to 8 eV, can raise many molecules to reactive states. On a metal or semiconductor surface, however, the time for competitive quenching (energy transfer) or charge transfer to the substrate is very fast, $10^{-13}$ to $10^{-15}$ seconds [2]. Thus only very fast photochemical processes like photodissociation via repulsive states are likely to occur on these surfaces. The surface can still perturb even such a fast process. A simple way is "kinematically": i.e., the surface or co-adsorbates can act as third bodies for energy or momentum transfer [3]. Also, through bonding, the surface can change the potentials along which the photochemistry proceeds. And finally, even if a surface photo-process can "out-run" competitive quenching or charge transfers, it still may be perturbed by them.

We studied the repulsive photofragmentation of $CH_3Br$ on a Br covered Ni(111) surface. We found the cross-section was not obviously different from the gas phase value. Velocity measurements of the ejected radicals indicate strong perturbation, which we attribute to resonant charge transfer processes, in analogy to those known to dominate electron and photon stimulated desorption at 20 to 100 eV energies [4].

## 2. EXPERIMENTAL RESULTS

The experiment was done in a UHV chamber, with Auger, sputter gun, time-resolved mass spectrometer, differentially pumped molecular beam for dosing, and cooling of the crystal to 45 K with a closed cycle refrigerator. The 193 nm laser was incident at $67^{\circ}$ to the surface normal, and the mass spectrometer observed at $13^{\circ}$ from the normal. The ionizer distance is 7 cm, ion flight time for $CH_3$ is 8.7 usecs. The laser intensity at the surface was uniform to about 15 %, and was held below .1 J/cm for 15 ns pulses. At this incidence angle, this causes negligible surface heating. The Ni(111) surface was prepared in the standard fashion.

$CH_3Br$ spontaneously decomposes on Ni(111) to produce ethane even at 45 K. The surface was "brominated" by exposing it to a monolayer of $CH_3Br$, annealing to 250 C, and repeating twice more. Auger showed about a monolayer of adsorbed Br with little carbon present. On this surface $CH_3Br$ physisorbed intact, showing a simple first monolayer thermal desorption peak near 125 K. All photolysis was done on this "brominated" surface.

Thermal desorption spectra (TDS) was used to monitor the decrease in the surface $CH_3Br$ after increasing doses of photons. A log plot of the TDS areas versus photon fluence (corrected for the surface light field using

Maxwell's equations) directly gives the cross section for photofragmentation. We found a value of $1.5X10^{-19}$ $cm^2$, about .3 to .5 of the gas phase value. We tentatively attribute the smaller value to an unfavorable geometry. We believe that this molecule adsorbs vertically on the surface. The light field is nearly parallel to the surface normal due to near cancellation of the other polarization by the reflected wave. But the electronic transition is primarily (83%) perpendicular at this wavelength [5]. So one would expect the measured cross section to be several times lower than the gas phase value. Polarization studies are now in progress.

The cross section results suggest that the photon absorption initially closely resembles that occurring for a gas phase molecule. However, time of flight (TOF) studies of photofragments ejected by the laser show that the evolution of the photofragmentation is severely perturbed. Figure 1(a) shows Br atom TOF data from $CH_3Br$ photolysis. The vertical scale gives the relative signal intensity. The dashed portion near time 0 is where the scattered photon signal disturbs the baseline. For less than a monolayer of $CH_3Br$, very little Br ejects from the surface. Instead, high-temperature TDS showed that most of the photo-generated Br atoms reacted with the surface to form a new, more weakly bound Br. Though the surface does not thermal decompose $CH_3Br$, it is not inert to atomic Br!

Above a monolayer of $CH_3Br$, substantial molecular ejection is observed. These molecules are ejected when a nearby molecule photolyses, the energetic fragments knocking it off the surface (similar to multilayer $CH_3Br$ on LiF [3]). This creates a Br signal due to the cracking pattern in the ionizer, and is responsible for the large right-hand component of the Br signal seen clearly at 15 monolayers. The remaining signal is very broad, and is both faster and slower than gas phase Br photofragments. This is understood to be due to collisional perturbations [6].

The $CH_3$ fragments show a roughly linear intensity dependence with coverage up to 3 monolayers, consistent with a vertically adsorbed molecule in the first layer, with the $CH_3$ up. At 4 monolayers coverage in Fig. 1(b), the cracking pattern of the molecules shows up as a distinct broad peak near 100 usec. At 4 monolayers, there is also seen an extremely narrow

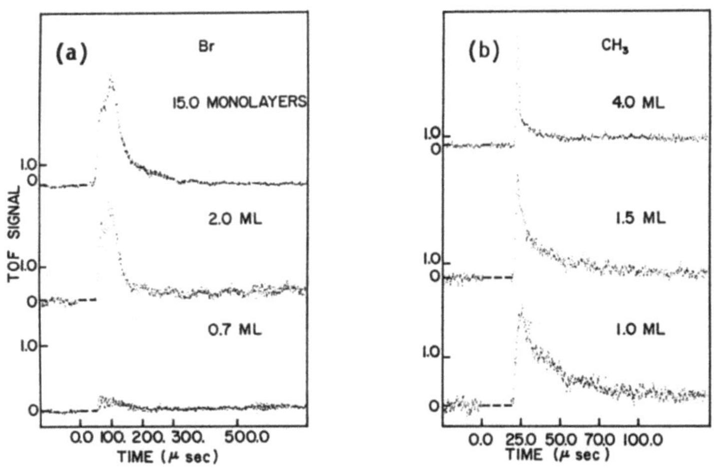


Fig. 1. TOF signals for $CH_3Br$ photolysis. Coverages as indicated

peak at about 25 usecs. This is probably due to molecules that have been ejected during the first part of the laser pulse, then photolyzed in the gas phase during the latter half. A similar component exists for the Br data, but it is masked, since it is small and not much narrower than the other two peaks in its vicinity.

The $CH_3$ TOF for 1 monolayer or less shows a steep rise at the gas-phase limit near 25 usec, and shows a very broad and generally slowed (compared to gas-phase) distribution. This slowing is an indication of some strong perturbations, particularly since the Br-surface bond formation might provide more kinetic energy to the products, by stabilizing the final state.

## 3. DISCUSSION

### 3.1 Quenching

Excitation of substrate electron-hole pairs or plasmons can remove all or part of the energy from an excited adsorbate. The coupling rate is proportional to the spontaneous photon emission rate $A_s$, and terms in $z^{-3}$ or $z^{-4}$ [2,7]. For Ni, the $z^{-3}$ term should dominate down to 1 Å [7], giving a rate of $A_s*[(\lambda/2\pi z)^3 nk/[1+\epsilon]^2]$, where $\lambda$ is the wavelength, n and k are the real and imaginary refractive indices, and $\epsilon$ is the complex dielectric constant of Ni at 198 nm [8]. This gives a rate of $2X10^5*A_s$ at $z = 3$ Å. $A_s$ is obtained from the integrated $^1Q$ cross section of about $2X10^{-15}$ cm [5], to be $4X10^6$/s. This gives a quenching rate of 1.5 ps at 3 Å, or 50 fs at 1 Å. This rate is simply not fast enough to dominate the 15 fs photofragmentation. For other systems, it may be important.

### 3.2 Charge Transfer ( CT )

It is typically energetically uphill to transfer an electron between a substrate and an adsorbed molecule. To transfer an electron from a ground state molecule at a distance z from the surface, to the lowest empty orbital of the substrate requires a change in energy of $E_{pos}$. The opposite direction of transfer requires $E_{neg}$. These are given approximately by

$$E_{pos} = IP_{mol} - EA_{sub} - (14.4 \text{ eV Å})/4z \quad (1a)$$

$$E_{neg} = IP_{sub} - EA_{mol} - (14.4 \text{ eV Å})/4z, \quad (1b)$$

where $IP_{mol}$ and $EA_{mol}$ are the molecular gas phase ionization potential and electron affinity, $IP_{sub}$ and $EA_{sub}$ are parallel quantities for the substrate, and the last term is the metallic image potential (less on a semiconductor). For $CH_3Br$ on a brominated Ni surface [9], these CT states are both about 3 eV above the ground state. These are energetically accessible, since the photolysis begins 6.4 eV above the ground state.

The rate of electron transfer should be highest for the one-electron resonant processes [2]. These are schematically shown in Fig. 2(a). The excited electron can hop to the Ni, or a Ni electron can tunnel to the hole state in the molecule, or both. Fig. 2(b) shows the ground, neutral-excited, and plus and minus CT states, versus distance from the surface [9]. The CT states are continuums, the curves are their lower edges. As the excited molecule accelerates down the upper potential, a sudden resonant electron hop can occur any time the excited curve is within the CT continuums, not just at the apparent crossing. The kinetic energy of the molecule cannot change instantaneously, so the difference between the excited and neutral curves at that particular distance represents the exci-

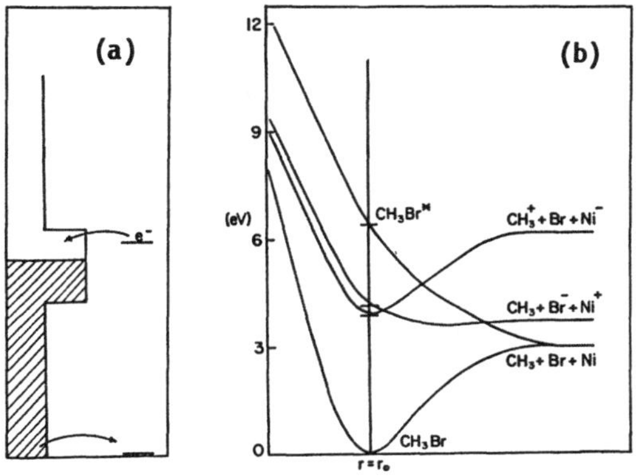


Fig. 2. a) The molecule-substrate resonant electron transfers are diagrammed. b) $CH_3$--Br potentials are shown, for ground, excited, and charge transfer states, versus bond distance, r.

tation energy of the substrate hole or electron. Charge transfer of either sign puts one on a potential whose dissociation limit is well above that of the neutral state. This and the excitation left in the substrate leaves less kinetic energy for the molecule, consistent with our finding slow methyls. A double electron hop leads back to the ground state. But the kinetic energy developed on the upper curves would be conserved, and this can lead to 1 to 2 eV of vibrational excitation of the ground state molecule [9]. This might allow surmounting the barrier preventing thermal decomposition, leading to adsorbed Br, and possibly ejected $CH_3$.

Recent works have examined CT rates for molecules physisorbed on metals. Peak linewidths from EELS, UPS, and inverse photoemission data, at 1 to 10 eV excitation energies, and theory, both suggest that rates no slower than 1 to 5 femtoseconds should be the rule for CT [2]. Since the time for photofragmentation is about 10 fs, these CT processes would be expected to dominate the photodynamical process, anytime they are energetically accessible. They are also slow enough to allow acceleration on the potentials, and reasonable probabilities for single versus double CT.

### 3.3 General Case

The results (and conjectures) outlined above certainly suggest an important role for charge transfer in the photochemistry of this system. Detailed analysis [6], and work now in progress in our lab will help firmly establish and characterize the phenomena. The fact that independent estimates of the CT rates [2] are generally much faster than the repulsive state fragmentation time of any molecule due to a UV photon (5 fs to 50 fs), means that any UV photochemistry on a metal or semiconductor should be dominated by resonant CT, if it is accessible. It is possible to construct a graph like Fig. 2 for most any combination of molecule and surface, but there is a much easier, versatile way to decide which cases should have strong CT.

The right side of Fig. 3 shows the filled and empty density of states for several representative metals and semiconductors. It is possible to compare these to the lines on the left half, which represent for each molecule where substrate density-of-states would have to exist for resonant charge transfer to occur. A resonant charge transfer requires the change in energy for electron hopping from the excited state to be zero, via an equation in analogy with (1). Then $EA_{sub}$ and $IP_{sub}$ refer to the specific filled and empty orbitals of the substrate that are resonant, via

$$EA_{sub}(r) = IP^*_{mol}(r) - (14.4 \text{ eV Å})/4z \quad (2a)$$

$$IP_{sub}(r) = EA^*_{mol}(r) + (14.4 \text{ eV Å})/4z. \quad (2b)$$

Note that the quantities depend on the intramolecular bond length r. This is obvious, since at the beginning of the fragmentation, with $r = r_0$, the appropriate $IP^*_{mol}$ is the (vertical) ionization potential of the ground state molecule minus the photon energy, about 4 eV. By the end, with r = infinity, the $IP^*_{mol}$ is the lesser of the ionization potentials of $CH_3$ or Br, 9.8 eV. This allows one to calculate the required empty density of state energies for resonant transfer as laying along the solid arrow in Fig. 3: the arrow's tail indicates the beginning of the fragmentation, and the head, the end. A similar analysis shows that for resonant transfer <u>to</u> the molecule, filled density-of-states must exist along the dashed arrow. Consistent with the conclusion from Fig. 2. CT of both signs can occur between $CH_3Br$ and Ni during the first part of the fragmentation. Fig. 3 presents similar curves for several other molecules: together they are fairly representative of just about any UV surface photolysis candidate. Note that for this figure, z has been fixed at 3 Å.

The format of Fig. 3 makes some very important generalizations obvious: CT of both signs should be important for most molecules to any "typical" metal, with work function between 4 and 6 eV. Exceptions include $Br_2$ (and $I_2$), which by virtue of a high molecular ionization potential and low photon energy, typically only would be perturbed by electron transfer <u>to</u> the molecule. Things are not changed very much by going to semiconductors like GaAs or Si. Wide gap semiconductors, however, like $CdS_2$, $SnO_2$, or $TiO_2$, would effectively shut down one or both of the CT processes, since the necessary states fall within the band gap.

An interesting case occurs when an electronegative molecule like $Br_2$ is paired with a very electropositive metal, like K. As one would expect, resonant electron transfer from the halogen to the metal is not allowed. But also <u>resonant</u> electron transfer <u>to</u> the halogen is also not allowed. This is because the filled density-of-states of K lies completely above the hole to be filled in the excited $Br_2$. Thus only two electron processes can fill the hole. If somehow, at a low enough temperature, $Br_2$ could be laid down intact on a K surface, UV photochemistry might proceed unperturbed.

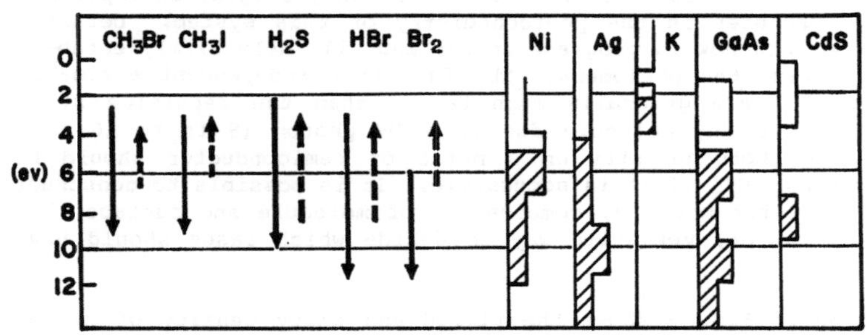


Fig. 3. Left: Density-of-states sampled during photolysis for various molecules. Dashed is for filled, solid is for the empty density-of-states. Right: The approximate density-of-states for various substrates

## 4. CONCLUSIONS

We have shown for this system that the initial photon absorption was not strongly perturbed by the surface. At one monolayer, the photolysis result was a reaction between the Br formed and the Ni surface, and the ejection of a $CH_3$ radical. The $CH_3$ velocities indicate strong perturbations, which we ascribe to charge transfers. We discuss the general case of UV surface photochemistry, and conclude that charge transfer should dominate most combinations of non-insulator substrates and molecules.


## Acknowledgements

Thanks are due to W. Meier, T. Gilton, G. Bach, and Y. Kim for assistance in this work, and to Dr. H. Metiu for many discussions. We also appreciate the financial support from the National Science Foundation (CHE 8415343, CHE 8411302, and the donors of the Petroleum Research Fund, administered by the Amer. Chem. Soc. One of the authors (F.L.T.) was supported by a Fulbright Fellowship from Spain.


## References


1. T.C. Chuang, Surf. Sci. Rep. 3, 1 (1983)
2. see P. Avouris, B.N.J. Persson, J. Phys. Chem. 88, 837 (1984), and references therein.
3. F.L. Tabares, E.P. Marsh, G. Bach, J.P. Cowin, J. Chem. Phys., 86, 738 (1987)
4. see other articles in these proceedings.
5. G.N.A. Van Veen, T. Baller, A.E. DeVries, Chem. Phys. 92, 59 (1985)
6. M.R. Schneider, E.P. Marsh, W. Meier, J.P. Cowin, (to be published)
7. B.N.J. Persson, N.D. Lang, Phys. Rev. B26 5409 (1982)
8. D.W. Lynch, W.R. Hunter, in Handbook of Optical Constants of Solids, ed. E. Palik (Academic Press, N.Y., 1985), p. 322
9. Detailed sources and discussion are in Ref. 6

# Chemically Induced Desorption of Nickel Oxide and Carbide by 5-200 eV $C^+$ and $O^+$ Ion Bombardment

*H. Kang, S.R. Kasi, O. Grizzi, and J.W. Rabalais*

Department of Chemistry, University of Houston,
Houston, TX 77004, USA

## 1 INTRODUCTION

Adsorbate desorption induced by reactive ion bombardment can occur via several mechanisms including collisional momentum transfer, electronic excitation originating from charge transfer, or chemical reactions. When the ion kinetic energy (Ek) is in the hyperthermal range (1-100 eV), which is comparable to chemical bond energies, the ion-surface interaction is greatly influenced by the chemical nature of the reactive projectile [1-3]. Therefore, by employing a suitable projectile-adsorbate combination, selective induction of ion-surface reactions and subsequent product desorption can be achieved. For this reason, we have undertaken experimental studies of reactive ion-induced desorption by employing mass and Ek selected single reactant species. These experiments probe the reactions of 5-200 eV $C^+$ and $O^+$ ion beams with nickel oxide (NiO) and nickel carbide (NiC) surfaces. These reactions are modeled using a Monte Carlo computer simulation to compare with the experimental results.

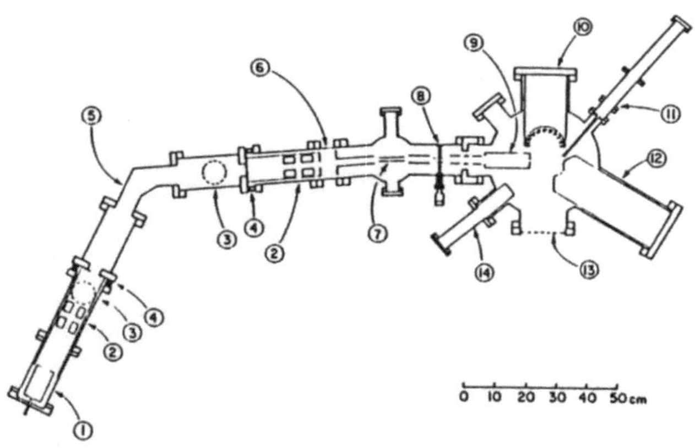


Fig.1: Schematic drawing of ion beam line and analysis chamber. The components include (1) ion source, (2) electrostatic quadrupole lenses, (3) vacuum pumping ports, (4) ceramic insulator, (5) 60° sector electromagnet, (6) gate valve, (7) 6° deflector plates to eliminate fast neutrals and vacuum pump, (8) beam aperture, (9) decelerator lens, (10) LEED, (11) UPS source, (12) CMA analyzer, (13) view port, (14) RGA mass spectrometer. The X-ray source is located out of plane above the UV source. The electron gun for AES is on the CMA axis.

## 2 EXPERIMENTAL

The spectrometer system used herein is shown schematically in Fig. 1 and has been described in detail elsewhere [2-4]. Briefly, it consists of a low-$E_k$, mass-selected, differentially pumped ion beam line attached to a UHV chamber which contains instrumentation for Auger electron spectroscopy (AES), X-ray and UV photoelectron spectroscopy (XPS and UPS), low-energy electron diffraction (LEED), and quadrupole mass spectrometry. Electron impact ionization of CO gas in a Colutron source is used to generate $C^+$ and $O^+$ beams. Efficient differential pumping of the beam line allows maintenance of the system pressure in the low $10^{-10}$ torr range during exposure to reactive ion beams. The sample is a Ni(111) single crystal which is cleaned by standard techniques as described previously [4]. NiO and NiC surfaces were prepared <u>in situ</u> as follows. Chemisorption of 8L $O_2$ on clean Ni(111) produces a p(2x2)NiO layer with an oxygen concentration of $6.3\times10^{14}$ atoms $cm^{-2}$ [5]. The NiC surface was prepared by reacting clean Ni with the low-energy $C^+$ beam. The carbon coverage of the NiC surface was estimated from the reaction probability data of low-$E_k$ $C^+$ ions with Ni(111) [4]. We used $C^+$ doses typically less than $1\times10^{15}$ ions $cm^{-2}$ in order to avoid the formation of "diamond-like" carbon films [6] observed at higher doses. These different carbon species can be identified from their AES lineshapes [6,7]. The sample was maintained at room temperature and the incidence angle of the beam was along the surface normal.

## 3 RESULTS

Reactions of active ions with adsorbate covered surfaces can generally be divided into two categories depending on the surface reactant species; one is the reaction with surface adsorbates X and the other with atoms of the substrate lattice M. The term <u>reaction probability</u> $P_r$, the probability of producing a specific product for each reactant ion impinging on the surface, will be used to represent the branching ratio. The dominant processes for these low-$E_k$ ions are either reaction with M or X, with probability $P_r^{M,X}$, or scattering from M or X, with probability $P_s^{M,X}$, so that we can write

$$P_r^M + P_r^X + P_s^M + P_s^X = 1. \tag{1}$$

Since the ratio of ions reacting with M or X depends on adsorbate coverage, it is desirable to define the reaction yield Y so that it will be independent of adsorbate coverage. We use the notation Y(i,j), where <u>i</u> represents the impinging particle ($C^+$, $O^+$, $Ne^+$, or $O_2$) and <u>j</u> represents the reactant surface species M or X (M = Ni and X = C or O). Y(i,j) is then

$$Y(i,j) = P_r^j/f^j, \tag{2}$$

where $f^j$ is the fraction of the primary reactant ion flux that interacts with species <u>j</u>. The $P_r$ for the reactions examined in this study are summarized in Table I. These results are expounded in this section in the same order as they appear in Table I.

The reaction of $C^+$ ions with a p(2x2)NiO surface in the range 5-200 eV was studied by monitoring the changes in adsorbate concentrations after $C^+$ exposure using UPS and AES. The variations in oxygen and carbon coverages as a function of dose for a 5 eV $C^+$ beam are shown in Fig. 2. The oxide desorption and carbide accumulation occur simultaneously. After extended $C^+$ exposures, the oxide desorption rate slows down while the rate of carbide formation increases. Extrapolation of the initial slopes of the

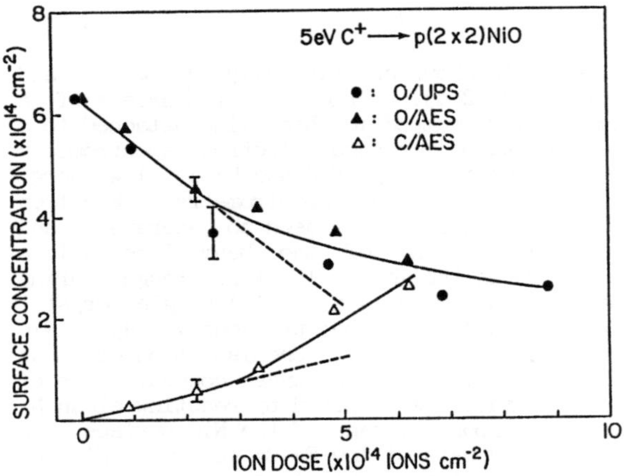


Fig.2: Dose dependence plot of the reaction of 5 eV $C^+$ with a p(2x2)NiO surface. Dotted lines are extrapolation of the initial slopes.

Table I: Reaction Probabilities, or Branching Ratios, for the Surface Reactions Investigated.

| | Reaction Probability | |
|---|---|---|
| Reaction | Adsorbate Reaction Channel | Ni Lattice Atom Reaction Channel |
| $C^+$(5 eV) + NiO[a] | 0.80 ± 0.1 | 0.24 ± 0.1 |
| $O^+$(5 eV) + NiC | 0.10 ± 0.1 | 0.84 ± 0.1 |
| $O^+$(20 eV) + NiC | 0.27 ± 0.1 | 0.66 ± 0.1 |
| $Ne^+$(20 eV) + NiO[a,b] | < 0.05 | ------ |
| $Ne^+$(20 eV) + NiC | < 0.05 | ------ |
| $C^+$(5 eV) + Ni[c] | ------ | 0.85 ± 0.1 |
| $O^+$(5 eV) + Ni[c] | ------ | 0.88 ± 0.1 |
| $O_2$(thermal) + NiC[d] | 0.05 ± 0.02 | 0.80 ± 0.2 |
| $O_2$(thermal) + Ni[f] | ------ | ~1 |

[a]p(2x2) NiO layer.
[b]Ref. 8
[c]Ref. 4
[d]Pressure reading in chemisorption studies is not calibrated.
[f]Ref. 9

data in Fig. 2 provides estimates of the initial $P_r$ into oxide desorption and carbide formation channels of 0.80 and 0.24, respectively. Excluding the portion of the $C^+$ beam that is channeled into NiC formation, the oxide desorption yield is calculated to be $Y(C^+,O) = 1 \pm 0.1$. This stoichiometry for oxide sputtering will be important in understanding the product formation mechanism. No molecular species, including CO, were detected on the surface.

The oxygen desorption yields of 5-200 eV $C^+$ and $Ne^+$ ions on the p(2x2) oxide layer are shown in Fig. 3. Typical ion doses used in this measurement are less than $3x10^{14}$ ions $cm^{-2}$ in order to obtain <u>initial</u> values of Y. The desorption behavior is remarkably different for $C^+$ and $Ne^+$ ions. Desorption by $C^+$ is most efficient at low $E_k$ and the yield $Y(C^+,O)$ decreases with increasing $E_k$. $Y(Ne^+,O)$, which occurs by a momentum transfer or physical sputtering mechanism, is not detectable in our system for $E_k<25$ eV and increases with $E_k$ [8]. At low energies $Y(C^+,O) >> Y(Ne^+,O)$ and the two curves approach each other near 200 eV.

NiC surfaces, formed by exposure of Ni to low energy $C^+$ ions, were exposed to $O^+$ beams. The results indicate a decrease in surface carbon concentration as well as formation of NiO with increasing $O^+$ dose. No molecular CO species were observed on the surface. Exposure of the NiC to $Ne^+$ beams in the same $E_k$ range (5-20 eV) indicated negligible carbon desorption, i.e., $Y(Ne^+,C) < 0.05$. Dose dependence studies monitored by AES are shown in Fig. 4 for the reactions of 5 and 20 eV $O^+$ beams with NiC surfaces prepared by 5 eV $C^+$ (dose = $8.2x10^{14}$ ions $cm^{-2}$) and 20 eV $C^+$ exposures (dose = $8.4x10^{14}$ ions $cm^{-2}$), respectively. The rate of oxide formation is faster than the rate of carbide desorption; this appears more profound (Table I) for the 5 eV $O^+$ beam than the 20 eV beam. At 5 eV, ~84% of the $O^+$ beam forms NiO while ~ 10% contributes to carbide desorption. At 20 eV the $P_r$'s for these channels change to ~66% and ~27%, respectively. Excluding the beam fraction leading to NiO formation, $Y(O^+,C) = 0.7 + 0.2$.

In order to compare reactivities of low energy active ions with those of thermal energy molecules, chemisorption of thermal $O_2$ on the NiC surface was examined. A NiC surface, prepared by 20 eV $C^+$ exposure (dose $8x10^{14}$ ions $cm^{-2}$), showed a 5-10% decrease in the carbon concentration after 2L $O_2$ chemisorption along with simultaneous formation of NiO. The UPS detected no molecular CO species on this surface. When the $O_2$ exposure is increased to 30L, the carbon concentration decreases continuously to 40% of its initial value. Using the rate of thermal $O_2$ collisions with the surface and the carbide concentration, $Y(O_2,C)$ is estimated to be ca. 0.05. $P_r$ for oxide formation on this surface (ca. 0.8) is comparable to the value on clean Ni [9].

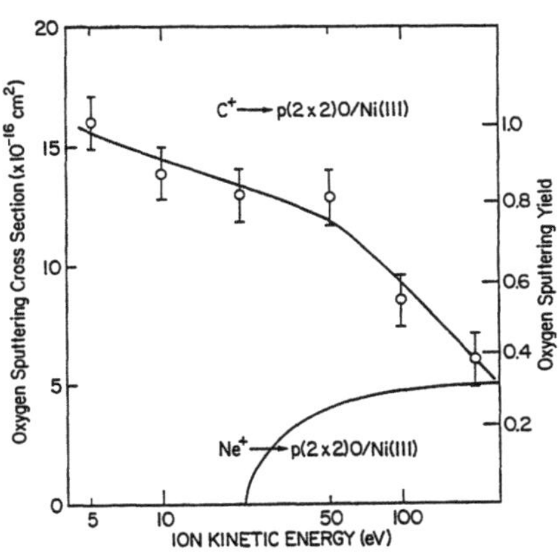


Fig. 3: Oxygen desorption yield as a function of $C^+$ kinetic energy for the reaction of $C^+$ with NiO and $Ne^+$ sputtering yield as a function of $Ne^+$ kinetic energy.

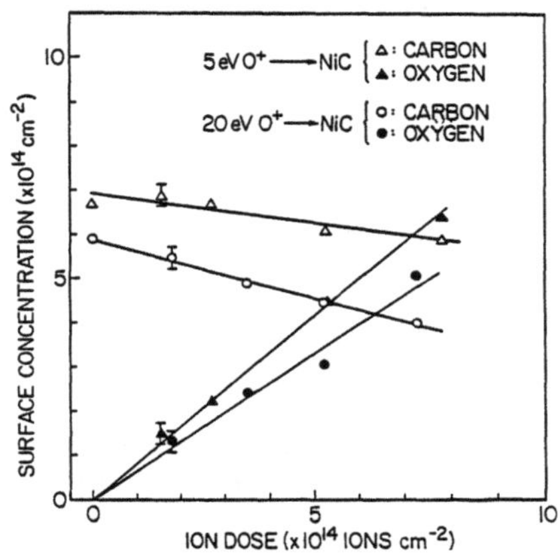


Fig. 4: Dose dependence plot of the reaction of and 20 eV $O^+$ with NiC surfaces. Surface concentrations are monitored by AES.

## 4 DISCUSSION

### 4.1 Energetics of Reactive Ion Induced Desorption

The remarkable difference (Fig. 3) in $C^+$ and $Ne^+$ desorption yields for the p(2x2) oxide is due to chemical reactions induced by reactive ions. Oxygen removal by $Ne^+$ occurs by physical sputtering involving a momentum transfer mechanism. Since momentum transfer is determined by the projectile/target mass difference, the amount of energy transferred in Ne and C collisions with O is similar. Therefore the difference between the $Y(C^+,O)$ and $Y(Ne^+,O)$ curves of Fig. 3 is attributed to the attractive potential between carbon and oxygen atoms which leads to chemical reaction. A plausible reaction is recombination of $C^+$ with surface oxygen to form a CO molecule. In formation of the strong CO bond, the recombination energy can be deposited as internal energy in the molecular CO product, thereby inducing CO desorption. $CO_2$ formation is another plausible explanation for desorption of oxides. However, $CO_2$ formation requires an encounter between C and two of the dispersed surface O atoms, which is not as likely to occur as CO formation. The measured oxide sputtering yield $Y(C^+,O)$ of near unity at low $E_K$ also implies CO formation with 1:1 ratio for C and O recombination.

The reaction of $O^+$ with a NiC surface results in both NiO formation and carbon desorption. The $Y(O^+,C)$ measurement has a large error due to the low intensity of the $O^+$ beam and the fact that oxide formation is the major reaction channel. However, the value $Y(O^+,C) = 0.7 \pm 0.2$ favors molecular CO formation as the major desorption mechanism. The reactivity of a carbide layer (deposited by high-temperature ethylene decomposition) towards $O_2$ has been the subject of recent kinetic studies [10,11] which indicate that CO desorption is the major product channel; gaseous CO evolution has been detected by mass spectrometry in one of these studies [11]. The efficient NiO formation may be due to the fact that the carbide has a more open structure than the p(2x2) oxide layer, exposing a larger fraction of Ni lattice atoms in the uppermost surface layer. It has been shown [4] that $C^+$ penetrates deeper into Ni, forming a more diffuse carbide layer than the NiO layer formed by either $O^+$ or $O_2$ exposure. As a result, the C concentration in the topmost carbide layer is lower than the O concentration in the topmost oxide layer when the same ion doses are used for NiC and NiO formation, providing greater possibilities for the competing NiO formation reaction.

The CO formation reactions on NiO and NiC surfaces are highly exothermic [12,13]:

$$C(g) + NiO(s) \longrightarrow NiCO(s) \qquad \Delta H = -8.43\ eV \qquad \text{(reaction 1)}$$

$$O(g) + NiC(s) \longrightarrow NiCO(s) \qquad \Delta H = -5.31\ eV\,. \qquad \text{(reaction 2)}$$

When reactants are in the form of a beam, their translational energy $E_K$ as well as internal excitation energy gained in surface neutralization [2,14] can be accommodated in the reaction coordinate. Therefore, the overall reaction exothermicity can be higher than the above values. Consider desorption of CO from a Ni surface [15]:

$$NiCO(s) \longrightarrow Ni(s) + CO(g) \qquad \Delta H = 1.55\ eV\,. \qquad \text{(reaction 3)}$$

Since the desorption energy is small relative to the overall exothermicities of the above recombination reactions, the CO molecule will desorb unless efficient surface quenching dissipates most of the excess

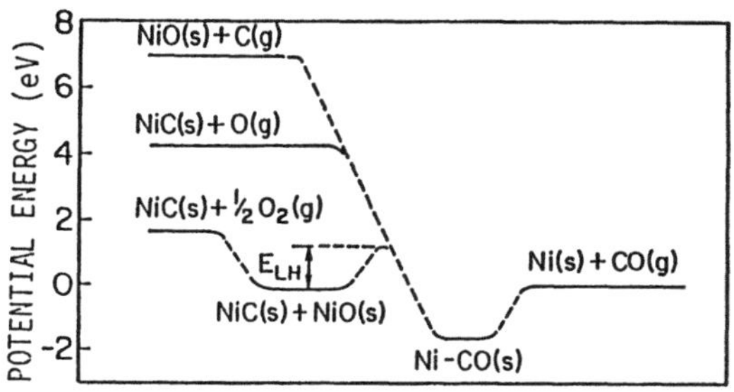


Fig. 5: Reaction coordinate diagrams for C and O recombination processes. Potential energies of reactants and products are drawn to scale, however, the $E_{LH}$ barrier heights are arbitrary. The products Ni(s) + CO(g) are chosen as the zero potential energy reference.

energy into the solid. The present results indicate that energy dissipation into the Ni lattice is very inefficient in these processes. This is in agreement with theoretical calculations of Tully [16] which indicate that surface quenching is not efficient for recombination of C and O on a Pt surface.

Comparing the reactivities of $O^+$ ions and thermal $O_2$ molecules toward NiC layers, carbide desorption occurs ca. 15 times more efficiently for $O^+$ than $O_2$. It is generally recognized [10,11] that the $O_2$ reaction proceeds via a Langmuir-Hinshelwood (LH) mechanism, i.e., dissociative chemisorption of $O_2$ is a prerequisite to induce C(s) + O(s) recombination. The slow rate of the $O_2$ reaction, despite the fact that formation of a NiO intermediate is as facile for $O_2$ exposure as for $O^+$ exposure (Table I), is indicative of an activation energy barrier ($E_{LH}$) in the recombination. Consider CO formation as a result of $O^+$ and $O_2$ reactions as depicted by the reaction coordinate diagrams of Fig. 5. Atomic oxygen, being a highly reactive radical species with high potential energy, can react with surface carbon atoms directly without involving surface oxide intermediates, proceeding via a pathway similar to the Eley-Rideal (ER) mechanism [17]. When this direct recombination occurs, the potential energy of the oxygen atom can be efficiently incorporated into the transition state because the reaction does not go through the NiO(s) intermediate by which energy can be dissipated into bulk. For this reason, the rate of reaction and CO desorption from a direct recombination reaction is expected to be fast. On the other hand, when the LH mechanism is involved, the potential energy differences among various reactants are largely equalized by the oxide formation and consequential energy dissipation; surmounting the barrier $E_{LH}$ becomes more difficult, and therefore, the reaction rate is comparably slow. We consider the highly efficient adsorbate sputtering in the reactions $C^+$ + NiO and $O^+$ + NiC to be an indication of feasible ER processes.

## 4.2 Monte Carlo Simulation of the Desorption of Adsorbed O Atoms by $C^+$

Computer simulation of the desorption of oxygen from a p(2x2)NiO surface induced by $C^+$ ions was carried out using a modified version of the TRIMSP code [8,18]. The target is assumed to be amorphous Ni with a submonolayer of O on top of the Ni surface. Physical sputtering was simulated as the sum of two contributions: 1) Direct knock-off ($S_{p1}$) of adsorbed atoms in C-O collisions ocurring either during the incoming part or the reflected part of the primary ion trajectories. 2) Sputtered substrate contribution ($S_{p2}$), where the secondary sputtered flux of Ni atoms knocks off adsorbed O atoms. For this part of the simulation, $C^+$ can be treated as a nonreactive projectile such as $Ne^+$. Calculational procedures and parameters were similar to those used previously [8] for sputtering of NiO by $Ne^+$ ions.

These parameters are: surface binding energy = 4.25 eV, final cut-off $E_K$ for the trajectory = 4.25 eV, and displacement energy of lattice atoms = 20 eV.

Simulation of chemically induced desorption requires inclusion of a chemical bond formation mechanism in the computation. Since the present TRIMSP code uses a purely repulsive Moliere-type potential, we adopted a simplistic approach to simulate the chemical interaction originating from attractive components by exploiting the sputtering mechanism asserted in the previous sections. Chemically induced desorption $S_c$ along with C-O bond formation were simulated by assuming that 1) C atoms moving within the range of the attractive C-O potential, i.e., within 2.5 Å from the average position of the O atoms, with $E_K$ equal to or less than the CO interatomic potential well depth (11.1 eV), will recombine to form CO, and 2) the internal energy gained in this recombination induces CO desorption when the reaction occurs in the outermost atomic layer.

The sum of the three contributions, $S_{p1}$, $S_{p2}$, and $S_c$ for the three different values of final energy $E_f(C^+)$, 5, 10, and 20 eV are shown in Fig. 6 along with the experimental results. The reasonable agreement observed between the calculation and experiment supports the validity of this model. The lowest $E_f$, 5 eV, is a typical cut-off value [18,19] below which a particle moving in a lattice is considered to be trapped. The highest $E_f$, 20 eV, corresponds to a center-of-mass energy of about 11 eV, which is near the CO interatomic potential well depth. This parameter serves to represent the trapping efficiency of C(g) into the CO potential well, or the degree of inelastic energy loss in the C-O collision for molecular CO stabilization. The sum of $S_{p1}$ and $S_{p2}$ has a threshold energy of about 20 eV and increases with projectile energy due to the increasing yield of Ni substrate sputtering. $E_K$ dependence of this sum qualitatively tracks the $Ne^+$ sputtering curve. $S_c$ becomes the major contributor to oxygen desorption in the low $E_K$ region approaching unity. At these energies, projectile penetration is limited only within a few Angstroms from the surface and therefore, almost all of $C^+$ ions react to form CO. At higher $E_K$ the ion penetration depth increases, leaving a smaller fraction of C in the upper layer; as a result, $S_c$ decreases with $E_K$. It needs to be mentioned

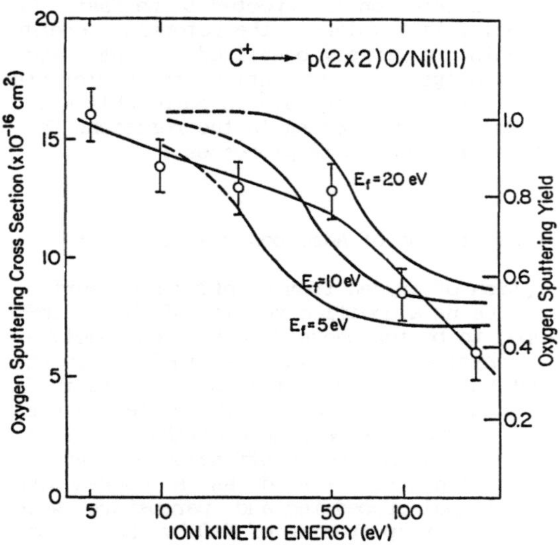


Fig. 6: Calculated total sputtering yield for final trapping energies $E_f$ of 5, 10, and 20 eV, along with the experimental curve from Fig. 3.

here that this simulation should be considered qualitative because it is based on a binary collision model which neglects the many-body aspects of low-energy atom-atom interactions.

## 5 CONCLUSIONS

(i) Low-energy reactive ion-surface collisions induce facile adsorbate desorption. The $P_r$ of the reactions $C^+$ + NiO and $O^+$ + NiC are generally orders of magnitude higher than the thermal $O_2$ reaction. Specifically, $C^+$ and $O^+$ chemically desorb surface oxide and carbide species, respectively, via C and O recombination with subsequent desorption of molecular CO.
(ii) The general mechanism governing adsorbate desorption processes can be altered by changing the chemical nature and $E_k$ of the reactive projectiles. For example, NiO sputtering by ion bombardment can be achieved via either a physical (by high $E_k$ $Ne^+$) or chemical (by low $E_k$ $C^+$) process. The reaction mechanism leading to chemical desorption of NiC can be controlled such that it occurs via a direct Eley-Rideal-type mechanism (using $O^+$ beams) or by an indirect Langmuir-Hinshelwood-type mechanism (using thermal $O_2$ molecules).
(iii) Monte Carlo simulation results, modified for inclusion of chemically induced desorption, attribute the high $Y(C^+,O)$ at low $E_k$ to the efficient trapping of $C^+$ into a C-O potential well to induce recombination and desorption. $Y(C^+,O)$ decreases with increasing $E_k$ due to ion penetration into bulk Ni.

## ACKNOWLEDGMENTS

This material is based on work supported by the National Science Foundation under Grant No. DMR-8610597.

## REFERENCES

1. H. Kang, T.R. Schuler, and J.W. Rabalais, Chem. Phys. Lett., 128 (1986) 348.
2. T. Darko, D.A. Baldwin, N. Shamir, J.W. Rabalais, and P. Hochmann, J. Chem. Phys., 76, 6408 (1982).
3. D.A. Baldwin, N. Shamir, and J.W. Rabalais, Surface Sci., 141 (1984) 617, and references therein.
4. H. Kang, S. Kasi, and J.W. Rabalais, J.Chem.Phys., submitted.
5. P.H. Holloway and J.B. Hudson, Surface Sci., 43 (1974) 141.
6. S. Kasi, H. Kang, and J.W. Rabalais, Phys.Rev.Lett., in press.
7. D.W. Goodman, R.D. Kelley, T.E. Madey, and J.T. Yates, Jr., J. Catal., 63 (1980) 226.
8. J.W. Rabalais, T.R. Schuler, and O. Grizzi, Nucl. Instru. Methods, in press.
9. P.R. Norton, R.L. Tapping, and J.W. Goodale, Surface Sci., 65 (1977) 13.
10. T.J. Vink, M.M.J. Vanzandvoort, O.L.J. Gijeman, and J.W. Geus, Appl. Surface Sci., 18 (1984) 255.
11. R. Sau and J.B. Hudson, Surface Sci., 102 (1981) 239.
12. $D^o(Ni\text{-}C)_s$ = 4.25 eV; B.E. Nieuwenhuys, Surface Sci., 126 (1983) 307.
13. $D^o(Ni\text{-}O)_s$ = 7.3 eV; L.C. Isett and J.M. Blakely, Surface Sci., 58 (1976) 397.
14. H.D. Hagstrum, in Electron and Ion Spectroscopy of Solids, eds. L. Fiermanns, J. Vennik, and W. Dekeyser (Plenum Press, N.Y., 1978) p.273.
15. H. Ibach, W. Erley, and H. Wagner, Surface Sci., 92 (1980) 29.
16. J.C. Tully, J. Chem. Phys., 69 (1980) 6333.

17. G. Ertl, CRC Critical Reviews in Solid State and Materials, vol.10., p.349.
18. J.P. Biersack and W. Eckstein, Appl. Phys. 34, 73 (1984).
19. E. Taglauer, U. Beitat, G. Marin and W. Heiland, J. Nucl. Mat. 63, 193 (1976).

# PSID SEXAFS Studies of Silicon Surface Chemistry

*G. Thornton*[1,*], *D.R. Warburton*[1], *R. McGrath*[2], *I.T. McGovern*[2], *and D. Norman*[3]

[1]Department of Chemistry, Manchester University, Manchester M13 9PL, U.K.
[2]Department of Pure and Applied Physics, Trinity College, Dublin 2, Ireland
[3]SERC, Daresbury Laboratory, Warrington WA4 4AD, U.K.

## 1. Introduction

Photon stimulated ion desorption (PSID) surface EXAFS (SEXAFS) showed early promise, being successfully applied to the Mo(100)/O system [1]. The technique offers a means to determine the coordination geometry of a substrate atom to which the desorbed adsorbate ion was bonded. Information regarding the adsorbate coordination geometry can also be obtained, although electron and fluorescence yield techniques can usually be used for this purpose if a majority surface species is the subject of interest. The surface specific nature of the PSID yield is unique amongst probes of the surface absorption coefficient. This conveys the possibility of using PSID SEXAFS to determine the coordination geometry about a substrate atom bonded to hydrogen or, for instance, the bond geometry of -OH on an oxide surface. In addition, the bond geometry of minority surface species can be probed, as in the Mo(100)/O study. Clearly, PSID SEXAFS is potentially a very powerful tool in surface structure determination.

Substrate edge PSID SEXAFS is dependent on ions being desorbed following photon absorption by surface substrate atoms to which they were bonded. Hence, the variation in ion yield above an absorption edge of the substrate atom (the EXAFS) contains information on the substrate atoms coordination. In practice, this means that the ion yield arising from a direct, Auger decay mechanism of PSID [2] must dominate over other sources of ion desorption. This is not always the case; the ion yield arising from a competing mechanism has been shown to dominate the ion yield from condensed gases [3], rare earth oxides [4], and $SrTiO_3(100)-H_2O$ [5]. This mechanism is X-ray induced electron stimulated desorption (XESD), which involves photon absorption in the bulk with transfer of energy to a surface complex via the outgoing photocurrent. The ion yield EXAFS resulting from this mechanism contains only bulk structural information. In general, it is not possible a priori to estimate the relative yields of direct PSID and XESD. Hence, positive proof of the surface specific nature of the ion yield EXAFS is a pre-requisite of PSID SEXAFS studies. Understandably, this has considerably reduced the perceived potential of PSID SEXAFS.

However, the perceived potential of PSID SEXAFS, although limited, still exists, especially in the areas where it has a unique capability. In order to further explore its possibilities in one of these areas, the geometry of substrate sites bonded to hydrogen, we have studied $H^+$ desorption from hydrogenated silicon surfaces at the Si K-edge. In this paper we will concentrate on the results obtained from Si(100)2x1 following dissociative $H_2O$ adsorption.

---

*Also at SERC Daresbury Laboratory, Warrington WA4 4AD, U.K.

## 2. Experimental

The experiments were carried out using the double-crystal monochromator [6] (1.75 $h\nu \leq 11 \leq$ keV) on station 6.3 at the SRS, Daresbury Laboratory. Ion yield measurements used the time structure of the SRS in single bunch mode (0.3 ns pulse width, 320 ns repetition period) in conjunction with a time-of-flight (TOF) mass spectrometer [5]. Photons were incident at 70° from the surface normal, with the ion yield about the surface normal being measured. Total electron yield (TEY) spectra were also recorded for comparison. All data were flux normalised. The sample was cleaned following standard procedures to form Si(100)2x1. A 2L (1L = 1.32 x $10^{-6}$ mbar s) saturation exposure to $H_2O$ was carried out with the sample at room temperature.

## 3. PSID:XESD

As noted above, the ion yield cannot be assumed to arise from surface absorption and it is necessary to determine the origin of the ion yield, in particular the PSID:XESD yield ratio. The XESD yield, $N_{XESD}(h\nu)$ is a function of the electron energy distribution of the outgoing photocurrent, $N(E,h\nu)$, and the ESD cross-section, $\phi(E)$:

$$N_{XESD}(h\nu) \;\alpha \int_{E_o}^{h\nu} N(E,h\nu)\, \phi(E)\, dE,$$

where E is the electron energy, and $E_o$ is the ESD threshold [3]. Because measurements of $\phi(E)$ are rather rare it is usually impossible to directly assess the XESD contribution to the photon-induced ion yield. However, it is worthwhile to consider some general guidelines. For condensed gases ion desorption following electron-stimulated ionisation of valence electrons is observed to have an appreciable cross-section, in contrast to that observed for sub-monolayer systems, where the effective $E_o$ value is typically 20 eV [2,7]. This allows low-energy secondary electrons, which form a substantial part of the photocurrent, to contribute to the XESD yield from condensed gases [3]. On this basis, the photon-induced ion yield from condensed gases is more likely to be dominated by XESD than that from sub-monolayer systems.

A second, and more obvious statement can be made regarding the PSID:XESD yield ratio: if an increase in the PSID ion yield is not expected at an absorption edge then any increase observed must arise from XESD. We can illustrate this point using ion yield [5] and photoemission [5,8] data obtained from $SrTiO_3(100)$-$H_2O$. Surface core-level chemical shift [8] and valence band photoemission [5] data are consistent with dissociative $H_2O$ adsorption on stepped $SrTiO_3(100)$, and with the formation of Ti-OH, Sr-OH, and O-H species. Hence, in the absence of β site desorption (i.e. desorption of an ion following ionisation of an atom separated from it by another atom), an increase in the $H^+$ PSID yield is not expected at Ti or Sr absorption edges. The observed increase in $H^+$ yield at the Ti K and Sr $L_1$ edges [5] must therefore arise from XESD. This argument can be pursued, admittedly in a more speculative vein, in considering the ion yield data obtained from rare earth oxides [4]. Here, as for $SrTiO_3(100)$-$H_2O$, the $H^+$ yield probably arises from surface hydroxyl species so that an increase in $H^+$ yield at cation absorption edges is again expected to be XESD dominated. Evidence for the XESD domination of the ion yield from sub-monolayer adsorption systems where PSID is expected is much more restricted. For instance, we deduced that the $OH^+$ yield from $SrTiO_3(100)$-$H_2O$ at the Ti K-edge arises mainly from XESD [5].

One means of assessing whether the ion yield arises chiefly from XESD, and that employed here, is to compare the absorption edge jumps in ion yield and total electron yield (TEY) [3]. If it is assumed N(E, hν) does not change markedly through the edge then the relative edge jump, ΔI/I, will be the same in ion yield and TEY if XESD dominates (I is the count rate below the edge, ΔI is the difference in count rate above and below the edge). It follows that PSID dominates if the edge jumps are substantially different. The assumption regarding N(E,hν), and therefore the use of relative edge jumps in estimating the PSID:XESD ratio, has been questioned [9]. However, we note that ion and electron yield edge jumps have been found to be the same where XESD dominates [3,5], which gives us some confidence in its validity.

## 4. Si(100)2x1-$H_2O$

The relative edge-jump in the $H^+$ yield at the Si K-edge (2.6) is widely different from that observed in TEY (9.1). With the assumption of an invariant N(E,hν), the relative edge-jumps can be used to estimate the upper limit of the XESD contribution to the ion yield, ca. 28%. On this basis we assume that PSID is dominant, and proceed with EXAFS analysis of the Si K-edge $H^+$ yield data.

The Si K-edge $H^+$ ion yield SEXAFS data are shown in fig. 1. The data were analysed, after background subtraction and normalisation to the edge jump, using a curve-fitting routine based on the rapid curved wave computational scheme [10]. The filtered Fourier transform of the data is shown in fig. 2 (solid line), peaks at 2 and 4 A, and the shoulder at 4.7 A corresponding to the first, second, and third nearest-neighbour shell distances. The feature at 3.25 A is an artifact of the analysis procedure,

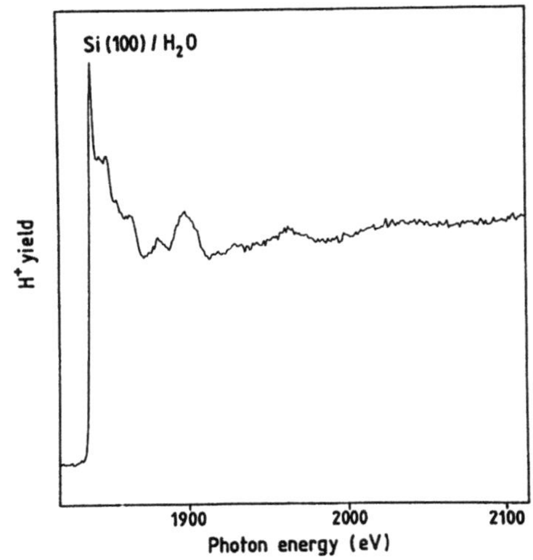


Fig. 1. $H^+$ ion yield spectrum from the 0.5 monolayer (ML) saturation coverage of $H_2O$ on Si(100), measured at the Si K-edge at 290 K; X-rays were incident at 70° off normal

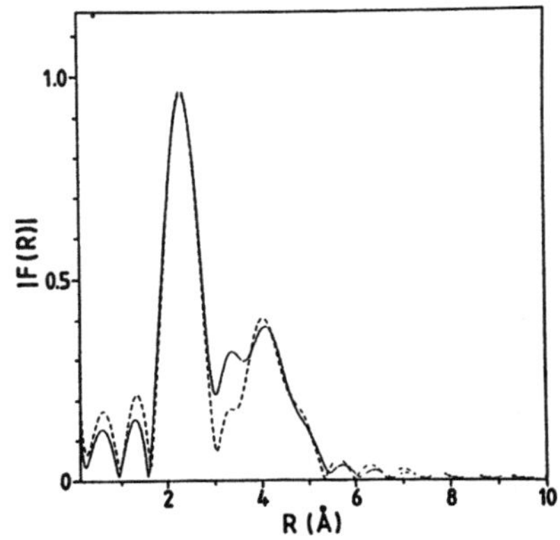


Fig. 2. The filtered Fourier transform of the $H^+$ yield SEXAFS data from Si(100)-$H_2O$ (solid line), with the corresponding 3-shell best fit (dashed line)

similar to features observed in many EXAFS experiments. The dashed line in fig. 2 represents the best theoretical fit to the data. Electron scattering phase shifts for Si were calculated using a "muffin tin" potential-based method and then refined to fit the measured TEY bulk EXAFS spectrum of Si. After phase-shift correction the distances for the first three nearest-neighbour shells are found to be 2.37±0.02 Å, 3.9±0.15 Å and 4.6±0.3 Å.

In this case the SEXAFS data reflect the silicon-silicon geometry in the presence of an "invisible" adsorbate, H. For our excitation geometry, backscattering from OH is also negligible; the effective coordination number is <<1 for an OH bonded to the second Si atom of the dimer. β site desorption of $H^+$ from Si-OH, evidenced in an ESDIAD study of Si(100)-$H_2O$ [11], appears to have a relatively small cross-section at the Si K-edge. It would give rise to an easily distinguished peak in the Fourier transformed spectrum at a position corresponding to the Si-O distance (ca. 1.6Å), and no such peak is observed. As $H_2O$ does not perturb the symmetry of the two-dimensional surface mesh, this situation invites comparison with dimer models of the clean 2x1 reconstructed surface of silicon (see fig. 3). Several different dimer models have been proposed, the symmetric dimer model of Levine [12] and the asymmetric dimer model of Chadi [13] being depicted in figs. 3c and d. The maintenance of 2x1 symmetry up to saturation indicates that each surface dimer forms an Si-Si(H)-Si(OH)-Si complex, without disturbance of the substrate geometry (fig. 3e). For comparison with the experimental data, we have calculated average first shell distances for six dimer model geometries of the Si(100)2x1 reconstructed surface, assuming that hydrogen is selectively bonded to the upper atom of asymmetric dimers which should be favoured since this has a higher electron density. Only the models of Chadi and Levine are consistent with the data, as shown in table 1. Further justification for our choice of asymmetric dimer geometry is gained by considering the effective coordination numbers. The calculated first shell value for the "up" site of the asymmetric dimers is ca. 50% higher than that for the corresponding "down" site and therefore closer to the experimentally determined value.

We conclude with a cautionary note. Because the surface dimer distance derived in this work is, within experimental error, the same as the bulk Si-Si separation, it is legitimate to pursue the question of the PSID

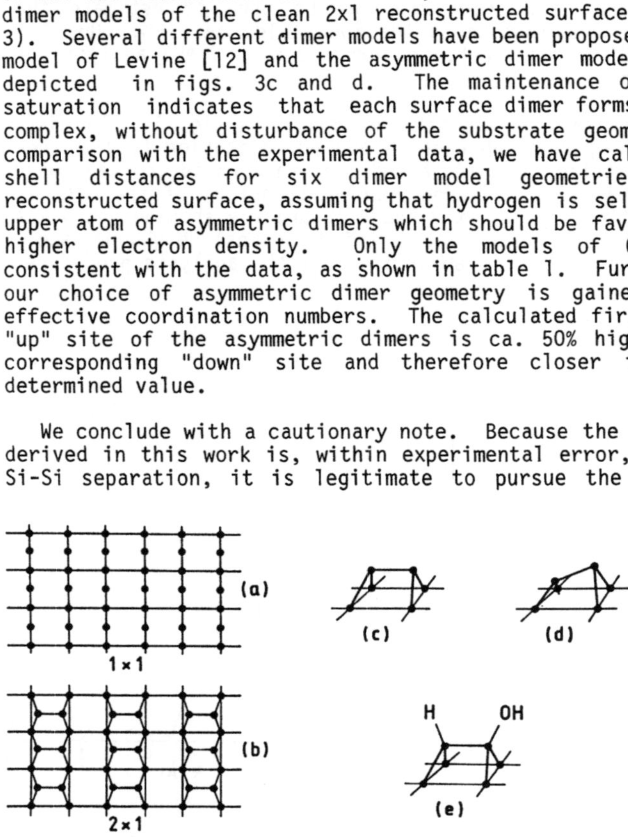


Fig. 3 (a) The unreconstructed Si(100)1x1 surface (top view). (b) The formation of dimers leading to the 2x1 surface net (top view). (c) A symmetric dimer. (d) An asymmetric dimer. (e) The adsorption sites for H and OH on the Si-Si symmetric dimer.

Table 1. A comparison of our experimental with theoretical nearest-neighbour shell distances (in A) assuming upper atom hydrogen bonding in the Chadi model

| Shell | Bulk (TEY) | Surface ($H^+$ yield) | Levine [12] | Chadi [13] |
|---|---|---|---|---|
| 1 | 2.35 | 2.37±0.02 | 2.35 | 2.39 |
| 2 | 3.84 | 3.9 ±0.15 | 3.7 | 3.9 |
| 3 | 4.50 | 4.6 ±0.3 | 4.6 | 4.6 |

dominance of the ion yield. While the available evidence, outlined above, suggests that PSID does dominate, Si K-edge N(E,hν) data would provide valuable additional information regarding the relationship between the ion and electron yield edge jumps and the PSID:XESD ratio.

## 5. Acknowledgements

This work was supported by the Science and Engineering Research Council U.K. Additional support was received from Vacuum Science Workshop Ltd., and from NATO. RMcG is grateful to the Dublin University Trinity Trust for financial support.

## 6. References

1. R. Jaeger, J. Feldhaus, J. Haase, J. Stöhr, Z. Hussain, D. Menzel, D. Norman: Phys. Rev., Lett., 45, 1870 (1980)
2. M.L. Knotek: Rep. Prog. Phys., 47, 1499 (1984)
3. e.g. R. Jaeger, J. Stöhr, T. Kendelewicz: Phys. Rev. B, 28, 1145 (1983); and Surf. Sci. 134, 547 (1983)
4. J. Schmidt-May, F. Senf, J. Voss, C. Kunz, A. Flodström, R. Nyholm, R. Stockbauer: Surf. Sci. 163, 303 (1985)
5. I.W. Owen, N.B. Brookes, C.H. Richardson, D.R. Warburton, F.M. Quinn, D. Norman, G. Thornton: Surf. Sci. 178, 897 (1986)
6. A.A. MacDowell, D. Norman, J.B. West: Rev. Sci. Instrum. 57, 2667 (1986)
7. M.L. Knotek and J.E. Houston: J. Vac. Sci. Technol. 20, 544 (1982)
8. N.B. Brookes, G. Thornton, F.M. Quinn: Solid State Commun. in press
9. D. Menzel: Proceedings of ICSOS-II, Springer Ser. Surface Sci., to be published
10. S.J. Gurman, N. Binsted, I. Ross: J. Phys. C 17, 143 (1984)
11. C.U.S. Larsson, A.L. Johnson, A. Floodström, T.E. Madey: J. Vac.Sci. Technol. A 5, 842 (1987)
12. J.D. Levine: Surf. Sci. 34, 90 (1973)
13. D.J. Chadi: Phys. Rev. Lett. 43, 43 (1979)

# Photon Stimulated Desorption from Oxidized Al(110)

*E.D. Johnson*[1], *R.F. Garrett*[1], *M.L. Knotek*[1], *and F. Sette*[2]

[1]National Synchrotron Light Source, Brookhaven National Laboratory, Upton, NY 11973, USA
[2]AT&T Bell Laboratories, Murray Hill, NJ 07974, USA

We have studied oxide films on Al(110) by photon stimulated desorption (PSD) on the Bell Labs U4 PGM at the National Synchrotron Light Source. Utilizing a time-of-flight technique [1] we have obtained ion energy distribution (IED) and relative ion yield (RIY) data at the Al 2p and O 1s edges for oxides prepared at various temperatures. These initial studies suggest that different sites for the surface hydroxyls exist, that they can be selectively prepared, and examined by PSD.

## I Introduction:

Aluminas are among the most technologically and economically significant materials in use today. They are known and extensively employed in fields ranging from heterogeneous catalysis to micro electronics, with far too many applications to enumerate here. One important aspect of aluminas is the wide range of surface acidity which can be obtained [2]. In the catalytic literature distinct sites have long been postulated for the surface hydroxyls which are generally accepted to be responsible for the acidity [3,4]. The chemistry of these sites can be related not only to the coordination of surface hydroxyls with respect to the underlying lattice, for example octahedral vs. tetrahedral, but also their proximity to other hydroxyls or anion vacancies on the surface. It is the complexity and number of possible sites of this type which makes this system so difficult to fully understand.

Planar aluminas on polycrystalline and single crystal aluminum substrates have been characterized in previous work by other surface spectroscopic methods [5-8] so a great deal is already known about how to prepare them in ultra-high vacuum (UHV). PSD may provide the ideal probe for examining surface acidity, and so is the technique utilized in this study.

## II Experimental:

These experiments were conducted at the Brookhaven National Laboratory National Synchrotron Light Source on the Bell Labs plane grating monochromator beamline U4 [9]. To prepare the oxides discussed here, an Al(110) crystal is sputtered clean at room temperature with 2 keV argon ions, annealed at 500C, and exposed to 500 Langmuirs of oxygen at either room temperature or 400C. These two aluminas exhibit the lowest and highest measured surface acidity respectively, and should provide a good test of PSDs sensitivity to local chemical environment and specifically to hydrogen containing acid sites. Oxides produced under different conditions are described elsewhere [5,6,10].

The ion spectrometer is a simple time of flight detector [1] utilizing a drift tube of ca. 4.5 cm which is biased negative with respect to the sample. The ions extracted from the sample region are detected by a dual channel plate electron multiplier. The detector is mounted at an angle of 45 degrees with respect to the incident beam in a UHV chamber which is also fitted with a CMA for electron energy analysis and Auger spectroscopy, a quadrupole mass spectrometer, and an ion gun for sputter cleaning the sample.

The signal from the detector is used to start a time to amplitude converter (TAC) which uses a timing signal from the ring as its stop pulse. The output of the TAC is collected by a pulse height analyzer (PHA) in an arrangement shown schematically in figure 1. The NSLS UV ring operates at a frequency of 52.88 MHz with a single bunch turn time of 170.2 ns, and a bunch width of 300 ps. A typical spectrum is shown in figure 2 from an oxide at a photon energy of 84 eV, just above the Al 2p edge. The 1.4 ns wide prompt peak arises from fluorescent photons, and may be used as an indication of the instrumental broadening. The breadth of the hydrogen ion peak is additionally affected by the drift tube voltage and the initial ion kinetic energy distribution. By acquiring spectra of this kind as a function of photon energy, characteristic edge jump spectra can be obtained. The respective ion yield data and a signal from a beam flux monitor are stored at each photon energy simultaneously, which facilitates their direct and unambiguous comparison.

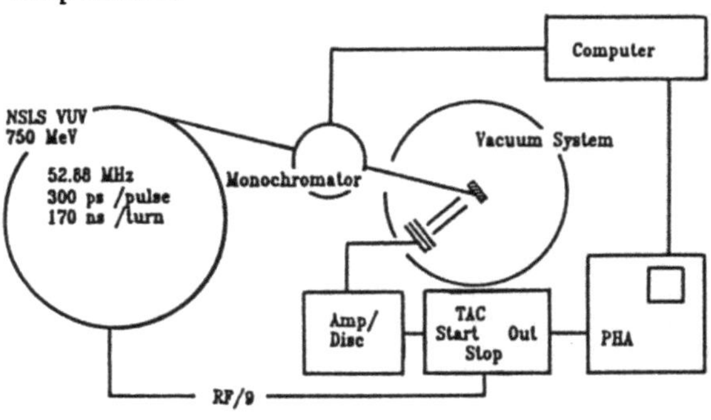


Figure 1 Experimental schematic. See text for the full description.

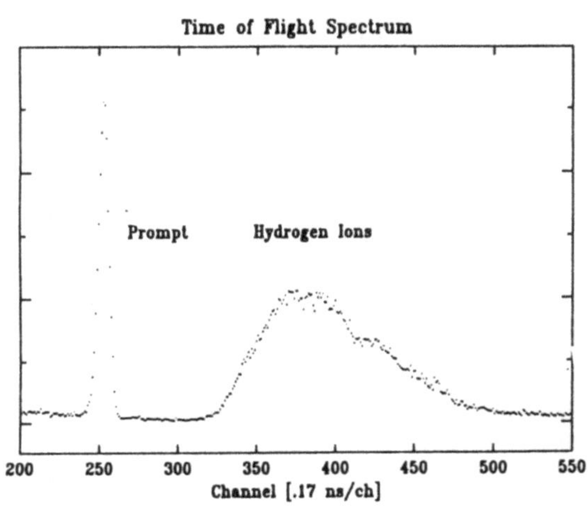


Figure 2 Time-of-flight spec trum for drift tube bias of 250 eV. The prompt is 1.4 ns fwhm.

III Results and Discussion:

One question which must be posed in examining PSD data at characteristic edges is the mechanism responsible for the production of the ions. The concern is always present that the ions are desorbed by a secondary electron cascade, or x-ray excited electron stimulated desorption (XESD), and not by a direct electronic excitation of the species of interest. A comparison of the total ion and total electron yields at the aluminum 2p edge as shown in figure 3 indicates that XESD does not contribute significantly to the ion yield since there is virtually no similarity between it and the electron yield.

In contrast, the Al 2p near-edge spectral dependences of the hydrogen ion yield and partial electron yield at 35 eV (an aluminum oxide intervalent Auger transition [11]) which are shown in figure 4, have qualitative similarities, indicating that both occur at oxide sites, but differ in several important respects. The PEY data are in general representative of the bulk oxide, while the ions which are observed can only come from the surface. The variation in intensity and energy of the features in these spectra can be ascribed to the different chemistry of these two environments, ie. the proton yield data contains information specific to the H bonding site [12]. The features of these spectra are usually interpreted in terms of the inner well resonances observed in x-ray absorption (XANES) [13,14,15]. The transitions labeled in figure 4 as A, B, and E are attributed to excitations into $a_1$(s-like), $t_2$(p-like), e and $t_2$(d-like) inner well states [13]. The A state is a core exciton corresponding to a 2p-3s transition in the $Al^{+3}$ ion and as such is essentially atomic in character, however, our initial results show that the exciton has a different threshold for each of the species observed and in fact for hydrogen appears to have two or more components related to the ion kinetic energies as discussed below (see inset to figure 4). The B, D, and E features are more like excited molecular state transitions and so exhibit sensitivity to their structural environment. The E state appears at roughly 100eV for tetrahedrally coordinated aluminum, while in octahedral samples it occurs at 90eV. For example the PEY data in figure 4 represent aluminum atoms primarily in octahedral sites, which might be expected in the bulk, while the ion yield data show

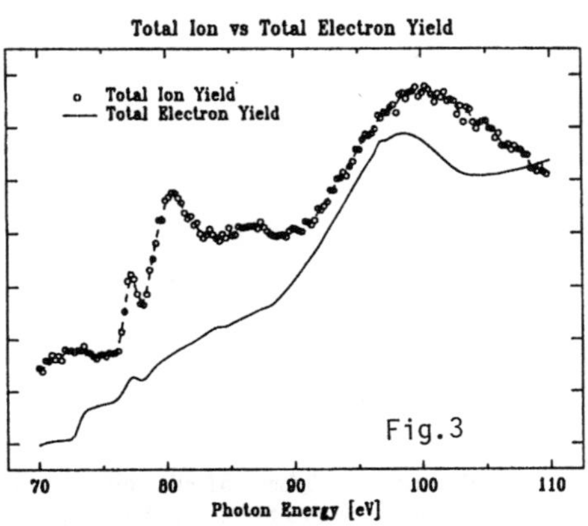


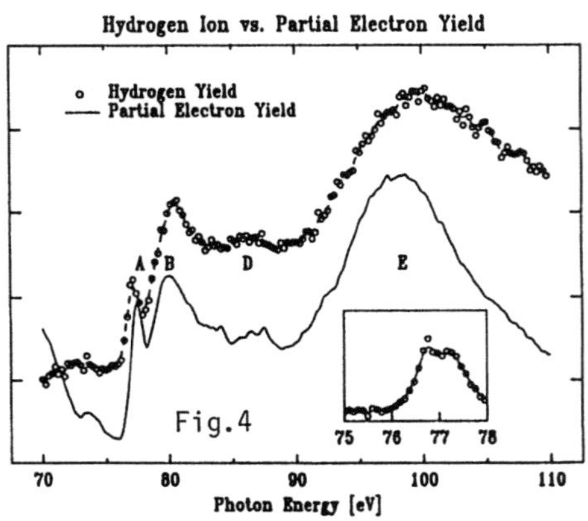


Figures 3 and 4 Comparisons of the total electron and ion yields (3), and 35 eV partial electron yield with hydrogen ion yield (4) for a high temperature oxide at the Al 2p edge.

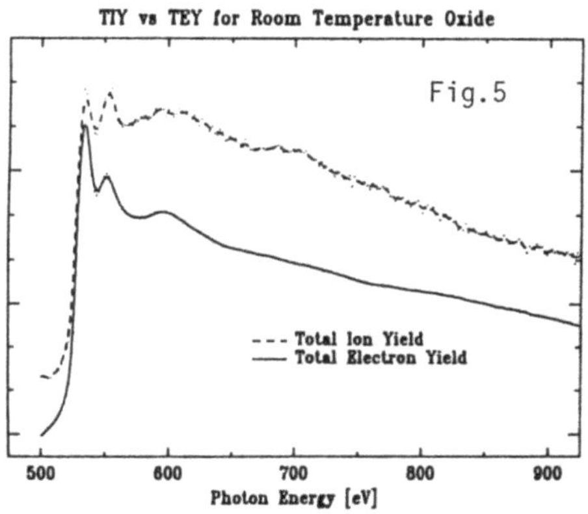


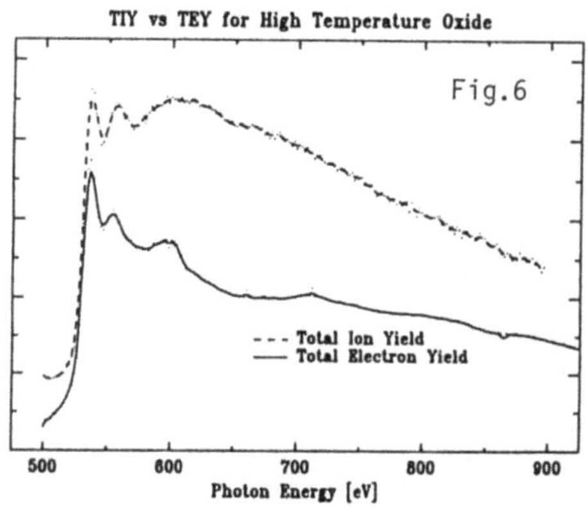


Figures 5 and 6 Comparison of total electron and total ion yields at the O 1s edge for room temperature and high temperature oxides respectively.

mixed octahedral and tetrahedral character. Other variations in the energy and intensity of these features can be used as fingerprints of the environment of the aluminum atoms and, when coupled with desorption spectra of hydrogen or hydroxyls can provide insight as to the nature of the surface species present on reactive aluminas.

Spectra taken at the oxygen edge are similar to the aluminum edge data just discussed in that the comparison of the total electron and ion yields strongly indicate that XESD does not contribute significantly to the desorption yield, as shown in figures 5 and 6. Also, by examining the ion yields as a function of photon energy as shown in figure 7, distinctly different spectral dependences can be seen for the various ions. The hydrogen yield exhibits a sharp onset with structural features more than 100 eV above the edge, the hydroxyl yield has a similar onset but dramatically drops in intensity only 50 eV above the edge, while the oxygen yield has a delayed onset with structure on the slowly increasing edge and does not achieve its maximum until almost 200 eV above the O 1s edge. As was the case at the aluminum edge, the oxygen edge data given by figures 5 and 6 show clear differences for oxides prepared at low and high temperatures.

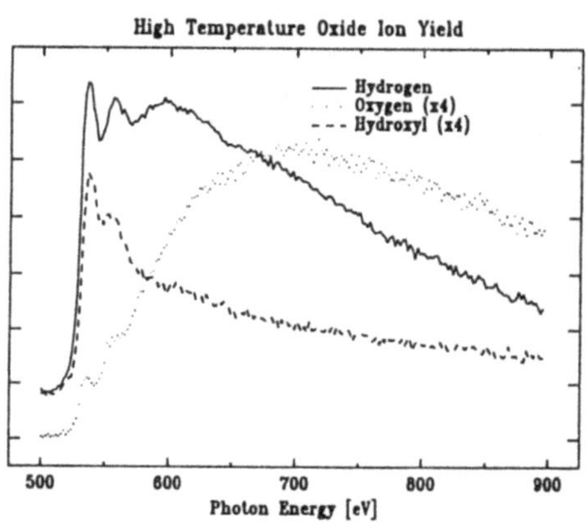


Figure 7 Hydrogen, hydroxyl and oxygen ion yields at the O 1s edge for a high temperature oxide.

An examination of full time-of-flight spectra at photon energies around the oxygen edge, shown in figure 8 for an oxide grown at 400C, reveals a narrow hydroxyl ion peak indicating a single desorption state for photon energies in this range. The oxygen peak, in contrast, is quite broad with its weight shifting from long to short arrival times as the photon energy is increased. These spectra can be fitted by assuming that oxygen ions are photodesorbed with three distinct kinetic energies. Although the data available at this time are sparse it appears that the highest kinetic energy species has a maximum intensity at relatively low photon energy, akin to the $OH^+$ yield, while the lower kinetic energy species exhibit a delayed onset and are similar in appearance to the total oxygen yield data of figure 7.

Similar spectra taken at photon energies around the aluminum edge reveal a hydrogen peak which has several different kinetic energy components. Interestingly enough, room temperature oxides exhibit only two components while high temperature films have three as shown in figure 9. This may be a significant observation with respect to the question of model alumina surfaces since the high temperature oxide is known to have a more acidic surface than its low temperature counterpart. Hydroxyl ions are observed with low intensity and a relatively narrow energy distribution, suggesting a single desorption state, while oxygen ions occur with still less intensity and a greater spread in kinetic energy. Signal to noise for these ions was too poor to allow a precise determination of their energies at the aluminum edge. The energies of the components to the various desorption peaks were determined by least squares fitting to Gaussian peak shapes. The ions are observed with a complex distribution of kinetic energies which is summarized below.

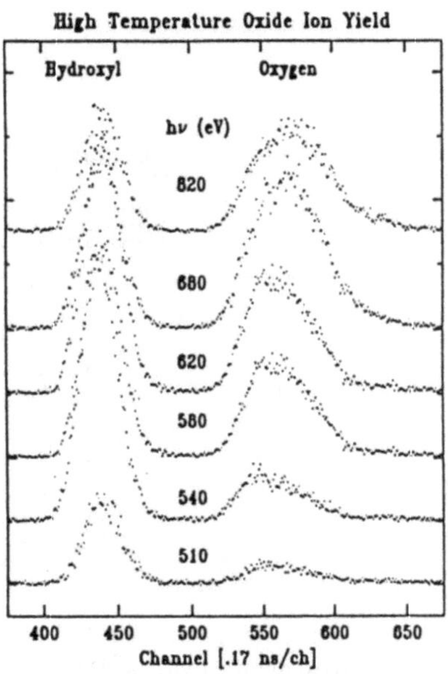


Figure 8 Time-of-flight spectra at the oxygen edge for various photon energies around the O 1s edge.

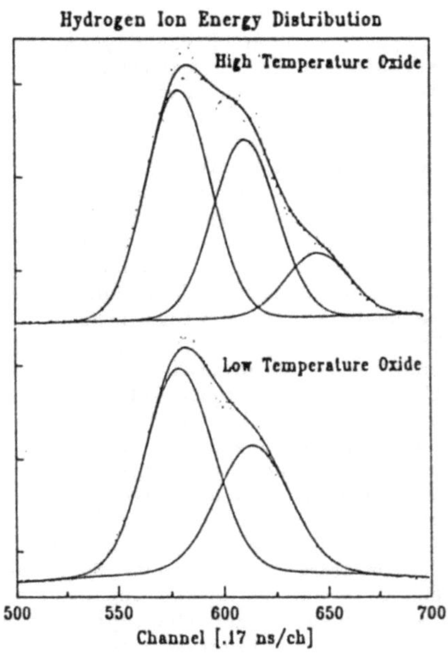


Figure 9 Hydrogen desorption spectra fitted with two and three components for room temp. and high temp. oxides respectively.

| Ion(+) | Ion energy (eV) for: Aluminum edge | Oxygen edge |
|---|---|---|
| H | .6, 1.5, 4.2 | 23.4 |
| O | (weak, broad) | 3.6, 7.0, 11.2 |
| OH | (weak, narrow) | 2.6 |

It is interesting to speculate on the causes for these observations. Because the only hydrogen expected to be native to these surfaces is in the form of hydroxyls, the relatively low kinetic energy of the hydrogen ions observed at the aluminum edge and very low oxygen ion yield at the aluminum 2p edge suggests that the desorbed protons are coming from a process which leaves the oxygen at the surface or possibly desorbing it as a neutral. Ostensibly the most acidic protons would be at the greatest distance from the oxygen atoms and would therefore experience the smallest accelerating force and hence have the lowest kinetic energy.

At the oxygen edge considerably different energy distributions are observed. Only one very high kinetic energy proton is observed while oxygen ions are detected with three well-defined kinetic energies, each with its own spectral dependence. The unusually high kinetic energy of the protons suggest that they were derived from a compact, highly ionized environment. This could be due to poor screening of the oxygen core hole in the oxide, or possibly as a result of the desorption and subsequent free-space dissociation of a hydroxyl. These protons could then be attributed to a Coulomb explosion which occurs too far away from the surface to be mediated by its presence. From the available data a single unambiguous explanation for the oxygen ion kinetic energies is not possible. Based on the observed energies it is tempting to think of oxygen atoms coordinated with one, two and three aluminum atoms for the 3.6, 7.0, and 11.2 eV ions respectively. Another possible explanation could be decay of desorbed hydroxyls to different final states producing oxygen ions at different kinetic energies. Clearly more study is required to understand the desorption process at this edge. The use of isotope labeled oxides and surface titrants should shed some light on the problem.

## IV Perspective:

Our initial investigation of this system by PSD has provided us with a rich data set which shows different spectral dependences and desorbing species for surfaces prepared at high and low temperatures. Given the existing knowledge of this system we believe that these variations are attributed to modification of the bonding site distribution with preparation conditions. The complex behavior of the system will require additional study, particularly at the oxygen edge, to provide detailed descriptions of the desorption mechanisms and bonding sites associated with the different kinetic energy species observed. Further studies using isotope labeled oxides and probe molecules are anticipated which should lend insight to these problems. We have demonstrated however that in principle the acid sites can be quantitatively determined and their response to various reactions studied.

## V Acknowledgments:

This work was performed under the auspices of the U.S. Department of Energy under Contract No. DE-AC02-76CH00016.

## VI References:


1 E.D. Johnson, M.L. Knotek; 1986 NSLS Annual Report
2 D L. Cocke E.D. Johnson, R.P. Merrill; Catal. Rev. 25, 163 (1984)
3 H. Knozinger, P. Ratnasamy, Catal. Rev. 17, 31 (1978)
4 J.B. Peri, R.B. Hannan, J. Phys. Chem. 64, 1526 (1960)
5 E.D. Johnson; PhD Thesis, Cornell University (1985)
6 E.D. Johnson, R.P. Merrill; JVST A 3, 1313 (1985)
7 K. Czanderna; PhD Thesis, Cornell University (1984)
8 H.S. Shin; PhD Thesis, Cornell University (1985)
9 M.R. Howells, Nucl. Instrum. and Methods 177, 127 (1980)
10 E.D. Johnson, R.P. Merrill; To be published
11 P.H. Citrin, J.E. Rowe, S.B. Christman, Phys. Rev. B 14, 2642 (1976)
12 M.L. Knotek, V.O. Jones, V. Rehn, Phys. Rev. Lett. 43, 300 (1979)
13 A. Bianconi, R.Z. Bachrach, S.B.M. Hagstrom, S.A. Flodstrom, Phys. Rev. B 19, 2837 (1979)
14 M.L. Knotek, R.H. Stulen, G.M. Loubriel, V. Rehn, R.A. Rosenberg, C.C. Parks, Surf. Sci. 133, 291 (1983)
15 A. Balzarotti, F. Antonangeli, R. Girlanda, G. Martino, Phys. Rev. B. 29, 5903 (1984)

# Stimulated Desorption from CO Chemisorbed on Cr(110): Sensitivity to Bonding Changes

*N.D. Shinn*[1] *and T.E. Madey*[2]

[1]Sandia National Laboratories, Division 1134,
Albuquerque, NM 87185, USA
[2]Surface Science Division, National Bureau of Standards,
Gaithersburg, MD 20899, USA

## 1. Introduction

Electron stimulated desorption (ESD) experiments using a time-of-flight pulse counting method [1,2] have revealed a correlation between the binding states of carbon monoxide on the Cr(110) surface at 90K and the positive ion yield. Consistent with previous qualitative observations [3] using the electron stimulated desorption ion angular distribution (ESDIAD) method [4], negligible $CO^+$ and $O^+$ desorption signals were measured from the $\alpha_1$ CO molecular binding state which saturates at 1/4 monolayer [3]. For $\theta_{CO} > 0.25$, terminally-bonded molecular CO ( the $\alpha_2$ CO binding state) co-exists with the $\alpha_1$ CO state [3] and both $CO^+$ and $O^+$ ions are observed. In addition, a decrease in the $O^+$ yield near saturation coverages ($\theta_{CO} \cong 0.35$) of the mixed $\alpha_1$ CO + $\alpha_2$ CO adlayer is attributed to intermolecular interactions which reduce the ion desorption probability for the ESD-active $\alpha_2$ CO species.

## 2. Experimental

These experiments, described in detail elsewhere [5], were performed in an ultrahigh vacuum chamber located at the NBS Synchrotron Ultraviolet Radiation Facility (SURF-II). The chamber is equipped for photoemission, photon- and electron-stimulated desorption, ESD with time-of-flight (TOF) detection for simultaneous ion kinetic energy and mass-to-charge (m/e) analysis, ESDIAD, low energy electron diffraction (LEED), Auger electron spectroscopy (AES) and sample cleaning capabilities [1]. Repeated cycles of argon ion bombardment and annealing were used to clean the Cr(110) crystal [3]. CO was adsorbed at 90K by using either a movable microcapillary-array effusive-doser [1] or by ambient exposures measured using the corrected sensitivity of a nude ionization gauge. CO coverages were determined by cross-correlating LEED, AES, photoemission and work function measurements to obtain a CO coverage vs. exposure calibration curve.

A cylindrical mirror analyzer operated for positive ion detection at a constant pass energy of 50 eV and a pulsed, co-axial electron gun operating at ~10 kHz, 100eV, $10^{-5}$ $A{\cdot}cm^{-2}$ beam current density were used to obtain ESD TOF data. The [110] sample normal was aligned with the CMA acceptance cone; i.e., inclined 42.5° with respect to the CMA axis. A +15V crystal bias was sufficient to compress the ion desorption cone (spatially predetermined by ESDIAD [3]) to within the CMA acceptance solid angle [5]. Details of the TOF experiment have been discussed previously [1,5].

## 3. Results

In initial experiments, the ion kinetic energy distributions of the clean and CO-dosed Cr(110) surface were measured to determine the most probable kinetic energies for $O^+$and $CO^+$ desorption. Three observations were made: (1) a sizable $H^+$ and trace $F^+$ ESD signals were found for the nominally clean Cr(110) surface, (2) the total ion desorption current ($H^+ + F^+ + O^+ + CO^+$) initially increases then decreases as $\theta_{CO}$ approaches saturation, and (3) a major fraction of the total ion yield has low ion kinetic energies ($E_k < 3$ eV) at high $\theta_{CO}$. The presence of the $H^+$ ESD signal on the clean and CO-dosed surfaces is attributed to trace quantities of adsorbed impurities (e.g., OH, $H_2O$) with high ESD cross sections but at levels undetectable via photoemission or AES [5]. The increase in total ion yield is due to the contribution of $CO^+$ and $O^+$ from adsorbed CO, specifically $\alpha_2$ CO (see below), whereas the decrease as $\theta_{CO} \longrightarrow 0.35$ is due to a reduction in both the proton yield and the $O^+$ yield. TIme-of-flight data at ion kinetic energies of 1-9 eV show that the ion signal in the range $1\ eV < E_k < 3\ eV$ is due to $CO^+$ ions and that the $O^+$ ions desorb with $E_k \cong 5$ eV.

Figure 1 shows a series of TOF spectra at 5 eV kinetic energy with increasing CO exposure. The calculated position of the $O^+$ ion with respect to the proton signal is indicated by a vertical arrow ($\Delta t$ = 13 μsec). For CO exposures up to 1L (1L = $10^{-6}$ torr sec) only the $\alpha_1$CO binding state is populated [5]. This chemisorbed molecular state is of interest chemically since it exhibits a greatly reduced C-O stretching frequency (1150 - 1330 $cm^{-1}$) relative to terminally-bonded CO ligands (1700 - 2000 $cm^{-1}$) and was proposed to be "lying down" on Cr(110) or significantly rehybridized [3,9]. The spectra in Fig. 1 are quantified in Fig. 2 by plotting the $O^+$ absolute counts, averaged over the five channels nearest to the calculated peak position, vs. CO exposure and derived CO coverage.

In ESD experiments of terminally-bonded CO on other transition metals [6-8], it has been found that the $O^+$ ( and $CO^+$) ion yield typically increases proportionally with $\theta_{CO}$ at low coverages but decreases as $\theta_{CO}$ nears saturation coverage. The behavior shown in Fig. 2 demonstrates that the $\alpha_1$ CO state on Cr(110) is not typical; the lack of any ion signal above the noise level at coverages approaching ~4 X $10^{14}$ $cm^{-2}$ can be used to estimate [5] that the $\alpha_1$ CO ESD ion cross section is reduced by at least a factor of 50 relative to terminally-bonded $\alpha_2$ CO on Cr(110). The onset of the terminally bonded $\alpha_2$ CO on Cr(110) is quite evident with the sharp increase of ESD signal after a 1L CO exposure. This onset correlates well with vibrational [3] and photoemission [9] data which also exhibit features indicative of a new CO binding state after a ~1 L CO dose.

Moreover, Fig. 2 shows that at high total coverages, a decrease in $O^+$ desorption signal is observed even though the CO adlayer consists of mixed binding states. This coverage-dependent decreasing probability for ion desorption has been attributed to adsorbate-

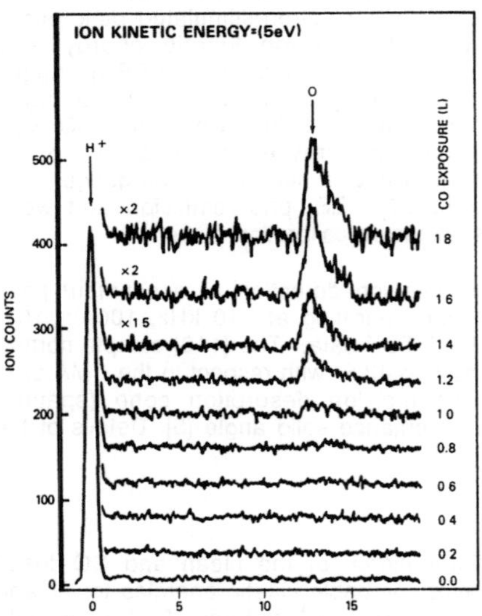


Fig. 1: $O^+$ yield as a function of CO exposure. Expanded vertical scales are used to show the $O^+$ onset

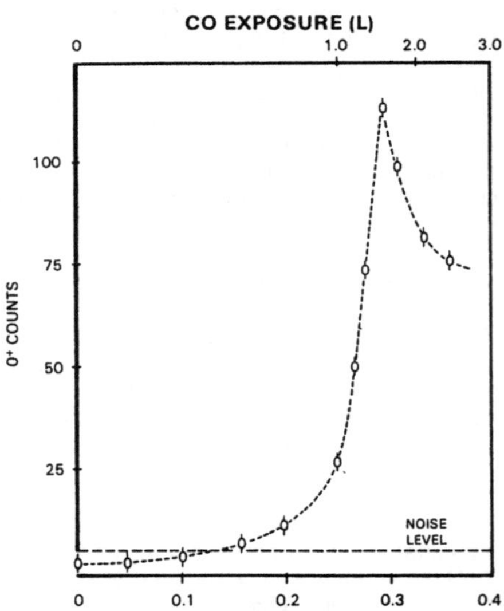


Fig. 2: Quantitative $O^+$ yield as function of exposure and coverage

adsorbate interactions which result in the delocalization of the desorption-inducing electronic state [8, 10, 11]. Our results show that if resonant through-space charge exchange is the cause of this decreasing $O^+$ signal, then this mechanism is not sensitive to the specifics of how CO is bonded to Cr(110). This could be explored further by considering various electronic configurations for CO(ads) in calculations of resonant through-space charge-exchange processes.

## 4. Conclusions

These experiments have shown that (1) the cross section for positive ion desorption ($O^+$, $CO^+$) from the $\alpha_1$ CO binding state on Cr(110) is at least a factor of 50 less than that for the terminally-bonded $\alpha_2$ CO state on the same surface, (2) the $CO^+$ ions desorb with characteristically lower kinetic energies (1-3 eV) than the $O^+$ ions (5-7 eV) for ESD from Cr(110)/$\alpha_2$CO, and (3) the $O^+$ ion yield from $\alpha_2$ CO decreases as the total CO coverage approaches saturation. Thus the positive ion ESD signal from chemisorbed CO on Cr(110) is sensitive to the binding state because of either the molecular orientation or rehybridization of the CO frontier orbitals. In addition, intermolecular interactions that are independent of CO binding state reduce the probability of positive ion desorption at high CO coverages.

## 5. Acknowledgements

This work was supported, in part, by the Office of Naval Research.

## References

1. D. M. Hanson, R. Stockbauer and T. E. Madey, Phys. Rev. B24, 5513 (1981).
2. M. M. Traum and D. P. Woodruff, J. Vac. Sci. Technol. 17, 1202 (1980).
3. N.D. Shinn and T. E. Madey, Phys. Rev. Lett. 53, 2481 (1984); J. Chem. Phys. 83, 5928 (1985).
4. See, for example: Desorption Induced by Electronic Transitions, DIET II, ed. by W. Brenig and D. Menzel, Springer Ser. in Surface Sci., Vol. 4 (Springer-Verlag, Berlin, Heidelberg, 1985), pp. 104-141.
5. N. D. Shinn and T.E. Madey, Surface Sci. 180, 615 (1987).
6. T.E. Madey, Surface Sci. 79, 575 (1979).
7. P. Feulner, H.A. Engelhardt, and D. Menzel, Appl. Phys. 15, 355 (1978).
8. F. P. Netzer and T. E. Madey, J. Chem. Phys. 76, 710 (1982).
9. N. D. Shinn and T. E. Madey, Phys. Rev. B33, 1464 (1986).
10. D. E. Ramaker, J. Vacuum Sci. Technol. A1, 1137 (1983).
11. E. B. Stechel and M. L. Knotek, Surface Sci. 139, 29 (1984).

# Temperature-Dependent PSD of $H^+$ from $H_2O/Si(111)$ (7×7)

*R.A. Rosenberg*[1], *C.-R. Wen*[2], *and D.C. Mancini*[3]

[1]Synchrotron Radiation Center, University of Wisconsin-Madison, 3731 Schneider Drive, Stoughton, WI 53589, USA
[2]Synchrotron Radiation Center and Department of Physics, University of Wisconsin-Madison, Stoughton, WI 53589, USA
[3]Materials Science Program, University of Wisconsin-Madison, 1500 Johnson Drive, Madison, WI 53706, USA

## 1. Introduction

The adsorption of water on the Si(111)(7X7) surface has been studied by a variety of techniques in recent years and the exact nature of the surface bonding and structure remains in doubt. Photoemission studies by Fujiwara [1] and Schmeisser and coworkers [2,3] were interpreted in terms of molecular adsorption. Electron-energy loss (EELS) spectra revealed the presence of Si-H and Si-OH bonds [3,4] which indicated that water dissociated on the surface. The situation is further complicated by the fact that photoemission spectra show differences depending whether the adsorption is done at room temperature (RT) or done at low temperature (LT<100 K) and then annealed to RT [2,3]. However, EELS spectra show no substantial differences between the two methods of preparation. In the only reported study to combine both techniqes, Schmeisser and Demuth concluded that on the the LT-dosed, RT-annealed surface, there existed both molecular and dissociated water, whereas on the RT-dosed surface there existed chemisorbed oxygen, OH radicals, and hydrogen [2,3]. In addition, the photoemission results of Ranke and Schmeisser indicate that on the low-temperature surface, in the low coverage regime, only molecular water exists [2].

We have applied the technique of photon-stimulated desorption (PSD) of ions to study this system. Results from this work not only give further credence to the previously mentioned adsorption models, but also shed new light on mechanisms of ion desorption.

## 2. Experimental

The experimental apparatus used in these studies is similar to one used previously [5]. It consists of a UHV chamber (base pressure = $1 \times 10^{-10}$ torr) equipped with LEED, a time-of-flight ion detector, a total-electron yield (TEY) detector, and a gas-doser system. During TEY measurements the microchannel plates on the ion detector were left on to maintain stability, resulting in a repulsive electric field in the vicinity of the sample. This field prevented low-energy electrons from reaching the TEY detector; consequently, most of the TEY signal is due to higher energy (~92 eV), Si(LVV) Auger electrons. These electrons have a shorter mean-free path than the low-energy, inelastically scattered electrons.

The Si(111) sample was cleaned by heating to T>1400 K, which produced a 7X7 surface as indicated by LEED. The gas-dosing was done in two ways. For coverages in the 0 to 6 Langmuir (L) regime, at LT, the dosing was done in a dynamic fashion by leaking $H_2O$ into the chamber, with the ion pump running, to maintain a nominal pressure of $8 \times 10^{-10}$ torr for an appropriate time. Due to an unknown contribution of ambient water, the low-coverage exposures given here represent a lower limit to the actual exposure. We estimate the actual exposure may be as much as a factor of 2 greater than those given here. The Si(111) surface has a very low sticking coefficient (~.01) at RT. Therefore, the RT dosing was done using a doser tube, to minimize the exposure times necessary to produce monolayer coverage (~100 L). The experiments were performed at the Synchrotron Radiation Center, University of Wisconsin, Madison. For experiments involving low-energy photons (15 to 40 eV), a Seya-Namioka monochromator was used, while the higher-energy experiments were done on both the UW extended-range grasshopper (ERG) beam line and the MARK V grasshopper beamline.

## 3. Results

Results of experiments performed in the energy range 15 to 40 eV are shown in Figs. 1 and 2, and data from higher-energy experiments are shown in Figs. 3 and 4. Figure 1 shows $H^+$ PSD spectra from a thick ice layer condensed at 95 K and then annealed slowly to higher temperatures. The spectrum at T=147 K has a threshold at 21 eV and peaks at 32 eV, which is characteristic of $H_2O$ ice [5]. At T=153 K, the ice starts to sublime, the overall intensity drops by a factor of 3, and the spectrum starts to shift toward lower energies. As the temperature is raised further, the intensity continues to decrease and the shift to lower energies becomes more pronounced.

Figure 2 shows spectra taken under three different exposure conditions. In Fig. 2(a) is the ion yield from the LT surface exposed to approximately 0.4 L $H_2O$; Fig. 2(b) shows the yield from a surface which was covered with an ice layer at 95 K, and then annealed to RT; Fig. 2(c) shows the ion yield after saturation (~100 L) exposure of the RT surface to water. The spectra in 2(b) and 2(c) are similar, with thresholds around 18 eV and relatively broad structure. The LT-dose, RT-anneal spectrum has more structure than the RT-dose spectrum, with shoulders at 22 and 27 eV. The spectrum in 2(c) is similar to that of a thick ice layer [5].

Figure 3 shows two sets of PSD and TEY spectra taken under similar conditions to those in Fig. 2 at photon energies near the Si(2p) threshold (99.5 eV in bulk Si). The $H^+$ yield from the RT-dosed surface has a higher

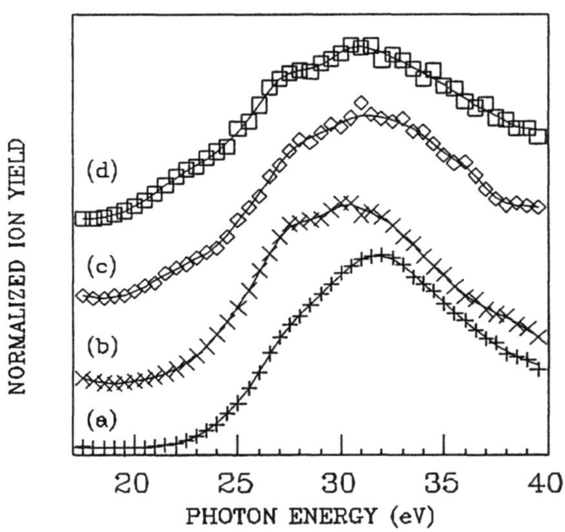


Fig. 1. $H^+$ PSD from a thick ice/Si(111) surface as the temperature is slowly raised: (a) T=147 (X1); (b) T=153 (X3); (c) T=160 (X7); (d) T=290 (X10)

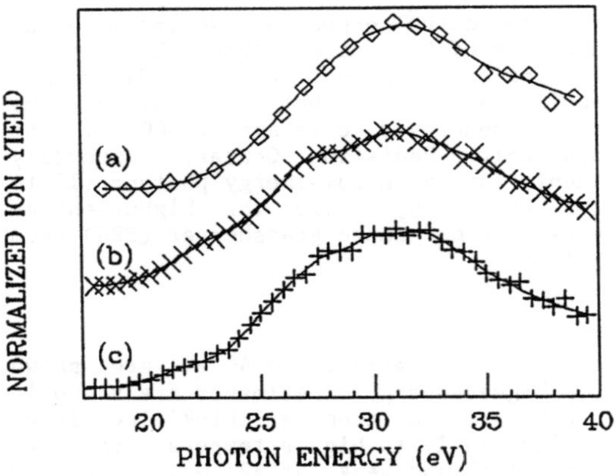


**Fig. 2.** $H^+$ PSD from $H_2O/Si(111)$ after different adsorption treatments: (a) 0.4 L at 95 K; (b) LT-dose, anneal to RT; (c) RT-dose. The vertical scale is the same in all three spectra.

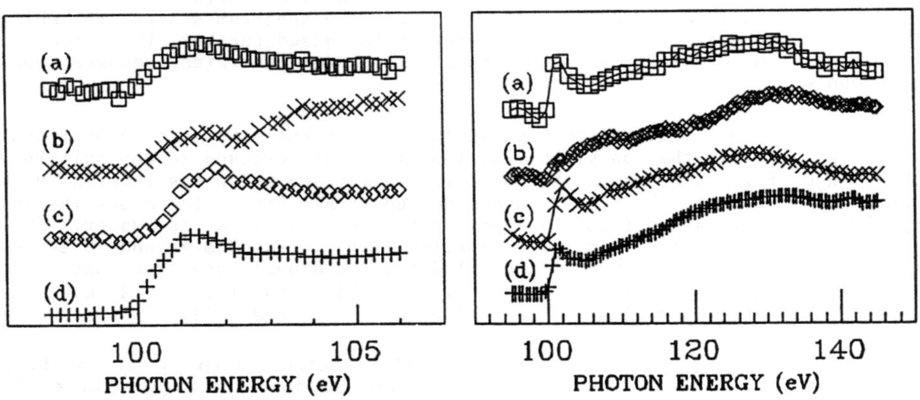


**Fig. 3.** $H^+$ PSD and TEY from $H_2O/Si(111)$ after different adsorption treatments: (a) 0.2 L at 95 K; (b) LT-dose, RT-anneal; (c) RT-dose; (d) TEY from clean surface. The spectra on the right cover an expanded energy range. The vertical scale is the same in all the PSD spectra.

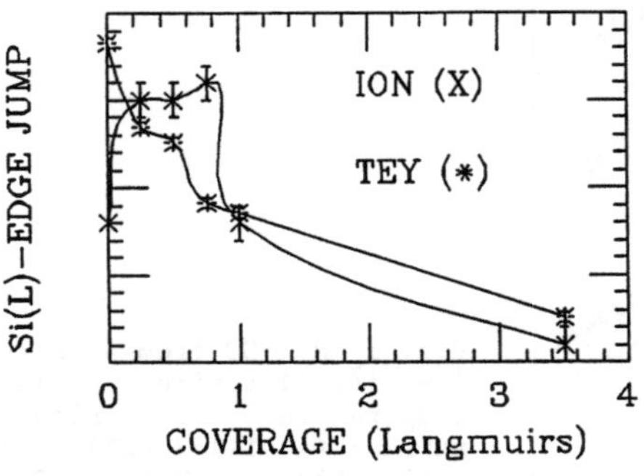


**Fig. 4.** $H^+$ PSD and TEY edge-jumps at the Si(L) edge as a function of water exposure at 95 K. As stated in the text the primary source of the TEY signal is probably Si(LVV) Auger electrons, which have a very short (~5 Å) mean free path.

energy threshold (100.4 eV) than the bulk Si (TEY) spectrum (99.8 eV) in agreement with earlier results [7]. The spectra in Fig. 2(a) and 2(b) each have a threshold commensurate with the bulk Si threshold, but the LT-dosed, RT-anneal surface shows an additional threshold at 102.6 eV.

In Fig. 4 are plotted the TEY and PSD edge-jumps (peak intensity at 101.5 eV minus background intensity at 99 eV) versus coverage. The TEY edge-jump slowly decreases with increasing coverage, while the PSD edge-jump saturates quickly, at approximately 0.2 L and then drops off rapidly after approximately 1 L exposure.

## 4. Interpretation

Our interpretation of the PSD spectra is based on the previous EELS and photoemission results, and may be summarized as follows:

1. At low temperatures and low coverages, water is adsorbed molecularly.

2. After a thick ice layer is adsorbed at low temperatures and then annealed to room temperature, there exists a mixed phase of both molecular and dissociated $H_2O$.

3. Room-temperature dosing results primarily in dissociated $H_2O$ and perhaps some oxide formation.

Using this model, we may interpret the spectra of Figs. 1-4. In Figs. 1 and 2, the low energy threshold (~18 eV), which is close to the threshold observed in $H^+$ PSD from condensed $SiH_4$ [6], is due to excitation of a Si(3s) electron which results in Si-H bond breaking. The shoulder at 27 eV in Fig. 2(b) could be due to O-H bond breaking via excitation of a repulsive state of the OH complex. The spectrum resulting from a 0.4 L exposure at 95 K is similar to that of ice (Fig. 1, T=147) and is due to molecular $H_2O$.

The spectra in Fig. 3 may also be interpreted in terms of this model. For the RT-dosed surface, the 100.4 eV threshold is due to excitation of a Si(2p) electron in a Si-H surface complex. The energy threshold is higher than bulk Si, probably because hydrogen has a higher electronegativity (2.1) than silicon (1.8) [7]. The PSD spectrum observed after exposure of the 95 K surface to 0.2 L $H_2O$ is very similar to the bulk Si (TEY) spectrum. The physisorbed water has very little influence on the neighboring Si atom and therefore the Si(2p) photoabsorption energy is not perturbed. The mechanism by which $H^+$ in the $Si(H_2O)$ complex desorbs could be either an interatomic Auger decay or dissociation of $H_2O^+$ in a metastable state which then predissociates to give $H^+$. On the LT-dosed, RT-annealed surface, the lower-energy threshold is due to photoexcitation of a Si(2p) electron in a $Si(H_2O)$ and/or a SiH surface complex, while the higher energy (102.4 eV) threshold results from excitation of a SiOH surface species and is quite close to the threshold observed by Bianconi and Bauer in SiO X-ray photoabsorption [8].

Finally, the edge-jump data of Fig. 4 demonstrates two important points. First, it indicates that the PSD-active site saturates at relatively low coverage (~0.2-0.4 L). This could be due to adsorption at defects or step edges. Second, it shows that there is little contribution of an indirect, X-ray-excited, electron-stimulated desorption (XESD) mechanism to the PSD spectra of Fig. 3. If the XESD mechanism was an important consideration for this system, the TEY and PSD edge-jumps would show the same behavior as a function of coverage. This is clearly not the case in Fig. 4 and is in contrast to previous work by Jaeger and coworkers who interpreted $H^+$ PSD data from the $NH_3$/Ni system in terms of a XESD mechanism [9].

## 5. Summary

1. The data presented here are consistent with the model developed by Schmeisser and coworkers [2,3] for water adsorption on Si(111)(7X7). At LT, water adsorbs molecularly, even at submonolayer coverages. After an ice layer is annealed to RT, water exists in both molecular (Si-$H_2O$) and dissociated (Si-H, Si-OH) forms. On the RT-dosed surface water dissociates to form Si-H bonds; we see no evidence for Si-OH.

2. The Si(L)-edge threshold excitation energies and their assignments are:

| Energy (eV) | Si bonding site |
|---|---|
| 99.8 | Si-$H_2O$, bulk Si |
| 100.4 | Si-H |
| 102.6 | Si-OH |

3. Our data shows that core-level excitation of a Si atom may result in $H^+$ desorption, even when the hydrogen atom was not directly bound to the Si atom. Evidently, interatomic Auger decay between the oxygen atom bound to the Si atom and the Si atom results in a metastable surface Si-$OH^+$ or Si-$H_2O^+$ complex which then decays to produce $H^+$.

Acknowledgements: This work was supported by the National Science Foundation and a grant from the Research Corporation.

## References

1. K. Fujiwara, Surf. Sci. 108, 124 (1981).
2. W. Ranke and D. Schmeisser, Surf. Sci. 149, 485 (1985).
3. D. Schmeisser and J.E. Demuth, Phys. Rev. B 33, 4233 (1986).
4. H. Ibach, H. Wagner, and D. Bruchmann, Solid State Comm. 42, 457 (1982).
5. R.A. Rosenberg, V. Rehn, V.O. Jones, A.K. Green, C.C. Parks, and R.H. Stulen, Chem. Phys. Lett. 80, 488 (1981).
6. R.A. Rosenberg, J. Vac. Sci. Technol. A4, 1463 (1986).
7. R.A. Rosenberg, P.J. Love, V. Rehn, I. Owen, and G. Thornton, J. Vac. Sci. Technol. A4, 1451 (1986).
8. A. Bianconi and R. Bauer, Surf. Sci. 99, 76 (1980).
9. R. Jaeger, J. Stohr, and T. Kendelewicz, Surf. Sci. 134, 547 (1983).

# Electronic Sputtering and Its Application to SIMS Analysis of Biomolecular Monolayers

*B.U.R. Sundqvist, A. Hedin, P. Håkansson, G. Jonsson, and G. Säve*

Division of Ion Physics, Department of Radiation Sciences,
Uppsala University, Box 535, S-751 21 Uppsala, Sweden

## 1. Introduction

Fast ions, i.e. ions with a velocity larger than the Bohr velocity, interact primarily with the electrons in a medium. In a good conductor like a metal, no late effects of the interaction will be found. However if the medium is an insulator, where the relaxation of the deposited electronic energy is slow, fast ion impact will lead to atomic displacement effects, i.e. local bulk damage of atomic order and surface erosion. In 1974 MACFARLANE and coworkers found that fast ions like fission fragments can induce desorption of intact ions of fairly large fragile and thermally labile molecules like small biomolecules [1]. They called their method Plasma Desorption Mass Spectrometry (PDMS). During the last few years that method has been used to study larger organic molecular ions than ever in the history of mass spectrometry [2,3]. About seven years ago it was established that PDMS is based on electronic sputtering from a layer of biomolecules [4].

## 2. Electronic Sputtering of Biomolecules

So far experimental data in the field concern yields of secondary ions. The time-of-flight method for mass analysis of secondary ions has mainly been used [5]. Recently we were able, for the first time, to measure sputter yields of intact neutral molecules of the amino acid leucine when a multilayer film of leucine molecules was bombarded with various fast heavy ions from the Uppsala tandem accelerator [6,7]. The neutral yields were found to be 10 000 times larger than the yields of positive secondary ions for primary ions of 90 MeV $^{127}$I. In Fig. 1 the yields of intact neutrals and positive and negative molecular

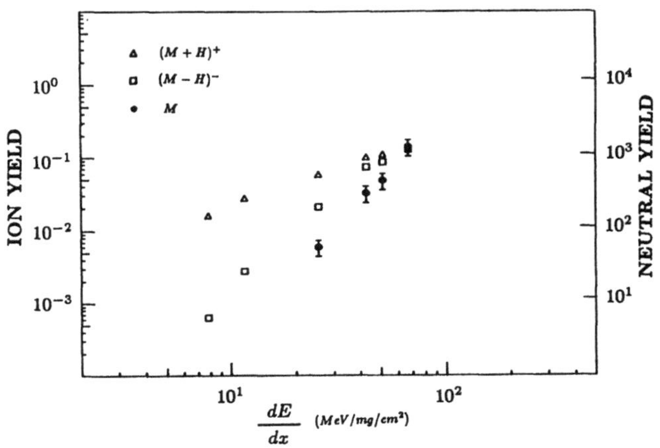


Fig. 1 The yields of intact ions and neutrals of the amino acid leucine (MW 131) as a function of electronic stopping power (see text) of different primary ions

ions are plotted as a function of electronic stopping power of the primary ions 78.2 MeV $^{127}I$, 48.7 MeV $^{79}Br$, 35.7 MeV $^{58}Ni$, 19.7 MeV $^{32}S$, 9.9 MeV $^{16}O$ and 7.4 MeV $^{12}C$. The energies of these primary ions were selected so as to keep the primary ion velocity constant. In this way the spatial distribution of the deposited electronic energy around the ion track is the same for all the different primary ions. The parameter varied is the energy density in the ion track. As can be seen in Fig. 1, the scaling of yields is different in the case of ions and neutrals. In this study it was also found that the absolute yields and the scaling were very reproducible for the neutrals while less so for secondary ions.

Secondary ion production from Langmuir-Blodgett films was studied with the aim to understand the influence of sample film thickness on secondary ion yields. In connection with those studies the idea came up to use marker molecules in the film in order to monitor the origin of molecular ions in the mass spectra. In Fig. 2 one set of results of these studies is shown [8]. The molecules used in the experiment were three different fatty acids, namely behenic acid (MW 340.6), arachidic acid (MW 312.5) and stearic acid (MW 284.5) [8]. In the inset in Fig. 2 the structure of the film is shown. As in the experiment described above different primary ions of the same velocity were used to bombard the sample. The primary ions used in this experiment were 78.2 MeV $^{127}I$, 19.7 MeV $^{32}S$, 9.9 MeV $^{16}O$ and 7.4 MeV $^{12}C$. As can be seen, the relative contribution of the different layers to the molecular ion intensities varies in a systematic way with the different primary ions. The integrated intensity is normalized in the four spectra. The interpretation is that the molecular ions can originate from layers below the surface and that the primary ion erodes a crater which is at least 150 Å deep in the case of 78.2 MeV $^{127}I$. In a third type of experiment a sample was irradiated and the yield of the molecular

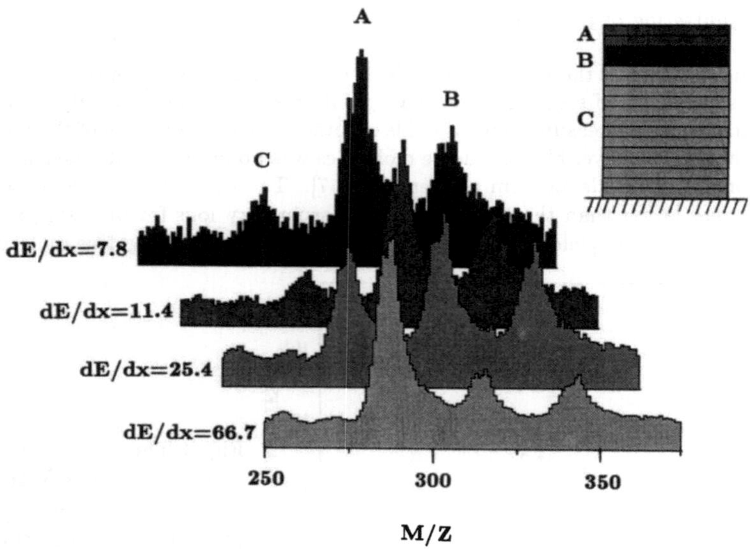


Fig. 2 Negative ion mass spectra of a mixed sample film, consisting of 2 layers of molecule A (arachidic acid, MW 312.5) on top of 2 layers of B (behenic acid, MW 340.6) on top of 13 layers of molecule C (stearic acid, MW 284.5) for four different primary ions. The four stopping powers namely 7.8, 11.4, 25.4 and 66.7 MeV/mg/cm$^2$ correspond to C, O, S and I ions

ion was studied as a function of dose (number of particles hitting the sample per $cm^2$). From the exponential slope of such a curve a damage cross section can be extracted [9] and if this microscopic area is assumed to be circular, a corresponding damage radius. Such an experiment was made with behenic acid in a surface layer position and with $^{127}I$ of 78.2 MeV as incident ions. The radius extracted was 60 Å. This radius may be related to the radius of a formed crater but might also be larger because this parameter may be influenced by late effects like "thermal" degradation.

If these briefly described findings are combined, the following picture of the process evolves. A fast heavy ion like a fission fragment impinging on a multilayer film of organic molecules forms a crater of about 50–100 Å radius at the surface and of at least 150 Å depth. This corresponds well to the total number of ejecta measured and the results of the Langmuir-Blodgett experiments. Molecular ions are formed from molecules originating from this volume of ejected material. The shape of the crater as well as the part of the volume from which the molecular ions originate varies with the energy density in the ion track. This is also the case for the ratios of neutral to positive ions to negative ions which in the case of 90 MeV $^{127}I$ on a multilayer leucine target is 10 000:10:1.

## 3. Biomolecular Mass Spectrometry with Fast Ion Impact and Time-of-Flight Analysis

The secondary ions produced in electronic sputtering are used for mass spectroscopic analysis of the sample. The secondary ions are mass analysed with the time-of-flight method which is particularly attractive when high mass analysis and monolayer samples are considered [10,11]. The PDMS technique can be regarded as a SIMS technique where the slow (keV) primaries are replaced by fast (MeV) ions. The applications in biomolecular mass spectrometry of electronic sputtering (PDMS) has been reviewed recently [12]. However the field develops rapidly and the singly most important development since that review was written is the replacement of multilayer samples prepared by the electrospray technique [13] with samples where the molecules are adsorbed to a polymer surface. Nitrocellulose [14] is so far the most useful backing in this sense. In Fig. 3 positive ion mass spectra of the molecule phospholipase $A_2$ (MW 13980) is shown. The three spectra correspond to: A. A sample prepared by the electrospray technique, i.e. a multilayer sample, B. A sample where a droplet of a sample solution is applied to a nitrocellulose backing and dried, i.e. it is probably also a multilayer sample, C. The sample in B has been rinsed with a water solution. In case C most likely a monolayer is left consisting of molecules bound to nitrocellulose. In the former cases the molecules "available" for desorption are mainly sample molecules bound to other sample molecules. In view of the finding discussed above, i.e. molecular ions may originate from molecules below the surface layer, the position of the molecule in the layer may also be relevant for the result. As can be seen in Fig. 3, the spectra in A and B show much broader molecular ion peaks,indicating that desorption and ionization occurs at higher energy densities than i C. Another difference in the spectra is the relative abundance of peaks corresponding to multiply charged molecular ions. The multiply charged ions are most abundant in case C and it may be connected to a lower energy density at which the molecular ions are formed. The underlying assumption is that the multiply charged ions are less stable than the singly charged ones due to the stored coulombic energy. It is important to remember that the multiply charged ions are much more efficiently detected due to their higher velocity at impact on the detector surface. The similarities of A and B as compared to C are probably associated with the multilayer and monolayer character of the samples.

In applying a sample solution to a nitrocellulose surface, which is hydrophobic and on a microscopic scale rather rough, it is important to lower the surface tension in the applied sample solution so that actually a large enough number of sample molecules reach the surface for adsorption to occur. This is an essential part of a new procedure for sample preparation by adsorbing large molecules to a nitrocellulose surface. With this method we have recently been able to observe larger intact molecular ions than before, namely molecular ions of the protein pepsin with a molecular weight of about 35 000 [15].

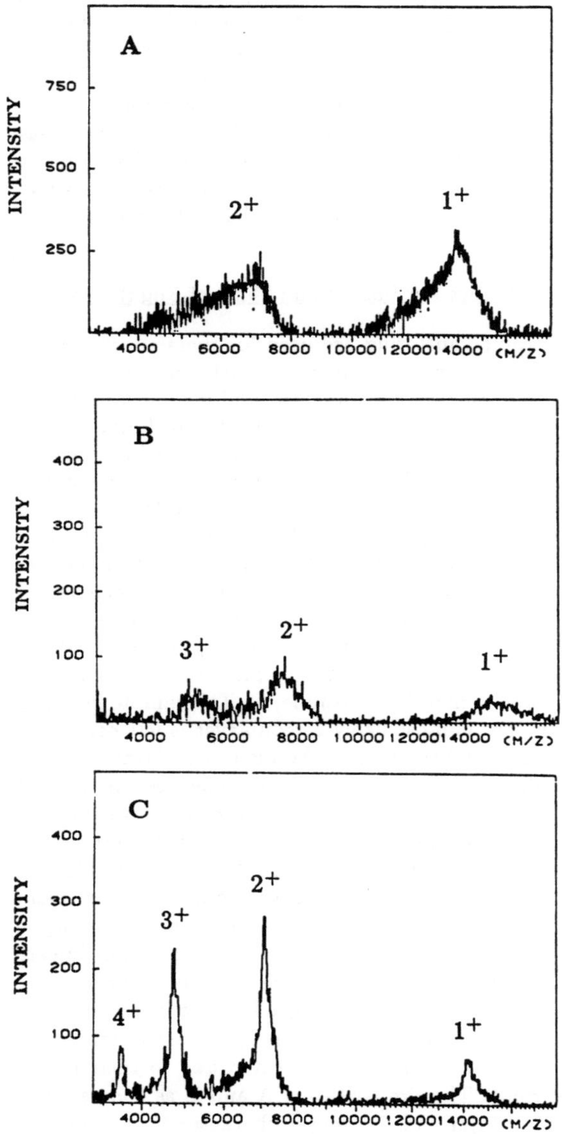


Fig. 3 Positive ion mass spectra of phospholipase $A_2$. (MW 13 980). Peaks labelled $n^+$ correspond to molecular ions in different charge states.

A. Electrosprayed, multilayer film.
B. A droplet of sample solution is deposited and dried on a nitrocellulose backing. This is most likely a multi-layer film.
C. The sample in B has been rinsed in water

4. References


1. D.F. Torgerson, R.P. Skowronski, R.D. Macfarlane: Biophys.Res.Commun. 60 (1974) 616
2. B. Sundqvist, I. Kamensky, P. Håkansson, J. Kjellberg, M. Salehpour, S. Widdiyasekera, J. Fohlman, P.A. Peterson and P. Roepstorff: Biomed.Mass Spectrom. 11 (1984) 242
3. B. Sundqvist, P. Roepstorff, J. Fohlman, A. Hedin, P. Håkansson, I. Kamensky, M. Lindberg, M. Salehpour and G. Säve: Science 226 (1984) 696
4. B. Sundqvist (ed.): in *Proceedings of the Nordic Symposium on Ion Induced Desorption of Molecules from Bioorganic Solids*, Nucl.Instr.Meth. 198 (1982)
5. P. Håkansson and B. Sundqvist: Radiat.Eff. 61 (1982) 179
6. M. Salehpour, P. Håkansson, B. Sundqvist and S. Widdiyasekera: Nucl.Instr.Meth. B13 (1986) 278
7. A. Hedin, P. Håkansson, M. Salehpour and B.U.R. Sundqvist: Phys.Rev. B35 (1987) 7377
8. G. Säve, P. Håkansson, B.U.R. Sundqvist, R.E. Johnson, E. Söderström, S.E. Lindquist and J. Berg: Submitted for publ. Appl.Phys.Lett., March 1987
9. M. Salehpour, P. Håkansson and B. Sundqvist: Nucl.Instr.Meth. B2 (1984) 752
10. B. Sundqvist, A. Hedin, P. Håkansson, I. Kamensky, M. Salehpour and G. Säve: Int.J.Mass Spectrom.Ion Proc. 65 (1985) 69
11. B. Sundqvist: in *Sputtering by Particle Bombardment III*, ed. by R. Behrisch and K. Wittmaack, Topics in Appl.Phys. (Springer Verlag, to be published 1987)
12. B. Sundqvist and R.D. Macfarlane: Mass.Spec.Rev. 4 (1985) 421
13. C.J. McNeal, R.D. Macfarlane and E.L. Thurston: Anal.Chem. 51 (1979) 2036
14. G. Jonsson, P. Håkansson, B. Sundqvist, P. Roepstorff, P. Nielsen, K.E. Johansson, I. Kamensky and M. Lindberg: Anal.Chem. 58 (1986) 1084
15. A.G. Craig, Å. Engström, H. Bennich, I. Kamensky and B.U.R. Sundqvist: (To be published)

# Limits of Applicability of the Popcorn Model for Desorption of Large Biomolecules by Fast Ion Impact

*P. Williams*[1] *and B.U.R. Sundqvist*[2]

[1]Department of Chemistry, Arizona State University, Tempe, AZ 85287, USA

[2]Division of Ion Physics, Department of Radiation Sciences, Uppsala University, S-751 21 Uppsala, Sweden

The popcorn model, introduced to explain fast ion desorption of intact, large biomolecules, is discussed with particular reference to desorption of low mass and very high mass molecules. In the model, desorption of a large biomolecule is shown to result from rapid expansion of the molecule, following excitation of most vibrational modes by the intense low-energy electron flux emanating from the primary ion track. It is shown that the popcorn effect can readily induce desorption of individual molecules as light as 500 u, and also that the high mass limit for desorption may lie beyond that presently attained. It is argued that desorption of molecules with molecular weights as high as $10^5$ might be possible by fast heavy ion impact.

## 1. Summary of the Popcorn Model

Recently Williams and Sundqvist /1/ proposed a new mechanism for the desorption of large biomolecular ions by fast (MeV/nucleon) ion impact. The model was designed to reconcile several features of the fast ion desorption process first demonstrated by MacFarlane and his colleagues /2/. Most important of these is the observation of desorption of intact large ions, as heavy as 34,000 u /3/, despite the intense radiolysis that must occur in the region of direct ionization within 5-10 Å radius of the incident ion trajectory -- the so-called "infratrack". Because electronic excitation, or ionization, events are efficient in breaking intramolecular bonds, it seemed clear that intact large ions must come from a region, the "ultratrack", sufficiently far from the primary ion trajectory that they have a low probability of experiencing either a direct ionization event arising from the primary ion passage or ionization by a high-energy secondary electron ejected laterally from the infratrack region. That desorption proceeds from the ultratrack is also apparent from the very high sputtering yields -- over 1,000 amino acid molecules/incident ion -- determined by Sundqvist et al /4/. But this presents the problem, first, of identifying an efficient excitation process far from the high-energy infratrack, and second, of explaining how a molecule containing several thousand atoms, if excited, can be set into coherent motion with sufficient kinetic energy to overcome a surface binding energy of several eV without rupturing any intramolecular bonds /1/.

The central suggestion of the Williams-Sundqvist model is that the primary excitation process leading to molecular desorption is that of vibrational excitation arising from the flood of low-energy secondary electrons that irradiates the ultratrack. Some 25% of the energy deposited by the primary ion into electronic excitation and ionization appears in the ultratrack as secondary electrons; the initial energy distribution of these electrons peaks at zero energy, and energy losses as they sweep through the ultratrack ensure that low energy electrons dominate at all locations. Electrons with energies too low to excite electronic transitions (a few eV) are known in the radiation chemistry community as "subexcitation" electrons. For a highly ionizing primary projectile such as a $^{252}$Cf fission fragment, typically a nuclide of mass 100-150u and energy ~ 1 MeV/nucleon, the energy loss is approximately $10^3$ eV/Å, so that over a ~ 20 Å depth corresponding to a layer of small spherical protein molecules each containing ~ 1000 atoms (mass ~ 10,000 u), secondary electrons carry some 2,500 eV into the ultratrack region. These electrons can be considered "hot" electrons moving in the conduction band of the protein layer. The electrons lose energy in inelastic scattering events, and are also decelerated in the electric field arising from the highly-ionized infratrack core. Electrons thermalized to the bottom of the conduction band are reaccelerated back to the core, inducing further excitations on the way, and either are captured in the core or pass through it to embark on a further transit of the ultratrack.

Cross-sections for vibrational excitation of polar bonds by subexcitation electrons are surprisingly high -- $\geq 10^{-16}$ cm$^2$ /5/ -- and a 1 eV electron can effectively excite up to 10 normal modes ($h\nu$ ~ 0.1 eV). Thus a few hundred low-energy electrons can excite all 3,000 vibrational modes of a 1,000-atom molecule. Williams and Sundqvist pointed out that this multiple excitation, occurring almost instantaneously on the time scale of molecular motion, must result in a rapid expansion of the molecule. The expanding molecule, pushing against the substrate, generates enough kinetic energy to push itself off the surface, much like an exploding grain of popcorn. A simple calculation yields velocities in the range 200-300 m/s, which correspond to kinetic energies 2-6 eV for a molecule of mass 10,000 u /1/. The model allows calculation of sputtering yields consistent with the high experimental values, explains the origin of the kinetic energy and momentum of the large desorbed molecules, and in particular provides a desorption mechanism which operates without any high-energy electronic excitation processes which would be expected to induce bond rupture and fragmentation.

The purpose of the present paper is to examine the range of applicability of the popcorn model. In particular we discuss the low mass and high mass limits of the process. The low mass limit for expulsion of a single molecule arises because there is a limit to the expulsion velocities that can be generated by the expansion, for a given degree of excitation, so that kinetic energy scales with molecular mass; the high mass limit must be dictated by the physical extent of the ultratrack and the availability of enough electrons for excitation. Finally, we will speculate on whether large molecule desorption by slow (keV/nucleon) ions, a process still not quantitatively understood, may be driven by expansion processes.

## 2. Surface Binding Energies and Surface Bond-breaking

For peptide or protein molecules, in multilayers or adsorbed on polymer films, bonding is predominantly due to hydrogen bonds and ion-pair interactions. Barlow and Thornton /6/ estimate that ion-pair interactions in proteins are similar in strength to hydrogen bonds -- ~ 0.2 eV -- and we assume the same here. Amino acids contain ~ 10-20 atoms and have masses ~ 100-200 u, so that a molecule of mass 10,000 u (say), will contain about 100 amino acid residues and something over 1,000 atoms. Assuming each residue can form at least one hydrogen bond or ion-pair bond with a neighbor, and roughly half the residues are located on the exterior of the molecule, we can estimate that perhaps 20-30 residues are involved in bonding to the substrate, giving a surface binding energy ~ 4-6 eV. To desorb the molecule, either it must acquire a kinetic energy in excess of this value, or these hydrogen bonds must be broken *selectively*, i.e. in a process that does not directly break the strong covalent intramolecular bonds. Such a selective bond-breaking process, induced by secondary electrons, was invoked in a model of Hedin et al./7/, but no mechanism was discussed. *Vibrational* excitation by subexcitation electrons provides such a mechanism, because a 0.2 eV bond cannot exist if each of the atoms involved in the bond is excited by ~ 0.3 eV (~ 0.1. eV/normal mode). Thus the desorption process need only supply sufficient kinetic energy to move the molecule out of bonding range of the surface before the local excitation is dissipated into the substrate.

## 3. Low-mass Limit

In their paper /1/, Williams and Sundqvist showed that the popcorn effect is easily capable of expelling entities with molecular weights ~ 10,000 u, and suggested that expulsion of smaller molecules may be understood if they are treated as assemblages; i.e. a roughly spherical collection of ~ 100 amino acid molecules (molecular weight ~ 100 u) should acquire velocities and thus kinetic energies similar to those acquired by a single large molecule of mass 10,000 u, and might therefore be desorbed as an assemblage, subsequently dissociating. However, it is interesting to consider how small a molecule may be desorbed individually by the popcorn effect. We assume, as in ref. 1, that all vibrational modes in the molecule are instantaneously excited to at least the (v=1) state. Then we estimate the desorption velocity from the degree of expansion of the molecule $\Delta x$ (which moves the center of mass a distance $\Delta x/2$) and the expansion time $\Delta t$, which is limited by the fundamental frequency of vibration of the entire molecule.

In ref. 1 the extent of expansion was estimated from thermal expansion coefficients for polymers to be ~ 3-5% for excitation of all bonds to the (v=1) level (equivalent to a temperature jump ~ 1000 K). For a discrete molecule we can arrive at very similar estimates by considering bond length changes due to anharmonicity, which are the microscopic basis of thermal expansion. For bond energies ~ 3 eV, comparable to single bond energies in organic molecules, bond *lengths* in diatomic molecules increase by ~ 2-3% on excitation to the (v=1) level, but anharmonicity of the many "soft" modes in polyatomic molecules -- bending modes, for example -- is significantly greater, and it is these soft modes that predominantly account for thermal expansion. It therefore seems reasonable to use a figure of 3-5% again for the expansion of quite small molecules. If we restrict our discussion to roughly spherical molecules, then the displacement $\Delta x$ of the center of mass due to excitation, is $\alpha r$, where $r$ is the molecular radius and $\alpha$ the degree of expansion. For $\alpha$ roughly constant, $\Delta x$ scales with $r$.

The expansion time, $\Delta t$, is limited by the fundamental vibrational frequency of the molecule as a whole. For a sphere, this fundamental vibration is the breathing mode for which the frequency, $\nu$ is given as $1/2r(E/\rho)^{1/2}$ /8/, where $E$ is Young's modulus for the molecule as a whole and $\rho$ is the density (assumed equal to 1 $g/cm^3$ here). The expansion time, $\Delta t$, is taken as 1/4 of the vibrational period, or $r/2(\rho/E)^{1/2}$ (an expansion time 1/2 the period was erroneously used in ref. 1). For spherical molecules, the expansion velocity ($\propto \Delta x/\Delta t$) is then roughly independent of r, but does scale with $E^{1/2}$. For large proteins, in ref. 1, a value for E ~ $10^{10}$ $N/m^2$ was used following the work of Suezaki and Go /8/, who chose this value to be intermediate between the value of $10^9$ $N/m^2$ typical of bulk polymers, and the value of $10^{11}$ $N/m^2$ determined experimentally for the $\alpha$-helix of poly-l-alanine. For small molecules, a value for E of $10^{11}$ $N/m^2$ seems appropriate. For small molecules, then, expansion velocities a factor of 3 greater than for large molecules are attainable by the popcorn mechanism -- 600-900 m/s for expansion by 3-5%-- and thus a kinetic energy ~ 1 eV could be attained by molecules as light as 250 - 500 u. This energy should be comparable to, or greater than, binding energies for adsorbed small molecules. The model applies equally well to linear molecules if we replace the molecular diameter, 2r, by the length, l, and if the long axis of the molecule is perpendicular to the surface as is the case in desorption of Langmuir-Blodgett films /9/. Clearly for molecules this small the continuum elastic theory used for larger molecules is questionable, and detailed calculations using molecular dynamics must be employed. Such calculations are in progress /10/. But the present discussion strongly suggests that the popcorn model can explain desorption of light, as well as very heavy, molecules.

## 4. High-mass Limit

Turning to the high-mass limit for desorption, we can make some simple arguments as follows. Assuming an irradiation density of low-energy electrons sufficient to excite every normal mode in the molecule, the ejection velocity arising from expansion is independent of molecular size for a fixed value of E (for large molecules a Young's modulus ~ $10^{10}$ $N/m^2$ should be used /8/). Therefore, if all surface bonds are ruptured due to the vibrational excitation, large molecules desorb as easily as small ones. The argument then hinges on the size of the molecule relative to the extent of the ultratrack. From the measured sputtering yield for the peptide LHRH -- 580 molecules of mass 1,182 u ejected per incident ion /11/ -- it may be deduced that ejection occurs efficiently from a volume ~ $10^3$ $nm^3$. There is strong evidence from studies of Langmuir-Blodgett films that ejection occurs from depths up to 10 nm /9/, hence from an area ~ 100 $nm^2$. If we assume that a central infratrack region ~ 1 nm in diameter must be excluded from discussion because bond rupture is most efficient in this region, then molecules some 4-5 nm in diameter, spanning the annular ultratrack, should be ejected intact. A spherical molecule 5 nm in diameter with a density ~ 1.5 $g/cm^3$, has a mass around 60,000 u.

As the molecular size increases, there is an increasing probability that the intramolecular bonds in that fraction of the molecule distant from the ion impact point will not be vibrationally excited. The fraction of the molecule that is not excited scales with the radius, r, and the velocity therefore decreases in proportion to $r^2$, but the mass increases as $r^3$, so kinetic energy continues to increase with molecular size (scaling as $M^{1/3}$). However, some fraction of the surface bonds may now remain intact; in addition, the lower velocity of the molecule may ultimately result in re-formation of surface bonds as the excitation in the ultratrack dissipates before the molecule moves out of bonding range. No good estimate for the lifetime of the excitation yet exists, but there is good reason to expect that it should be relatively long in protein multilayer samples, because intermolecular coupling should be weak. For proteins adsorbed on polymer films (the polymer substrate being similarly excited by the primary ion passage) the excitation lifetime may be shorter, particularly for extensively cross-linked (rigid) polymers.

The largest measured sputtering yield (~ 580 LHRH molecules/ion /11/) corresponds to a total ejected mass of ~ 685,000 u. 7 molecules of mass ~ 100,000 u (~ 6 nm diameter) close-packed in a plane would have a similar mass. It seems arguable therefore that molecules as heavy as 100,000 u might be ejected by fast heavy ions. The difficulties in detecting such ions to date may well lie in the known inefficiency with which such heavy projectiles eject secondary electrons from an electron multiplier surface.

A final comment on very large molecules is that they should in fact be rather resistant to fragmentation resulting from bond cleavage in electronic excitation/ionization events. In part, this is because many of the functional groups in the molecule are subject to extensive tertiary interactions, for example, interchain hydrogen bonding or non-bonded (repulsive) interactions with near neighbors, which can hold the atoms in place following a bond cleavage long enough for the bond to re-form. Secondly, the complexity of the potential energy surfaces in a very large molecule offers extensive possibilities for electronic excitation to be dissipated through level-crossing processes rather than causing bond cleavage. This means that very large molecules may be able to survive being in close proximity to the infratrack, and the high-energy processes occurring there, thus increasing the irradiation efficiency leading to desorption.

## 5. Expansion Phenomena in Other Desorption Processes

A final word concerns the question of whether the popcorn mechanism may be operating in other desorption processes, in particular, in desorption by keV ion impact. The greatest success at low energies has come in desorption from solutions in relatively volatile liquids such as glycerol. Although it is felt that the role of the glycerol may be in some way to buffer the radiation damage process, allowing extended sputtering, it is interesting to speculate whether the liquid plays a direct role in the ejection process. Michl and co-workers /12/ and Murray and Rabalais /13/ have suggested models for sputtering of condensed volatile substances, including water, that involve rapid expansion of a gasified region as a supersonic jet. Such an expansion, which should similarly occur in sputtering of glycerol, may well supply the coherent "push" that is required to get very large molecules off the surface without fragmentation arising from too violent impacts. The large molecule ejection process may thus be expansion-driven just as in the "popcorn" case, but in this case it is the matrix that is expanding and the large molecules are merely passengers. An obvious signature of such an effect would be a constant ejection velocity, i.e. an ejection kinetic energy that scales with molecular mass.

## 6. Conclusion

The popcorn model appears successfully to explain many of the experimentally observed features of fast ion desorption. A unique signature of the process is that it produces ejection velocities which are roughly independent of molecular size, so that kinetic energies increase with molecular mass for a given degree of excitation, making ejection of very large molecules a potentially efficient process. This article has attempted briefly to show that the process may work for relatively small molecules (molecular weight ~ 250-500 u), and has indicated that the high mass limit for the process may be as high as 100,000 u.

## 7. References


1. P. Williams and B. Sundqvist, Phys. Rev. Lett. 58 (1987) 1031
2. R.D. Macfarlane and D.F. Torgerson, Science, 191 (1976) 920
3. B. Sundqvist, Unpublished.
4. B. Sundqvist, A. Hedin, P. Hakansson, M. Salehpour, G. Save, S. Widdyasekera, and R.E. Johnson, Nucl. Instrum. Methods Phys. Res., B14 (1986) 429
5. F. Linder, in "Electronic and Atomic Collisions", edited by G. Watel (North-Holland, Amsterdam, 1978), p.51
6. D.J. Barlow and J.M. Thornton, J. Mol. Biol. 168 (1983) 867
7. A. Hedin, P. Hakansson, B. Sundqvist, and R.E. Johnson, Phys. Rev. B31 (1985) 1780
8. Y. Suezaki and N. Go, Int. J. Pept.Protein Res. 7 (1975) 333
9. G. Bolbach, S. Della-Negra, C. Deprun, Y. Le Beyec, and K.G. Standing, Rapid. Comm. Mass Spectrom. 1 (1987) 22
10. R. Lucchese, Texas A & M University. Private communication.
11. B. Sundqvist, A. Hedin, P. Hakansson, G. Jonsson, M. Salehour, G. Save, S. Widdiyasekera and T. Roepstorff, in "Proceedings of the Third International Conference on Ion Formation from Organic Solids", edited by A. Benninghoven, Springer Proceedings in Physics Vol 9, (Springer-Verlag, New York, 1986)
12. D.E. David, T.F. Magnera, R. Tian, J. Stulik and J. Michl, Nucl. Instrum. Methods Phys. Res. B14 (1986) 378
13. P.T. Murray and J.W. Rabalais, J. Am. Chem. Soc. 103 (1981) 1007

# Photon-Stimulated Desorption in High Electric Fields

*S. Jaenicke*[1], *A. Ciszewski*[1,*], *J. Dösselmann*[1], *W. Drachsel*[1], *J.H. Block*[1], *and D. Menzel*[2]

[1]Fritz-Haber-Institute of the Max-Planck-Society, Faradayweg 4–6, D-1000 Berlin 33, Germany
[2]Physics Department, Technical University Munich, James-Franck-Str., D-8046 Garching, Fed. Rep. of Germany

## 1. Abstract

**In photon-stimulated desorption (PSD) of molecules, the initial excitation can branch into several channels, leading to deactivation or to the desorption of neutrals, molecular ions and charged fragmentation products. The presence of a high electric field in particular influences the reaction path. Also, the field stabilizes adsorption states which are not observed in its absence. Examples are the adsorption of molecular hydrogen at temperatures as high as 120 K and the formation of whiskers grown from water vapour in a high electric field.**

## 2. Introduction

In photon-stimulated desorption (PSD), an initial electronic excitation either in the substrate or the adsorbate leads to desorption [1, 2]. The subsequent evolution of the system after the primary excitation contains a competition of processes which constitute very interesting physics. The transfer of electronic excitation into molecular motion which finally leads to desorption along a repulsive potential curve will only take place if the electronic excitation stays localized on the bond concerned for a sufficient time. One approach to further improve the understanding of these competing processes is to selectively influence part of them. Here this is done by the application of a high static electric field. To first order, the field acts only on ionic states by bending their potential down at larger distances [3]. Therefore, the lifetime of ions near the surface is considerably decreased which will lead to an enhanced desorption yield. On the other hand, the high field can also produce new adsorption states by a process known as field adsorption [4].

Schematically, the time evolution of the excitation follows this pattern [5]:

| | |
|---|---|
| initial electronic excitation | ($\approx 10^{-16}$ s) |
| fast redistribution of electronic energy | ($\approx 10^{-15}$ s) |
| a slower displacement of nuclear positions, resulting in ejection | ($10^{-14}$ – $10^{-13}$ s) |
| a modification of the escaping species as it recedes from the surface | ($\approx 10^{-13}$ s) |

During the last three steps, recapture and reneutralization of ionic states can take place. While the high field will change the transition probabilities for the first two steps to a certain extent due to a polarization-induced change of overlap, the main effects will arise on the time scale of the last two steps. The driving force for the expulsion of an

---

* On leave from: Institute of Experimental Physics, University of Wrocław, Pl 50-205 Wrocław, Poland

ionic state is the repulsion between the ion and the neighbouring atom cores which is partially screened by the electron cloud. These internuclear forces decay very rapidly with distance. A steady external field will help to remove ionic states from the surface. In order to be significant, such external fields have to be in the order of several V/Å. Fields of this magnitude can be generated if field emitters are used as the substrates. These are sharp needles where the radius at the tip is less than 500 nm.

## 3. Experimental

The apparatus used for photon-stimulated desorption from field emitters has been described elsewhere [6]. The field is generated by applying a high voltage to the sample, a field emitter, which is in close proximity of a grounded counter electrode. Detection and mass analysis of the desorbed ions is done by time-of-flight (TOF) analysis, utilizing the time structure of the synchrotron radiation. The measurements were performed during 'single bunch' operation at the storage rings DORIS at DESY in Hamburg and BESSY in Berlin. Monochromatic light in the 5 to 35 eV range was produced with a normal incidence grating monochromator, and a high intensity TGM was used between 30 and 120 eV.

## 4. Results

Hydrogen on a metal is perhaps the simplest system to study. Previous experiments with 'white' synchrotron light [7] had shown that it was possible to obtain PSD from field emitters even though the monitored area was less than $10^{-10}$ cm$^2$. It was hoped that the disadvantage of the very small target size would be partially offset by an increased desorption yield because the field was expected to suppress the reneutralization reaction and increase the desorption probability. Unfortunately, at field strengths high enough to significantly increase the photo-desorption yield, field ionization of the gases in the reaction chamber set in. This led to a high background ion count in the TOF spectra, again reducing the sensitivity. This problem was overcome by reducing the gas pressure. The best results were obtained when the gas under examination was dosed in continuously to a pressure of $2 \cdot 10^{-10}$ mbar. This low working pressure made a background pressure of less than $5 \cdot 10^{-11}$ mbar mandatory.

Spectra of field-assisted PSD of hydrogen from tungsten show three peaks, identified as $H^+$, $H_2^+$, and $H_3^+$. At low field strength, $H_2^+$ dominates; at higher field the amount of $H^+$ increases [Fig. 1]. $H_3^+$ is the product of a field-assisted surface reaction [8, 9]; it is observed only in a narrow range of field strengths. That $H_3^+$ is indeed a surface product is confirmed by the fact that its concentration depends critically on the crystallographic plane; also on certain (mostly fcc) metals like Ni, no $H_3^+$ is found [10]. The desorption signal decreases rapidly with increasing temperature and is virtually absent above 160 K. Since thermal desorption of atomically chemisorbed hydrogen takes place at higher temperatures, e.g., at $230 \pm 10$ K for the $\alpha$-state on Ni(110) and even higher for the $\beta$-states on nickel and tungsten [11, 12], it has to be concluded that the hydrogen ions observed in field-assisted PSD originate from a molecularly bound state. This idea was further confirmed by measurements on mixtures of $H_2$ and $D_2$. No isotope exchange was found in the desorbing ions. These measurements were performed at 80 K because the yield of molecular ions decreases dramatically at higher temperatures. Increased thermal dissociation of the desorbing ions leads to $H^+$ and $D^+$ so that the ion yield carries no information of the molecular state on the surface. In another experiment a Nb field emitter was used as the substrate for hydrogen desorption studies. It is known that Nb dissolves hydrogen very rapidly [13] so that surface hydrogen can be replenished from the bulk. Nevertheless, no enhancement of the hydrogen ion yield over that found with W or Ni tips was observed.

These results show that the desorption cross section of atomic hydrogen is smaller than $10^{-24}$ $cm^{-2}$, the detection limit of our apparatus. We therefore conclude that the observed signals stem from field-adsorbed hydrogen molecules. Thus, the high electric field appears to have a twofold influence in this case: it produces the adsorbed ground state, and possibly enhances its desorption yield after excitation. It is not possible to check the second effect experimentally since the corresponding state exists only in a very narrow range of field strength.

The assignment that the desorbing species is field-adsorbed $H_2$ and not H is independent of the nature of the initial excitation. In order to gain information on it, the ion yield was measured as a function of photon energy. In these measurements, rhodium was used as the substrate and the sample temperature kept at 80 K. The result is shown in Fig. 2. Increased desorption of $H^+$ and $H_2^+$ sets in around 50 eV. The ratio $H^+/H_2^+$ depends only on the field strength, indicating that both ions originate from a common source and are generated by a common primary excitation. Steps in the yield spectrum can be correlated with the $4p_{1/2}$ and $4p_{3/2}$ transitions in the rhodium substrate. This is expected both for a KF mechanism [14] (interatomic Auger decay of a substrate core level) and for XESD [15] (desorption induced by secondary electrons, e.g., Auger electrons and photoelectrons, which are generated in the metal bulk under

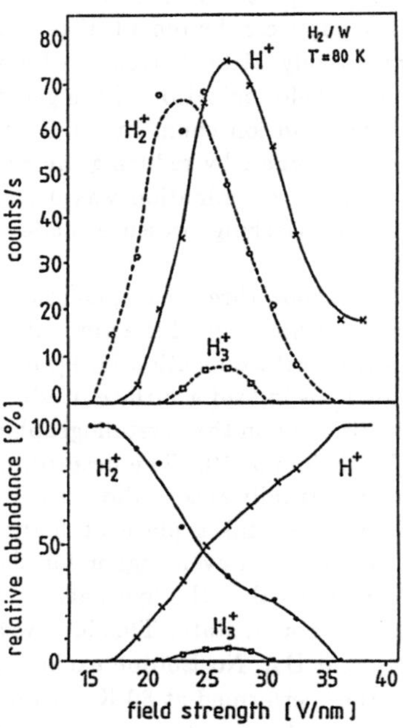


Fig. 1 Field strength dependence of the hydrogen species

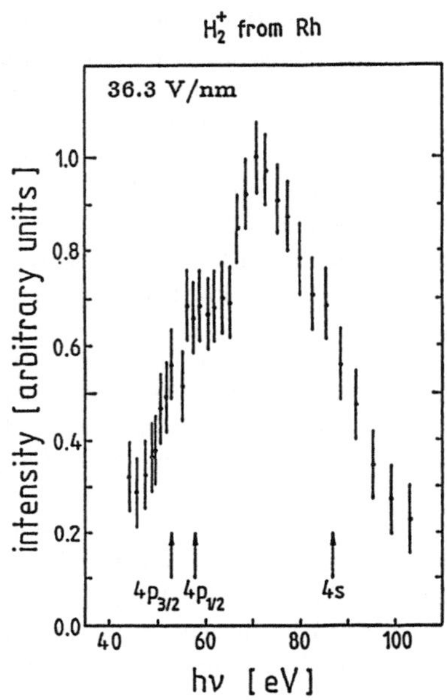


Fig. 2 PSD spectrum at 36.3 V/nm. The arrows indicate inner core energies of rhodium metal

photon irradiation). An interatomic Auger decay is important only in cases of negligible intraatomic decay channels. In desorption from clean metal surfaces, there are abundant valence band electrons available for filling the core hole. Therefore, an interatomic process seems unlikely to be responsible for desorption. Also, the KF mechanism depends on an overlap of the adsorbate valence states with the substrate core levels, which should be extremely small for molecular $H_2$. Even the action of the high field which strips the surface atoms of part of their charge and also pulls the electron density of the adsorbate towards the metal should not change this situation drastically. We therefore believe that XESD by rhodium Auger electrons is the most likely process leading to desorption. It is also possible that multiple excitation of $H_2$ leads to desorption, similar to the process observed recently by Auer et al. [16] for CO on ruthenium under zero field conditions. Such contributions cannot be excluded from the available data because of the poor resolution of the spectra. Future measurements on metals which do not possess core levels in this range (e.g. Ni) will help to resolve this question.

Another interesting system is water layers grown in the high electric field on a metal or semiconductor surface. Without the field, water is known to first form dimers and islands until at higher coverage, a monolayer and later, multilayers are formed [17]. The by far dominating desorption product observed is $H^+$ when photons or electrons of up to several hundred eV energy are used [18]. Even under ion bombardment, the predominant desorbing species are protons. If thick ice layers are sputtered with 6 keV $Ar^+$ or $Kr^+$ ions, protonated water clusters have been observed [19]. Similar clusters are also found if in the presence of a static field, PSD is performed with UV radiation. A typical mass spectrum for PSD from a blunt nickel tip is shown in Fig. 3. Clusters with up to 16 water molecules have been observed. Since the desorption yield for all these cluster ions depends on the wavelength in the same way, it is assumed that a common excitation step leads to desorption. The observed threshold for desorption lies below 7 eV. This value is considerably below the ionization energy of water in the

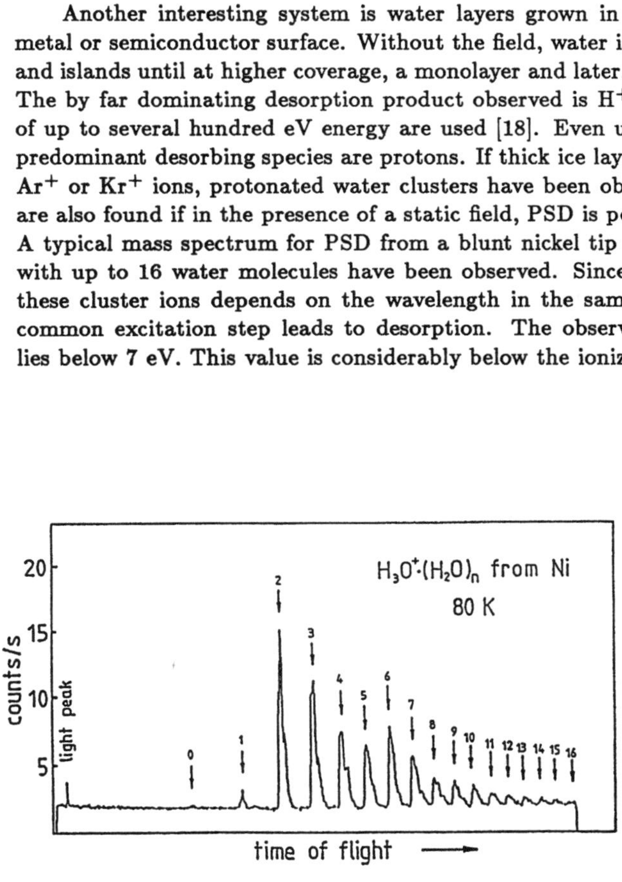


Fig. 3 TOF-mass spectrum of the products of field-assisted PSD from a water-covered Ni tip. The number of water molecules surrounding a central $H_3O^+$ is indicated.

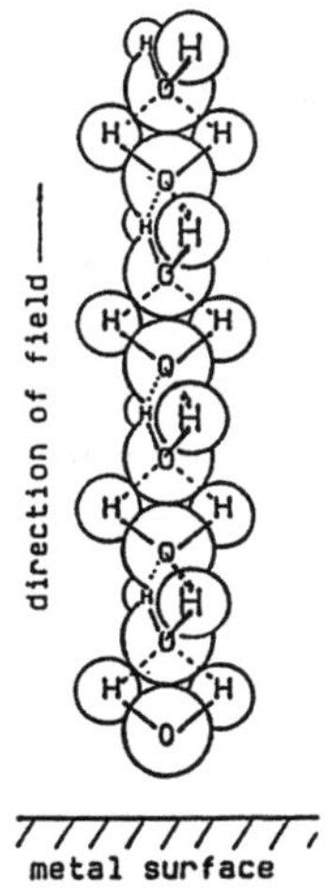


Fig. 4 Model of a water whisker grown in an electric field

gas phase [20] but close to the absorption maximum of liquid water [21]. This wide absorption band is thought to arise from exciton formation which have been shown to lead to desorption in rare gas solids.

It is well known that water tends to form clusters on metal surfaces [17]. Christianson et al. [19] calculated the size distribution of protonated water clusters desorbed by energetic ions. Their model uses the RRKM theory to calculate the energy redistribution into the various vibrational modes. Whereas their results fit well with their experiment, the size distribution observed in field-assisted PSD is entirely different. In particular, the molecular ion $H_3O^+$ and the dimer $H_3O^+ \cdot H_2O$ are almost absent. The maximum in the distribution is found for the trimer $H_3O^+ \cdot 2H_2O$; the distribution falls slowly to higher masses. Following a model proposed by Anway [22], we assume that under the influence of the field, water crystallites at the surface form more or less branched 'whiskers' [23]. A model of such a whisker is shown in Fig. 4. Starting from such a structure, the different desorbing ions are formed simply by breaking two O–H bonds. This model therefore explains that the cluster size distribution is independent of the excitation energy.

## 5. Conclusions

Photon-stimulated desorption in high electric fields has been observed for several systems. The desorption mechanisms are not yet fully understood. For the desorption of protonated water clusters from ice structures grown in the electric field the initial excitation lies below the ionization threshold of the water molecule. The spectra are independent of the substrate (Ir, $IrO_2$, Ni) and resemble the absorption spectrum of liquid water. The desorption of hydrogen ions from metals contains a minor contribution with a desorption threshold below 10 eV. The main contribution stems from more energetic excitations above 50 eV. In the desorption spectra, structures related to inner core transitions in the metal, rhodium, are visible. The most likely explanation is that desorption occurs via an excitation of field-adsorbed $H_2$ by electrons liberated in the bulk metal upon photon absorption, possibly superimposed on direct photoexcitation of the adsorbate.

## 6. Acknowledgements

A.C. thanks the Max-Planck-Society for a research fellowship. The support provided by the staff members of HASYLAB and BESSY is gratefully acknowledged. This work has been funded by the German Federal Minister for Research and Technology under contract number 05390 FX B2 TP3.

## 7. References

1. *DIET I*, N. H. Tolk, M. M. Traum, J. C. Tully, T. E. Madey, Eds., Springer Series in Chemical Physics, Vol. 24, (Springer, Heidelberg 1980)
2. *DIET II*, W. Brenig, D. Menzel, Eds., Springer Series in Surface Sciences, Vol. 4, (Springer, Heidelberg 1985)
3. R. Gomer: *Field Emission and Field Ionization: Harvard Monographs in Applied Science*, **9**, (Harvard University, Cambridge, Mass. 1961)
4. E. W. Müller, S. B. McLane, J. A. Panitz: Surf. Sci. **17**, 430 (1969)

5. J. C. Tully: In Ref. [1], p. 31
6. S. Jaenicke, U. Weigmann, J. R. Pitts, W. Drachsel, J. H. Block, D. Menzel: Chem. Phys., in press
7. W. Drachsel, U. Weigmann, S. Jaenicke, J. H. Block: In Ref. [2], p 245
8. N. Ernst, J. H. Block: Surf. Sci. **126**, 397 (1983)
9. N. Ernst, J. H. Block: Phys. Rev. **B29**, 7092 (1984)
10. C. F. Ai, T. T. Tsong: J. Chem. Phys. **81**, 2845 (1984)
11. K. Christmann, F. Chehab, V. Penka, G. Ertl: Surf. Sci. **152/153**, 356 (1985)
12. P. W. Tamm, D. L. Schmidt: J. Chem. Phys. **54**, 4775 (1971)
13. *Gmelins Handbuch der Anorganischen Chemie: Niobium*, B1, System No. 49, (Verlag Chemie, Weinheim 1970)
14. M. L. Knotek, P. J. Feibelman: Surf. Sci. **90**, 78 (1979)
15. R. Jaeger, J. Stöhr, T. Kendelewicz: Phys. Rev. **B28**, 1145 (1983)
16. S. Auer, P. Feulner, D. Menzel: Phys. Rev. **B35**, 7752 (1987)
17. P. A. Thiel, T. E. Madey: submitted for publication in Surface Science Reports
18. R. A. Rosenberg, V. Rehn, A. K. Green, P. R. LaRoe, C. C. Parks: In Ref. [1], p 247;
D. Coulman, A. Puschmann, H. P. Steinrück, P. Feulner, D. Menzel: ECOSS-9 (1987); to be published
19. J. W. Christiansen, I. S. T. Tsong, S. H. Lin: J. Chem. Phys. **86**, 4701 (1987)
20. L. Karlson, L. Mattsson, R. Jadrny, R. G. Albridge, S. Pinchas, T. Bergmark, K. Siegbahn: J. Chem. Phys. **62**, 4745 (1975);
A. M. Köhler: PhD-Thesis, University of Hamburg, 1987
21. L. R. Painter, R. D. Birkhoff, E. T. Arakawa: J. Chem. Phys. **51**, 243 (1969)
22. A. R. Anway: J. Chem. Phys. **50**, 2012 (1969)
23. S. Jaenicke, A. Ciszewski, W. Drachsel, U. Weigmann, T. T. Tsong, J. R. Pitts, J. H. Block, D. Menzel: J. Physique (Paris) **47**, C7-343 (1986)

# Observations of Photon-Induced Gas Desorption from Technological Materials Using Synchrotron Light

*E.M. Williams*

Department of Electrical Engineering and Electronics,
The University of Liverpool, P.O. Box 147, Liverpool, L69 3BX, U.K.

## 1. INTRODUCTION

The advent of intense light beams from synchrotron sources provides a ready opportunity to redress the dearth of reliable data on the efficiency of the process of photon-induced gas desorption from solid surfaces. Paradoxically, the operation of vacuum within synchrotron machines is itself greatly influenced by the process, as was recognised in the pioneering work of Fischer, Mack and co-workers during the 1960's /1/. With the widespread development of electron storage rings in recent years, there has been a great deal of interest in assessing the extent of self-cleaning within machines, and in examining the contributions of surface pre-treatments such as high temperature bakeout or plasma conditioning /2/. These considerations for the provision of vacuum walls with low desorption characteristics are equally relevant to the operation of controlled fusion machines, since the gaseous impurities can lead to serious plasma contamination. Indeed, the process can possibly lead to a runaway situation as successive waves of desorbed impurity act to stimulate further photon desorption. The present programme of experimental work is directed particularly towards elucidating the operation of the JET plasma experiment at Culham Laboratory where the radiation spectrum spans an energy range around 0.5 to 10 keV at a power level of a few watts per $cm^2$ /3/. These conditions have been simulated using the Soft X-ray beam line at the Synchrotron Radiation Source (SRS), Daresbury Laboratory.

## 2. EXPERIMENTAL

The strategy in the experiment was to employ the whole of the synchrotron spectrum emanating from the machine, within a 4 mrad horizontal spread, to irradiate a sample at normal incidence. By observing the changes in partial pressure accompanying irradiation, to proceed on the basis of known pumping speeds and photon fluxes to evaluate the neutral desorption efficiency expressed in terms of the number of desorbed molecules per incident photon /4/. It may be noted that following this practice, the desorption yields are characteristic of the sample material in equilibrium with its vacuum environment. The samples were contained in a stainless steel desorption cell, and a strip some 5 x 0.6 $cm^2$ of the total surface area (6 x 6 $cm^2$) was irradiated with synchrotron light. The desorption cell communicated with the storage ring via a carbon window, with a platinum mirror interposed between the window and the source. By altering the setting of glancing angle at the mirror, thereby moving the beam line and the desorption cell, it was possible to effect a high-energy, cut-off on the spectrum and so gain some insight of any energy dependence. The measurements were all obtained with glancing angles of $0.5^o$ and $2^o$ and the spectra corresponding to these settings are shown in Fig. 1. For clarity

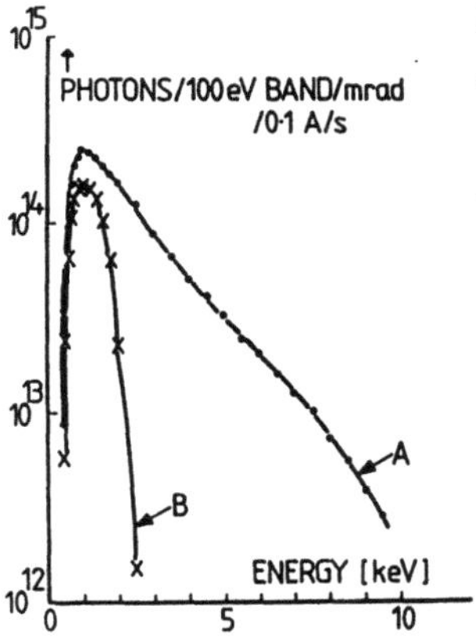


Fig. 1 The photon flux with mirror settings of (A) 0.5°, and (B) 2°. The estimated total photon fluxes, integrated over all energies, are 10.7 x $10^{15}$ and 5.0 x $10^{15}$ /mrad/0.1A/s, with mean energy 2.2 and 1.0 keV.

of presentation, the narrow component of response just below the carbon K edge at 282 eV is omitted in the display. The significant number of photons in this low-energy spike is included in the estimates of photon fluxes quoted in the figure caption. Typical values of the flux density and power density at the two mirror settings, taking 200 mA as a typical circulating current in the storage ring, are in the region of 2.9 x $10^{16}$ photons/$cm^2$/s, 10 W/$cm^2$ (mean energy 2.2 keV) and 1.3 x $10^{16}$ photons/$cm^2$/s, 2.1 W/$cm^2$ (mean energy 1 keV).

## 3. RESULTS AND DISCUSSION

Our experiments to date have been conducted with stainless steel 304L, inconel and graphite, with an ambient pressure ~ 1 x $10^{-7}$ torr reflecting an unbaked vacuum system. All samples were, however, pretreated by vacuum bakeout at 400°C prior to incorporation in the test. Preliminary inspections revealed an initial fast rise in pressure with the introduction of the photon flux in near step fashion at each sample. This was facilitated by the mechanical movement of the beam stopper in the immediate vicinity of the storage ring. In absolute terms this fast activity corresponded to a yield in the region $10^{-4}$ mol/photon. The evolution of the pressure during continuous irradiation pointed also to the occurrence of beam-induced thermal outgassing and beam-induced depletion of desorption signal. The exact scale of these additional, slower, processes was, of course, dependent on the input power level, but the influence of the thermal outgassing was markedly most pronounced for graphite (involving significant CO production).

The species of desorbed gas with the instantaneous response in direct photon desorption were $H_2$, CO, $CO_2$ and $CH_4$. This is in keeping with earlier experience involving both electron and photon stimuli at surfaces /4,5/. The findings of steady state desorption efficiencies with stainless steel are collected in Fig. 2 which includes data at the two mirror positions and contrasts the reponse of vacuum pre-bakeout (400°C) with a non-treated sample. The energy influence over the restricted mean-energy range in the test is not strong; at most there is seen a tendency for the efficiency to increase. CO and $CO_2$ contribute the largest desorption yields for the two sample conditions, with largely similar magnitudes in both cases. The main effect of the pretreatment would seem to advance the hydrogen desorption; this is in keeping with the

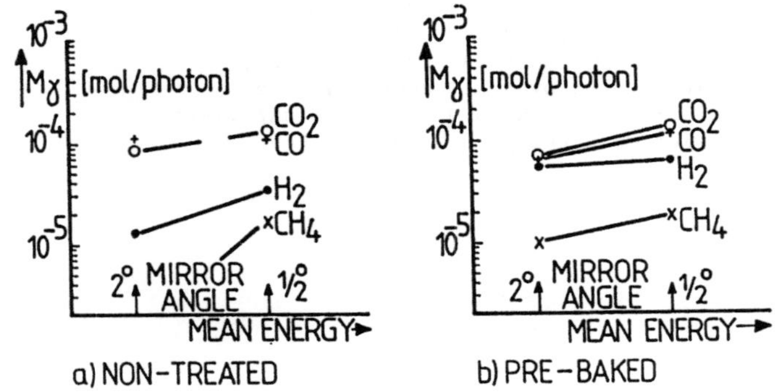


Fig. 2 Determinations of desorption efficency $M_\gamma$ with stainless steel; a) sample untreated, b) sample pre-treated by vacuum bakeout.

general finding of the increasing preponderance of hydrogen following bulk or surface treatment of the material. With an inconel sample (pre-baked) the results were quantitatively similar to stainless steel, while for pre-baked graphite the steady state desorption yields of the leading CO and $CO_2$ species were lower at around $2.5 \times 10^{-5}$ mol/photon.

The action of beam-induced depletion is illustrated in Fig.3a which shows a continuous recording of the $CO_2$ partial pressure at the untreated stainless steel sample. Here, following the initial rapid response of direct photon-activated desorption, the signal gradually decreases to the steady state level. The analysis of the depleting component (Fig.3b) identifies a desorption cross-section of $7 \times 10^{-19}$ cm$^2$ which is quite comparable with the findings in electron-stimulated desorption /6/. The coverage in the desorption state, determined using the earlier measured yield is $1.4 \times 10^{-14}$ cm$^2$, i.e. in the region of a few tenths of a monolayer.

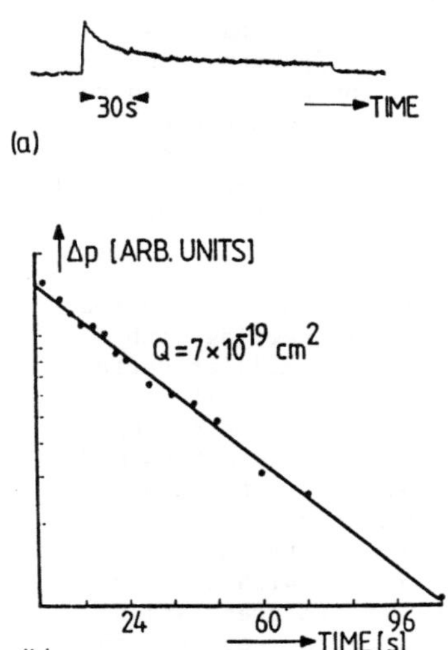


Fig. 3 Photon-induced gas depletion from stainless steel. a) Trace of $CO_2$ partial pressure with $4 \times 10^{16}$ photons/s. b) Semi-logarithmic plot of the depleting component.

The steady state levels of desorption efficiency are to be seen in terms of a dynamic balance between beam-induced depletion of gas sites, of both surface and sub-surface origin, with replenishment actions by adsorption from the ambient gas and by diffusion from within the bulk. With an base pressure ~ $10^{-7}$ torr there is a ready dominance from the gas phase component, but this will change with the provision of U.H.V. conditions which is now being pursued. The measured yields are lower than those that have been attained in recent simulations of storage ring operation /7/, which is a reflection on the significantly greater photon flux density arranged in the present tests. In terms of the mechanisms involved with photon-stimulated desorption, the values of desorption efficiency seen here, as with the magnitude of the desorption cross-section, are quite consistent with the two stage process of photoelectron generation followed by electron-stimulated desorption.

## 4. ACKNOWLEDGEMENTS

The work is a joint project between the University of Liverpool and Culham Laboratory, U.K.A.E.A. We acknowledge the support of S.E.R.C., Daresbury Laboratory and Staff in this work.

## REFERENCES

1. G.E. Fischer and R.A. Mack: J. Vac. Sci. Technol. 2, 123 (1965)
2. B.A. Trickett: Vacuum 37, in the press
3. G. Duesing and K.J. Dietz: J. Vac. Sci. Technol. A, 3, 1151 (1985)
4. E.M. Williams, F. Le Normand, N. Hilleret, and G. Dominichini: Vacuum 35, 141 (1985)
5. R.S. Vaughan-Watkins and E.M. Williams: Proc. 8th Vac. Cong. and 4th Int. Conf. Solid Surf. Vol. II, p.387
6. P.A. Redhead, J.P. Hobson and E.V. Kornelsen: The Physical Basis of Ultra-High Vacuum (Chapman and Hall, London 1968)
7. O. Gröbner, A.G. Mathewson, H. Störi, P. Strubin and R. Souchet: Vacuum 33, 397 (1983)

# Part VII

# Insulators

# The Origin of Positive Ions and Excited Neutrals in Electron Stimulated Desorption from Alkali Halides

*R.E. Walkup, Ph. Avouris, and A.P. Ghosh*

IBM Thomas J. Watson Research Center,
Yorktown Heights, NY 10598, USA

## Introduction

The interaction of electron and photon beams with insulating materials is of substantial current interest. Alkali halides are the most ionic of the ionic insulators, and they have comparatively simple geometric and electronic structures, thus they are useful prototypes for basic studies. When energetic electrons or photons are incident on alkali halides, particles are desorbed with high efficiency. The flux of particles leaving the bombarded region includes ground-state neutrals, excited-state neutrals, and positive ions. The vast majority of the desorbed particles are ground-state neutral atoms and molecules, which are ejected via bulk-defect mediated processes [1]. The ground-state yields can be remarkably large - for 1 keV electrons incident on NaCl at T=300 C, roughly 5 Na atoms and 5 Cl atoms are desorbed per incident electron [2,3]. The desorption mechanisms for the ground-state particles have been qualitatively understood for some time. The initial excitation caused by the incident radiation is electron-hole generation in the bulk. The holes (neutral halogen atoms) are incorporated into stable defects called H centers which can diffuse to the surface, where neutral halogen atom desorption can occur. Neutral alkali atoms can be formed on the surface by recombination of electrons and alkali ions, and these neutral atoms can then thermally desorb. This process appears to be limited by the diffusion of electrons (as F centers) to the surface [4]. The very large yields of ground state neutrals are related to the fact that the initial excitation results in a stable defect in the bulk, and desorption can occur much later after defect diffusion to the surface. The time delay due to the diffusion of defects to the surface has been directly observed in experiments using pulsed electron beams [4,5].

In order to gain a more complete picture of stimulated desorption from these materials, it is also important to understand the origin of the excited-state neutrals and positive ions. The mechanism most often invoked to explain positive ion desorption from ionic insulators is the Knotek-Feibelman mechanism [6], in which the formation of a long-lived multi-hole state on an anion causes ion expulsion by repulsive Coulomb interactions. In NaCl for example, $Cl^+$ can be formed by core ionization of $Cl^-$, followed by Auger decay. The $Cl^+$ ion can be regarded as a localized two-hole state, which is expected to be sufficiently long-lived for desorption to be possible [7,8]. The repulsive forces which would drive the desorption process arise from the reversal of the Madelung potential. However, for the alkali halides, there are a number of experimental observations which are difficult to explain by this mechanism. For example, the yield of positive alkali ions is generally greater than that of positive halogen ions, and the mass distributions contain clusters such as $NaCl^+$ and $Na_2Cl^+$. Difficulties have also been suggested by theoretical investigations [9]. Trajectory calculations indicate that the Coulomb repulsion mechanism fails to expel positive ions from normal lattice sites of alkali halides. That is, even if re-neutralization does not occur, positive ions are not expected to be expelled from the surface of a perfect crystal. Instead they become trapped, and the repulsive potential energy goes into defect formation and heat [9]. For the excited-state neutral atoms, no microscopic mechanism has been developed to explain how such particles could be formed at the surface and expelled into vacuum.

Recently, we have experimentally investigated the formation of positive ions [10] and excited neutrals [3,11] produced by electron bombardment of alkali halides. We have found that both the positive ions and excited neutrals are produced by gas-phase collisions between electrons (including both the incident beam and secondary electrons) and ground-state neutral species. This mechanism accurately accounts for: the velocity distribution of the excited neutrals, the absolute yield of excited neutrals, the dependence of excited-state yield on incident electron current, the distribution of population among the various excited states, the mass distribution of positive ions, the absolute positive ion yields, and the dependence of ion yield on incident electron current. Moreover this mechanism is consistent with all previous data on these systems (to the best of our knowledge). In this article we summarize our experimental measurements on positive ion and excited neutral production by electron bombardment of alkali halides.

## Experiment

The experimental apparatus is outlined below. A 200 eV - 2000 eV electron beam with currents of 0.1 - 10 $\mu$A was used to bombard alkali halide crystals in ultra-high vacuum ($\simeq 5 \times 10^{-10}$ Torr). The crystals were cleaved in air then cleaned by pumping in UHV at room temperature overnight followed by heating to the desired temperature. This procedure has been shown to yield atomically clean alkali halide surfaces [12]. The electron beam was incident at 45° from the normal to the (100) surface. The bombarded region had dimensions of $\simeq$1mm × 2mm. Sample temperatures from 290 K to 700 K were studied. Ground-state neutral species and positive ions were monitored with a quadrupole mass-spectrometer (QMS). The axis of the QMS was along the surface normal, with direct line-of-sight to the bombarded region. For neutral detection, the ionizer of the QMS was turned on, and for ion detection, the ionizer was turned off and biases were applied to collect the positive ions. Ground-state alkali atoms were probed by laser-induced fluorescence and laser absorption spectroscopy, with the laser propagating parallel to, and in contact with the bombarded portion of the surface. Excited-state neutrals were detected by laser-induced fluorescence and optical emission spectroscopy. Both low resolution (grating spectrometer) and high resolution (Fabry-Perot interferometer) instruments were used to monitor emission from the excited atoms.

## Results and Discussion

The velocity distributions of the excited atoms were probed by laser-induced fluorescence (LIF) and high resolution optical emission spectroscopy. The laser technique was restricted to the detection of Na*(3p) produced by electron bombardment of NaCl [3]. High resolution emission profiles, made with a Fabry-Perot interferometer, were measured for the Na(3p→3s), and Na(4d→3p) transitions produced by electron bombardment of NaCl, and for the Li(2p→2s) transition produced by electron bombardment of LiF [11]. The profile of the Na $3P_{1/2} \rightarrow 4D_{3/2}$ transition, probed by LIF, is shown in Fig. 1. The LIF signal vs. laser frequency is directly related to the velocity distribution along the laser axis, by the Doppler effect. The experimental profile is accurately characterized by a Maxwell-Boltzmann velocity distribution with a best fit temperature of T = 565 ± 10 K (dashed curve). This agrees with the surface temperature as measured with a thermocouple ($T_S$ = 558 K) to within error. Identical results were obtained with the Fabry-Perot interferometer by measuring the Na(3p→3s) emission line shape. The emission profile for the Li(2p→2s) transition produced by electron bombardment of LiF is shown in Fig. 2. Two peaks are observed due to fine structure of the 2p level. Again, the line shape is consistent with thermal kinetic energies for the excited atoms. For all temperatures where we were able to record LIF or emission profiles, the line shapes implied thermal velocity distributions for the excited atoms, with temperatures in equilibrium with the surface temperature to within error.

If the electronically excited atoms were formed on the surface and directly expelled into vacuum, one would generally expect to find kinetic energies greater than the surface temperature. Direct expulsion of an excited particle would presumably be due to excitation to some repulsive potential curve - thus complete thermalization would seem unlikely. Indeed, in every known example of direct desorption, the desorbed particles are hyperthermal. However, the ground-state neutral alkali atoms are known to thermally desorb, with a kinetic temperature in equilibrium with the surface temperature. For the ground-state neutrals, desorption is de-coupled from the initial excitation process - defect diffusion to the surface occurs as an intermediate step. The neutral ground-state atoms once formed are weakly bound to the solid and can be thermally desorbed. Thus the observation that the excited atoms are thermal suggests a link between the ground-state and excited-state desorption mechanisms.

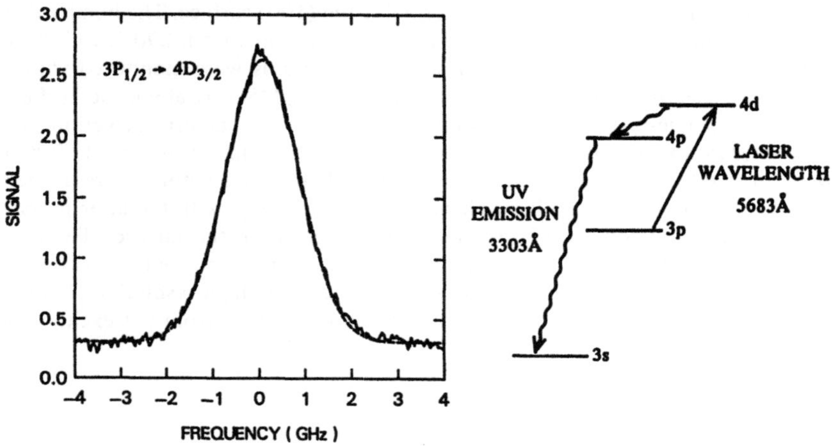


Figure 1. Experimental Doppler-broadened profile of the sodium $3P_{1/2} \rightarrow 4D_{3/2}$ transition measured by LIF (solid curve). The best-fit Maxwell-Boltzmann distribution (dashed curve) has T = 565 ± 10 K, in agreement with the surface temperature $T_S$ = 558 K.

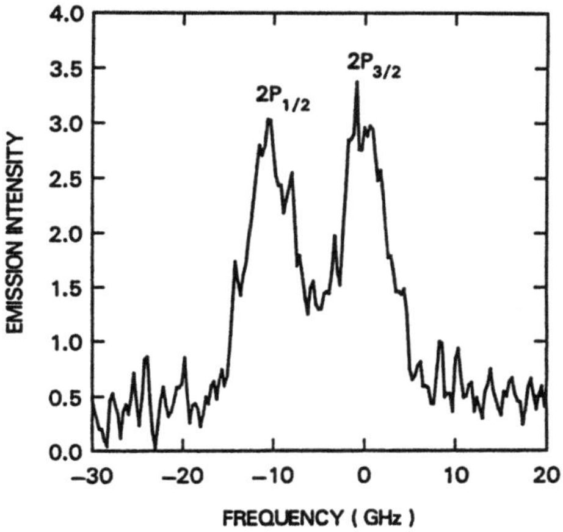


Figure 2. Experimental emission profile of the Li(2p→2s) transition produced by electron bombardment of LiF. Two peaks are are observed due to fine structure in the excited state. The width of the peaks (≃5.5 GHz FWHM) is consistent with the combined Doppler width at the sample temperature (≃3.0 GHz FWHM) and the instrument resolution (≃2.8 GHz FWHM).

We have made careful measurements of the yield of neutral atoms and molecules, excited atoms, and positive ions as a function of incident electron current, for a wide range of sample temperatures. It is important to understand that the surfaces of alkali halide crystals can be altered by electron bombardment, and this can affect yield measurements. In order to obtain results which relate to stoichiometric alkali halides, it is necessary to choose experimental conditions which minimize radiation damage. At low temperatures, the bombarded material is observed to develop an alkali enriched stoichiometry [2]. However, for NaCl, it has been shown that at sufficiently high temperatures (T>570 K), the material remains stoichiometric even for very large doses [2]. Thus there are two possible methods of minimizing effects due to radiation damage: (1) use very small electron doses, $<< 10^{13}$ electrons $cm^{-2}$, and (2) use temperatures which are high enough to prevent modification of the stoichiometry. Because of limitations in experimental sensitivity, the first method is impractical; thus we must resort to the second technique. As an additional measure, we restrict the bombardment time to 2 to 3 minutes, and we make comparisons with results obtained by translating the sample to bring unexposed material under the beam.

The yields of excited-state atoms, Na*(3p), and ground-state atoms, Na(3s), are shown in Fig. 3 as a function of electron current for bombardment of NaCl at T=570 K by 1 keV electrons. The ground-state yield was mónitored by LIF, and the excited-state yield was monitored by optical emission spectroscopy. The ground state yield is approximately linear in current and the excited-state yield is approximately proportional to the current squared. Small deviations from purely quadratic behavior were observed due to mild radiation damage of the sample. These deviations could be minimized by increasing the sample temperature, reducing the electron dose, and translating the sam-

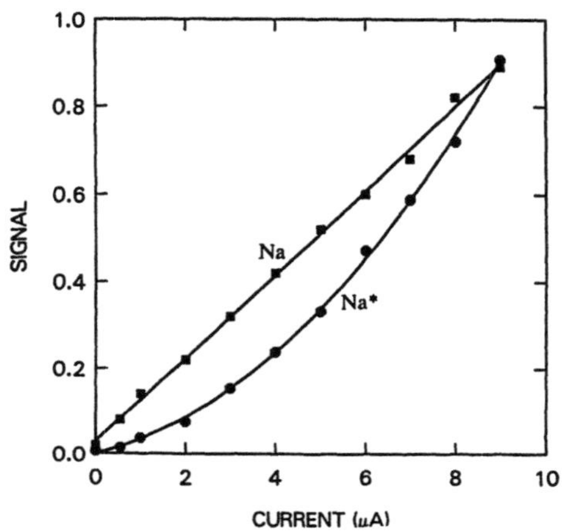


Figure 3. The Na*(3p) excited state yield (dots) and Na ground state yield (squares) are shown vs incident electron current for T=570 K.

ple to bring fresh material under the beam. The quadratic behavior clearly indicates a two-step process, where one electron is required to desorb ground-state Na, and a second electron is required to produce Na*. In previous studies [13], the current dependence was measured at room temperature, and the excited-state yield was found to be roughly linear vs. incident electron current. This behavior is a consequence of beam-induced surface modification. At low temperatures the surface region becomes alkali enriched - in fact alkali metal can be formed. Under these conditions the ground-state yield becomes nearly independent of current because it is determined by evaporation of alkali atoms from alkali metal. The excited-state yield then takes on an approximately linear current dependence. We have examined the emission profiles under these conditions and found that the excited atoms are always thermal. As the temperature is increased, the sample is more nearly stoichiometric, and the yield dependence tends toward purely quadratic.

Since the yield of ground-state atoms is quite large, there is a substantial probability that excited-state atoms will be produced by electron-impact excitation of the ground-state atoms in the gas phase. Both the incident and secondary electrons must be considered. For NaCl bombarded by 1 keV electrons, the secondary electron yield is quite large, ~ 14 [14]. In addition, the excitation cross section is a factor of ten larger for low energy electrons (2-10 eV) than at 1 keV [15]. About half of the secondary electrons are expected to have kinetic energies greater than the excitation threshold (2 eV) [16]. Thus for 1 keV incident electrons it is clear that secondary electrons are considerably more effective at excitation than the primary beam. It is also clear that the relative importance of secondary vs. incident electrons depends on beam energy since both the secondary yields and the excitation cross sections are energy dependent. It is straightforward to estimate the absolute yield of excited atoms produced by excitation in the gas phase because the absolute yields of ground-state neutrals, the secondary electron yields, and the atomic excitation cross sections are all known. The probability that a given ground-state alkali atom will be excited is $\sim n_e \sigma_{ex} v_e \tau$, where $n_e$ is the electron density, $\sigma_{ex}$ is the excitation cross section, $v_e$ is the electron speed, and $\tau$ is the transit time of the atom across the beam. For a 1 $\mu$A beam in a 1 mm × 2 mm area, the probability that a given Na atom will be excited is predicted to be $\sim 10^{-5}$, i.e. the number of excited atoms produced per second is smaller than the number of ground-state atoms which desorb per second by a factor of ~ $10^{-5}$. This agrees with the experimental estimate to within error [3].

If the excited atoms are produced by collisions between ground state atoms and electrons, one should observe a variety of excited states - with emission intensities determined by the excitation cross sections. Upon careful examination of the emission spectrum, a variety of atomic emission lines are observed. We determined the relative emission intensities using a calibrated monochromator and photomultiplier. Our experimental results and those of Szymonski et al. [2] are shown in Table 1 together with the excitation cross sections at 5 eV [15]. The two experiments are in good agreement, and the emission intensities are seen to scale with the excitation cross sections to within error (~30%). For alkali atoms, electrons excite the lowest resonance line much more efficiently than any other transition, thus this line stands out in the emission spectrum.

Table 1. Comparison of the experimental emission intensity ratios from [2] and this work with the cross sections for electron impact excitation of sodium atoms by 5 eV electrons [15]. The cross sections and emission intensities are scaled to 1.0 for the 3p → 3s transition.

| Transition | $\sigma_{ex}$[15] | Expt.[2] | This Work |
|---|---|---|---|
| 3p → 3s | 1.000 | 1.000 | 1.000 |
| 4p → 3s | 0.0095 | 0.011 | 0.008 |
| 3d → 3p | 0.103 | .... | 0.090 |
| 4d → 3p | 0.0090 | 0.005 | 0.008 |
| 5d → 3p | 0.0048 | 0.004 | 0.004 |
| 6s → 3p | 0.0022 | 0.002 | .... |

In Fig. 4 we compare the mass distribution of neutrals and positive ions produced by electron bombardment of NaCl at T=570 K. The neutral species which are stimulated to desorb are Na, Cl, $Na_2$, and $Cl_2$. The yield of each of these neutrals was observed to be approximately linear in incident current over the experimentally available range, 0.1 - 10 μA, as shown in Fig. 5 (upper curve) for neutral Na. In addition to the neutrals listed above, mass peaks due to residual gas, i.e. $H_2O$, CO, and $CO_2$, were detected. The neutral molecules NaCl and $Na_2Cl_2$ were also present in the residual gas, and there were trace amounts of organic contaminants. The observed signals from NaCl and $Na_2Cl_2$ are consistent with the equilibrium vapor pressures [17] at 570 K, i.e. ~ $1 \times 10^{-12}$ Torr for NaCl, and ~ $3 \times 10^{-14}$ Torr for $Na_2Cl_2$. Upon electron impact ionization, NaCl neutrals yield $Na^+$ and $NaCl^+$ ions, and $Na_2Cl_2$ neutrals primarily yield $Na_2Cl^+$ ions [18]. The yields of NaCl and $Na_2Cl_2$ neutrals were observed to be independent of incident electron current, as expected for molecules which thermally desorb from the surface. The neutral NaCl signal vs. incident electron current is shown in Fig. 5 (lower curve). The ion mass distribution shown in Fig. 4 is seen to closely parallel the neutral mass distribution. The ions include $Na^+$, $Na^{++}$, $Cl^+$, $Cl^{++}$, $Na_2^+$, $Cl_2^+$, $NaCl^+$, and $Na_2Cl^+$. The yields of $Na^+$, $Na^{++}$, $Cl^+$, $Cl^{++}$, $Na_2^+$, and $Cl_2^+$ were each observed to be approximately proportional to the square of the incident electron current. The $Na^+$ signal vs. current at T=570 K is shown in Fig. 5 as an example. As was observed for the excited neutrals, small deviations from purely quadratic behavior occur due to

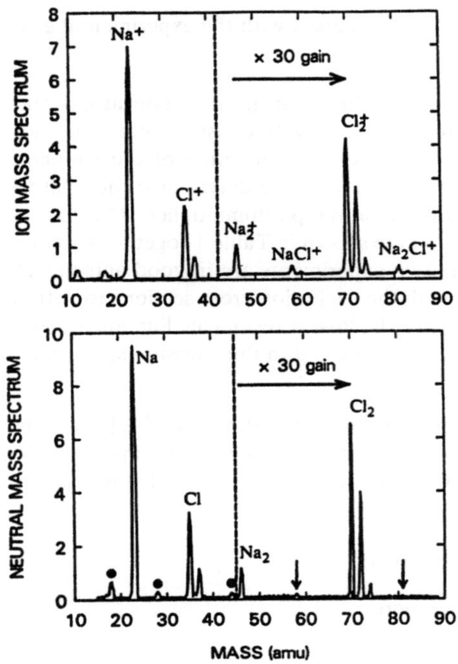


Figure 4. The upper curve shows the mass distribution of positive ions produced by bombardment of NaCl by 1 keV electrons at 570 K and $I_e = 1\mu A$ . The positive ions are labeled except for the peaks at m/q of 11.5, 17.5, and 18.5, which are due to $Na^{++}$ and $Cl^{++}$. The lower curve shows the mass distribution of neutrals under identical conditions except $I_e = 5\mu A$. Peaks at mass 18, 28, and 44 are due to $H_2O$, CO, and $CO_2$ in the residual gas.

mild radiation damage of the sample. At elevated temperatures and with limited doses, the deviations from quadratic behavior were small. In contrast with the other ionic species, the yields of $NaCl^+$ and $Na_2Cl^+$ were observed to be approximately linear in incident electron current, as shown in Fig. 5 for $NaCl^+$ at T=570 K. Order-of-magnitude estimates of the ratio of ion to neutral flux were made, based on the ratio of the experimental signals combined with estimates of the ionization efficiency of the QMS. For incident electron currents of $\simeq 1\ \mu A$, the estimated $Na^+$ ion to Na neutral flux is $\sim 10^{-6}$. This estimate is in good agreement with similar estimates by previous workers [19,20].

The ion mass distributions, absolute ion yield, and dependence of ion yield on incident current are all accounted for by a gas-phase ionization mechanism. The ions $Na^+$, $Cl^+$, $Na_2^+$, and $Cl_2^+$ are formed by single ionization of the corresponding neutral species; $Na^{++}$ and $Cl^{++}$ are formed by double ionization of neutral Na and Cl atoms; and $NaCl^+$ and $Na_2Cl^+$ are formed by ionization of neutral NaCl and $Na_2Cl_2$ molecules. The probability that a given neutral atom (or molecule) will be ionized is $\sim n_e \sigma_i v_e \tau$, where $n_e$ is the electron density, $\sigma_i$ is the ionization cross section, $v_e$ is the electron ve-

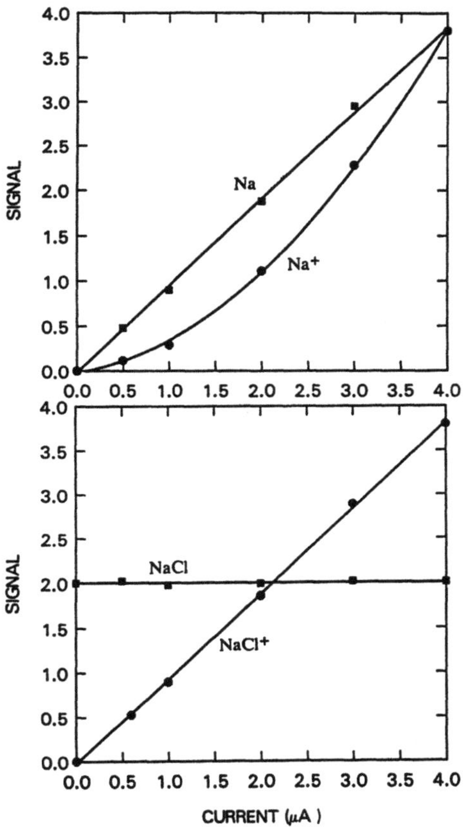


Figure 5. The upper panel shows the Na neutral yield (squares) and $Na^+$ ion yield (dots) vs. incident electron current at T=570 K. The lower panel shows the NaCl neutral yield (squares) and $NaCl^+$ ion yield (dots) vs. current at T=570 K.

locity, and $\tau$ is the interaction time. The ionization probability can be estimated using literature values for the cross sections [21], secondary yields [14], and energy distributions [16]. For a 1 $\mu$A incident beam in a 1 mm x 2 mm area, the net probability that a Na atom will be ionized is estimated to be $\sim 5 \times 10^{-7}$. This agrees with the experimental estimate of $\sim 10^{-6}$ to within error (a factor of $\sim$ 10). Similar estimates for Cl atoms indicate that the probability that a Cl atom will be ionized is $\sim 2 \times 10^{-7}$ for the conditions listed above. Secondary electrons are expected to produce most of the Na ionization and a good fraction of the Cl ionization. The other neutral species $Na_2$, $Cl_2$, NaCl, and $Na_2Cl_2$ are expected to be ionized to a similar extent.

Qualitative trends in the ion mass distribution can be understood in terms of the gas-phase ionization mechanism. For example, consider the relative abundances of $Na^+$ and $Cl^+$. The flux of neutral Na and Cl are equal (for our conditions), and the velocity distributions of both species are primarily thermal. The yield of $Na^+$ is somewhat larger than the yield of $Cl^+$ because Na is more efficiently ionized due to its low IP and large cross-section. Doubly-charged ions are also observed. For 1 keV incident electrons, $Na^{++}$ and $Cl^{++}$ are detected with ratios of $Na^{++}/Na^+ \sim 0.04$ and $Cl^{++}/Cl^+ \sim 0.1$ (uncorrected for QMS transmission). To within a factor of two, these ratios agree with estimates using cross sections for double-ionization of neutrals by the incident electrons [21]. The gas-phase mechanism also accounts for the presence of $NaCl^+$ and $Na_2Cl^+$: the neutral precursors are present in the gas phase, with densities sufficient to account for the ion yields observed.

## Conclusion

In conclusion, we have presented experimental data which indicate that the positive ions and excited neutrals produced by electron bombardment of alkali halides are formed by gas-phase collisions between desorbed ground-state neutrals and electrons. Both secondary electrons and the primary beam contribute to the excitation and ionization processes. For typical conditions, ~ 1$\mu$A electron beam with ~ 1 mm diameter, the probability that a neutral will be excited or ionized in the gas phase is $10^{-5}$ to $10^{-7}$. Although this probability is small, it completely accounts for the excited atoms and positive ions observed. Thus a unified picture of desorption from alkali halides is now obtained, in which ground-state neutrals are stimulated to desorb via bulk defect mediated processes with very large yields; and the positive ions and excited neutrals are formed by gas-phase processes.

## References


1. See M. Szymonski, Radiation Effects 52 , 9 (1980) and references therein
2. M. Szymonski, J. Ruthowski, A. Poradzisz, Z. Postawa, and B. Jorgensen, in Desorption Induced by Electronic Transitions, DIET 2 , ed. by W. Brenig and D. Menzel (Springer-Verlag, Berlin, 1984) p.160
3. R. E. Walkup, Ph. Avouris, and A. P. Ghosh, Phys. Rev. Lett. 57 , 2227 (1986)
4. G. M. Loubriel, T. A. Green, P. M. Richards, R. G. Albridge, D. W. Cherry, R. K. Cole, R. F. Haglund Jr., L. T. Hudson, M. H. Mendenhall, D. M. Newns, P. M. Savundararaj, K. J. Snowdon, and N. H. Tolk, Phys. Rev. Lett. 57 , 1781 (1986)
5. H. Overeijnder, R. R. Tol, and A. E. De Vries, Surf. Sci. 90 , 265 (1979)
6. M. L. Knotek and P. J. Feibelman, Phys. Rev. Lett. 40 , 964 (1978); P. J. Feibelman and M. L. Knotek, Phys. Rev. B 18 , 6531 (1978)
7. P. J. Feibelman, Surf. Sci. 102 , L51 (1981)
8. D. E. Ramaker, C. T. White, and J. S. Murday, Phys. Lett. 89A , 211 (1982)
9. R. E. Walkup and Ph. Avouris, Phys. Rev. Lett. 56 , 524 (1986)
10. R. E. Walkup, Ph. Avouris, and A. P. Ghosh, to appear in Phys. Rev. B15 , (1987)
11. R. E. Walkup, Ph. Avouris, and A. P. Ghosh, to appear in J. Vac. Sci. Tech., (1987)
12. J. Estel, H. Hoinkes, H. Kaurmann, H. Nahr, and H. Wilsch, Surf. Sci. 54 , 373 (1976)
13. N. H. Tolk, L. C. Feldman, J. S. Kraus, R. J. Morris, M. M. Traum, and J. C. Tully, Phys. Rev. Lett. 46 , 134 (1981); N. H. Tolk, L. C. Feldman, J. S. Kraus, R. J. Morris, T. R. Pian, M. M Traum, and J. C. Tully, in Inelastic Particle-Surface Collisions, ed. by E. Taglauer and W. Heiland (Springer-Verlag, Berlin 1981), p. 112
14. CRC Handbook of Chemistry and Physics , 66th edition, ed. by R. C. Weast (CRC Press, Boca Raton 1985)
15. J. O. Phelps and C. C. Lin, Phys. Rev. A 24 , 1299 (1981); and E. A. Enemark and A. Gallagher, Phys. Rev. A 6 , 192 (1972)

16. As a rough approximation, we used the best-fit analytic expression for the secondary electron energy distribution for x-ray bombardment reported by B. L. Henke, J. Liesegang, and S. D. Smith, Phys. Rev. B 19 , 3004 (1979)
17. G. M. Rothberg, M. Eisenstadt, and P. Kusch, J. Chem. Phys. 30 , 517 (1959)
18. J. Berkowitz and W. A. Chupka, J. Chem. Phys. 29 , 653 (1958)
19. A. Friedenberg and Y. Shapira, J. Phys. C 15 , 4763 (1982)
20. T. R. Pian, M. M. Traum, J. S. Kraus, N. H. Tolk, N. G. Stoffel, and G. Margaritondo, Surf. Sci. 128 , 13 (1983)
21. R. H. McFarland and J. D. Kinney, Phys. Rev., 137 , A 1058 (1965); T. R. Hayes, R. C. Wetzel, and R. S. Freund, Phys. Rev. A 35 , 578 (1987); F. A. Stevie and M. J. Vasile, J. Chem. Phys. 74 , 5106 (1981); L. J. Kieffer and G. H. Dunn, Rev. Mod. Phys. 38 , 1 (1966).

# Ion Desorption Induced by Core Exciton States in MgO

*R.L. Kurtz*[1], *R. Stockbauer*[1], *F. Senf*[2], *R. Nyholm*[3], *and S. A. Flodström*[4]

[1]National Bureau of Standards, Gaithersburg, MD 20899, USA
[2]HASYLAB DESY, Fed. Rep. of Germany
[3]MAX Laboratory, S-22100, Lund, Sweden
[4]Royal Institute of Technology, S-10044 Stockholm, Sweden

## 1. Introduction

Metal oxides, in particular those that are highly ionic, are ideal systems for the study of mechanisms of stimulated desorption [1]. Many of these ionic materials are also maximal-valency compounds and are useful for examining the Knotek-Feibelman mechanism of stimulated desorption [2]. MgO is one of the most ionic of these materials and, with the $Mg^{2+}$ cation in a $1s^22s^22p^6$ configuration, it is also maximal-valent. Calculations indicate that the (100) surface Madelung energy for a positive charge at a lattice $O^-$ site is quite large (>23 eV) [3]. From this perspective, one would expect an interatomic Auger decay process to be operative and, with the strong Madelung repulsion, intense ion desorption signals. Although this would appear to imply that MgO should give intense ion desorption signals, previous studies have shown it to be stable under electron irradiation [4]. Studies of desorption from this material therefore should be instrumental in assessing the validity of charge-transfer and lattice dynamical concepts.

In this investigation, we have studied photon stimulated ion desorption from MgO(100) and we report results for photon energies spanning the O $1s$ and Mg $1s$ levels. An excitonic state has been observed for excitations near the O $1s$ onset and the data imply that these excitons are formed preferentially in the near surface region; they are very efficient in producing $O^+$ and $H^+$ desorption. This is also the first observation of desorption induced by a core exciton that is formed on the ligand [5]. Excitations of the Mg $1s$ level, in contrast, show no evidence for excitonic states nor are these excitations efficient in producing $O^+$. This implies that either an interatomic Auger process is not efficient at neutralizing the Mg $1s$ core hole or that the resulting state does not live long enough to result in ion desorption.

## 2. Experimental Results

The studies of ion desorption from MgO surfaces were performed at HASYLAB on the DORIS synchrotron light source at DESY, Hamburg. The wiggler/undulator beam-line was used with the FLIPPER monochromator to permit excitations spanning all of the electronic levels in the material [6]. Mass analysis of the desorbing ions was achieved using a time-of-flight (TOF) technique that utilized the time structure of the storage ring. A 1.4 kV ion accelerating voltage collected all positive ions leaving the sample giving unit mass resolution at $m/e$=16, i.e. in the vicinity of $O^+$ ions. Photoexcitation of single-crystal MgO produced substantial charging that was dependent on the photon energy, however, the large accelerating potential mitigated this effect on the ion spectra. Electron spectroscopies could be performed reliably only on thin oxide layers that were formed on evaporated Mg films; these surfaces did not charge appreciably.

The MgO(100) surfaces were prepared in UHV by sputtering with 500 eV Ar ions followed by annealing. The thin-film MgO was prepared by evaporating Mg onto stainless steel and exposing to $O_2$. The most intense desorbing ions observed were $O^+$ and $H^+$, however, $OH^+$ and $F^+$ were also seen and were associated with surface contamination.

In Fig. 1, the $O^+$ and $H^+$ yields obtained for photon excitations of the O $1s$ level are displayed. Aside from the two features that are indicated by dashed curves, similar structures are observed in both curves at identical photon energies and are nearly equal in intensity. The

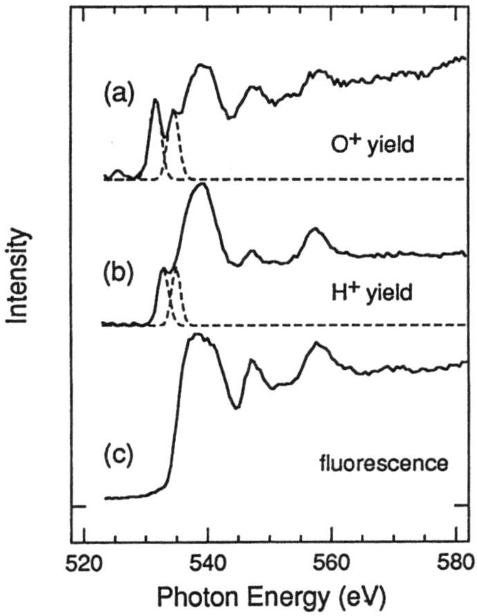


Fig.1. Ion desorption and fluorescence from MgO(100) with photon excitations spanning the O 1$s$ level: (a) $O^+$ ion yield; (b) $H^+$ ion yield; and (c) fluorescence. The onset for photoexcitation from the O 1$s$ to the first unoccupied conduction band state is at 534.9 eV. The dashed curves are Gaussian peaks that indicate the location and intensities of the exciton contributions.

fluorescence yield from this surface also shows the same structure above 535 eV, however, the features at the energies of the dashed peaks in Fig. 1 are completely absent. Bulk-sensitive energy loss spectra (ELS) from MgO microcrystals display structure at identical excitation energies with an additional small peak at 531.6 eV [7].

If the binding energy of the O 1$s$ level obtained from XPS is compared with the location of the O 2$p$ valence band [8] and the known 7.77 eV band-gap of MgO [9], one would predict that excitations from the O 1$s$ level to the first unoccupied conduction band state would begin with a photon energy of 534.9 eV. The core to conduction band transitions, therefore, give rise to the features observed for $h\upsilon$>535 eV.

As shown in Fig. 1, ion desorption is initiated with photon excitations that place the initial photoelectron into a normally disallowed (band-gap) state. However, such a state is allowed in the presence of the deep O 1$s$ core hole. We observe two such excitonic states in the $O^+$ yield with photoexcitation energies of 531.6 and 534.6 eV. If these features are fit with Gaussian curves and then subtracted from the experimental data, a photoexcitation onset at 534.9 eV is obtained which is in good agreement with the predicted core-to-conduction band onset.

Similarly, the pre-edge structure in the $H^+$ yield curve can be decomposed into Gaussian peaks at 533.4 and 535 eV and their subtraction from the experimental data yields an onset at 535.0 eV. The excitonic states giving rise to $H^+$ desorption are different than those producing $O^+$ since surface hydrogen is believed to be stabilized in the form of hydroxyl groups [10]. This results in a different chemical environment for the O and, hence, shifted excitonic levels.

Figure 2 shows the total electron yield (TEY) obtained from MgO formed by $O_2$-exposure of an evaporated Mg film. In curve (a), the Mg has been exposed to 10 L and in (b) the exposure was 30 L of $O_2$. As in Fig. 1, the data exhibit features at photon energies lower than the expected onset for core-to-conduction band excitations. As the oxide thickness in increased with the 30 L exposure, the bulk excitation structure observed for photon energies greater than 535 eV becomes increasingly well-defined and exhibits features at the same energies as observed in the $O^+$ and $H^+$yields. Also, with increasing oxide thickness, the relative intensity of the excitonic levels, indicated by the dashed Gaussian curves in Fig. 2 decreases. This suggests strongly that the excitonic levels are formed preferentially at the surface of the MgO. This implication is supported by the lack of excitonic structure in the fluorescence yield; a spectroscopy that is

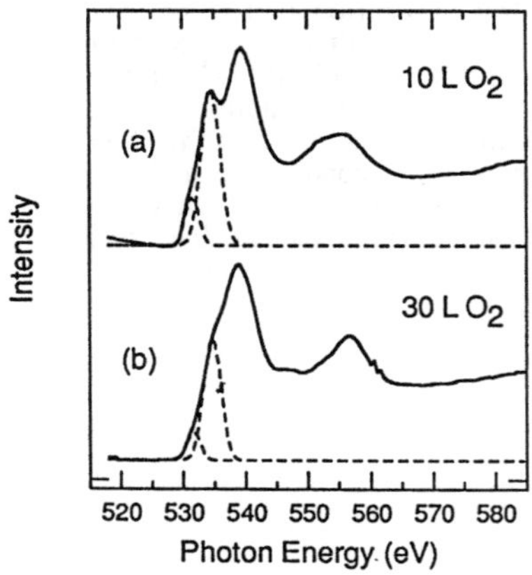


Fig.2. Total Electron Yield (TEY) from an evaporated Mg film that has been exposed to (a) 10 L and (b) 30 L $O_2$. The exciton contribution is clearly evident in both of these curves,however,as the oxide thickness increases in (b), the exciton contribution decreases in relative intensity.

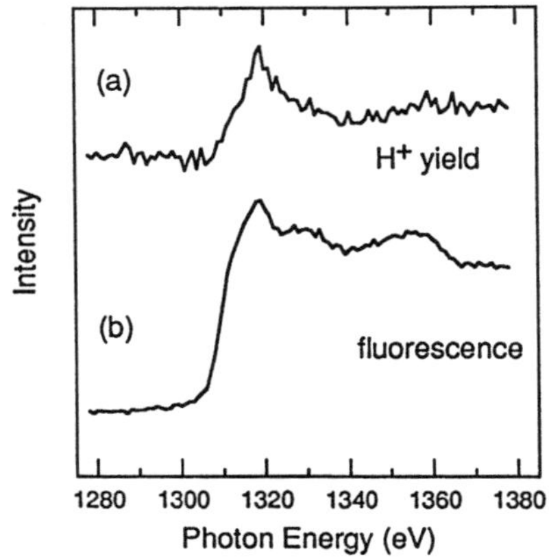


Fig.3. $H^+$ yield and fluorescence obtained from the MgO(100) surface for excitations spanning the Mg 1*s* level. The onset for excitations from the Mg 1*s* level to unoccupied conduction band states is at 1308 eV. Structure from core excitons is not evident in these curves.

primarily bulk sensitive. ELS measurements from a microcrystalline MgO powder exhibit a small feature at 531.6 eV whose intensity is that expected when a strong surface loss process is in effect [5,7].

The decay mechanism that occurs following the production of the core-excitonic state is most likely to involve an intraatomic Auger decay mechanism. Since the excitation involves an O 1*s* core hole and the overlap of O 1*s* core wavefunctions is small, it is not expected to be mobile. Therefore, the decay is likely to occur at the site where it was initially created. For this to result in a desorbing ion, the excitation must occur at the surface; this is consistent with the TEY from the thin oxide film described above.

For an excited lattice $O^{2-}$ species to decay to produce an $O^+$ ion, it is necessary that 3 electrons be transferred in the process. This implies that the excitonic electron is a *spectator* to the core neutralization from the occupied valence states. If two electrons are ejected in the Auger decay, the $O^{2-}$ would become $O^0$. When the O 1*s* core hole is filled, however, the excitonic electron state is no longer allowed. This requires the ejection of the spectator electron resulting in an $O^+$ final state. Although the details of the subsequent lattice dynamical response have yet to be studied in detail, preliminary calculations indicate that the lattice response is significant and that the initial 23 eV of Madelung repulsion results in only ~2 eV of kinetic energy transmitted to the departing $O^+$ [11]. This is due to the fact that the mass of the Mg cation is close to that of the O; much of the Madelung energy is absorbed in the recoil of neighboring atoms.

Fig. 3 shows the $H^+$ and the fluorescence yield for photon energies spanning the Mg 1*s* level. Exciton states are not observed, only an onset that is in good agreement with the predicted onset for core-to-conduction band excitations. In addition, the $O^+$ yield is extremely weak for excitations of this core level; the intensities were too low to be able to allow a measurement of yield versus photon energy. This implies that either an interatomic Auger decay is not an efficient mechanism for Mg 1*s* core-hole neutralization or that the resulting excited states do not produce an $O^+$ state that is long-lived enough to allow the ion to escape. Also, the lack of excitonic structure over the Mg 1*s* level indicates that the excitons reported here are associated only with O ligands.

## 3. Summary

Core-exciton-induced desorption of $O^+$ and $H^+$ from MgO(100) and oxidized thin-film Mg has been observed at the O 1*s* level. These excitons are localized and total-electron-yield measurements imply that they are produced in the near-surface region. The energies of the excitons that give rise to $O^+$ and $H^+$ are different and it is proposed that these energies reflect chemical differences between the lattice $O^{2-}$ and surface $OH^-$ species. The excitons are proposed to decay by a mechanism where the exciton electron is a spectator to an intraatomic Auger process. In contrast, no excitons are observed at the Mg 1*s* level implying that those excitations only occur on the O-ligand. As well, the $O^+$ yield is low at the Mg 1*s* edge indicating either that an interatomic Auger decay process is not an efficient core-hole neutralization mechanism or that the resulting excited states are not long-lived enough to result in $O^+$desorption.

## Acknowledgements

The authors would like to thank Professor C. Kunz and Hamburger Synchrotron-strahlungslabor, Deutsches Elektronen-Synchrotron (HASYLAB-DESY) for the use of the wiggler beamline and the Flipper monochromator. This work was supported in part by the Swedish Natural Science Research Council and the U. S. Office of Naval Research.

## References

1. M. L. Knotek and P. J. Feibelman, Surf. Sci. 90, 78 (1979).
2. M. L. Knotek and P. J. Feibelman, Phys. Rev. Lett. 40, 964 (1978).
3. J. Kendrick, E. A. Colbourn and W. C. Mackrodt, Radiat. Eff. 73, 259 (1983).
4. H. Tokutaka, M. Prutton, I. Higginbotham, and T. E. Gallon, Surf. Sci. 21, 223 (1970).
5. R. L. Kurtz, R. Stockbauer, R. Nyholm, S. A. Flodström, and F. Senf, Phys. Rev. B 35, 7794 (1987).
6. J. Barth, F. Gerken, C. Kunz, and J. Schmidt-May, Nucl. Instr. and Meth. 208, 307 (1983).
7. Th. Lindner, H. Sauer, W. Engle, and K. Kambe, Phys. Rev. B 33, 22 (1986).
8. J. C. Fuggle, Surf. Sci. 69, 581 (1977); J. C. Fuggle, L. M. Watson, D. J. Fabian, and S. Afrossman, *ibid.* 49, 61 (1975).
9. D. M. Roessler and W. C. Walker, Phys. Rev. 159, 733 (1967).
10. E. A. Colbourn and W. C. Mackrodt, Surf. Sci. 117, 571 (1982).
11. R. E. Walkup and Ph. Avouris, Phys. Rev. Lett. 56,524 (1986); R. E. Walkup, private communication.

# Studies of Sapphire Photoablation by Laser-Induced Fluorescence and Photothermal Deformation Measurement

*R.W. Dreyfus, F.A. McDonald, and R.J. von Gutfeld*

IBM T.J. Watson Research Center,
P.O. Box 218, Yorktown Heights, NY 10598, USA

Crystalline sapphire displays a low (0.6 J/cm$^2$) threshold for etching with 193 nm excimer pulses. Understanding of this low-threshold etching requires knowledge of the possible pathways by which laser energy can result in material removal. The different energy channels can in general lead to thermal or photochemical mechanisms as well as combinations of the two. Long lived electronically excited states usually will lead to bond breaking whereas very short lived states resulting in thermal phonons give rise to highly elevated local temperatures. A reasonable estimate of the maximum surface temperature attained during irradiation is sufficient to differentiate between classical thermal vaporization and electronic, i.e. photochemical, mechanisms for etching.

Experiments have been undertaken to distinguish between these two mechanisms. Initially, we utilized laser-induced fluorescence (LIF) measurements [1], see Fig. 1. The LIF technique makes it possible to infer the temperature of the laser-irradiated surface undergoing etching from the energies of AlO radicals in the etch plume (the ablated material moves rapidly away from the surface). The amplitude of the LIF signal as a function of excitation wavelength provides the necessary information from which the rotational/vibrational energy distribution of AlO is determined. From these data we infer an equivalent surface temperature of ~500K, c.f. Figs. 1(a) and (b). On the other hand, translational energy distributions for this same system have much higher energies which partly originate in plasma effects above the surface. Figure 2 shows the rapid increase in velocity which coincides with the onset of significant plasma-induced light emission. Both the above results, that is the rotational/vibrational as well as the translational energies for the AlO emanating from the sapphire surface imply a photochemical, i.e. direct electronic, etching mechanism.

Pulsed photothermal deformation (PPTD) results [2] help to complete the model for sapphire etching in two important ways: First, the results indicate that sapphire has a large surface absorption at 193 nm, amounting to about 3% of the incident light, see Fig. 3. This value implies a surface temperature increase which approaches ~130K in the region just *below* the experimentally observed threshold for etching. Figure 4 shows that the inferred lin-

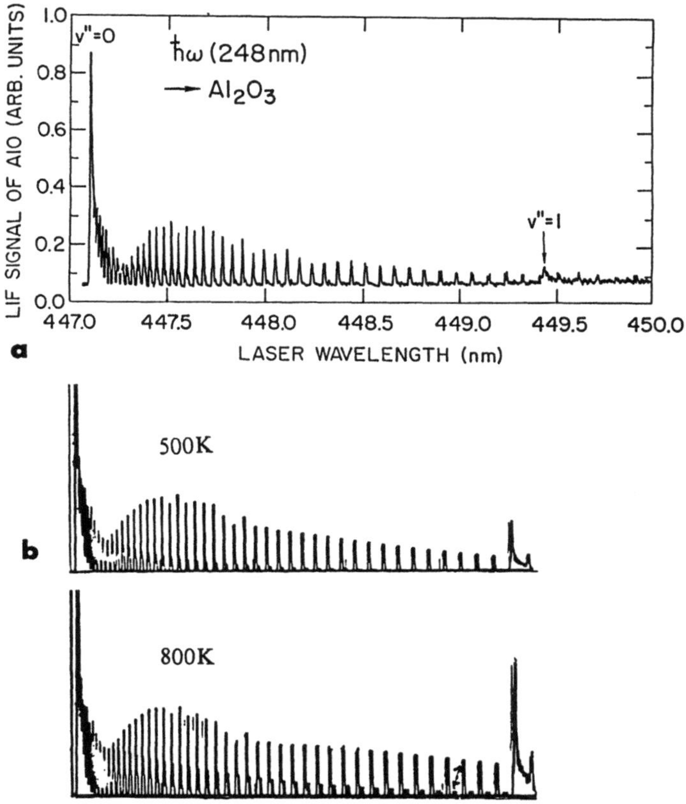


Fig.1(a). The LIF spectrum of AlO diatomics in the plume above excimer etched sapphire ($Al_2O_3$). Bands arising from both the v"=0 and 1 ground states are visible. (b). Calculated spectra for AlO at temperatures of 500K and 800K. The scale for the abscissa is identical to that of (a). Experimental results in (a) are evidently in better agreement with a temperature near 500K rather than 800K.

ear temperature increase with laser fluence extrapolates smoothly and linearly onto the earlier LIF data taken *above* the ablation threshold at 248 nm. Thus, these two entirely different experimental methods for measuring temperature both indicate ablation to occur at temperatures close to 200K above ambient, far below the temperature required for thermal vaporization. This comparison of temperatures above and below threshold also confirms the fact that plasma and gas kinetic (collisional) effects do not noticeably alter the observed AlO rotational/vibrational results (for surface temperatures as high as ~500K)shown for the region *above* threshold. The fact that the PPTD results are linear in fluence below the ablation threshold indicates that the surface energy absorption leading to ablation is due to a one

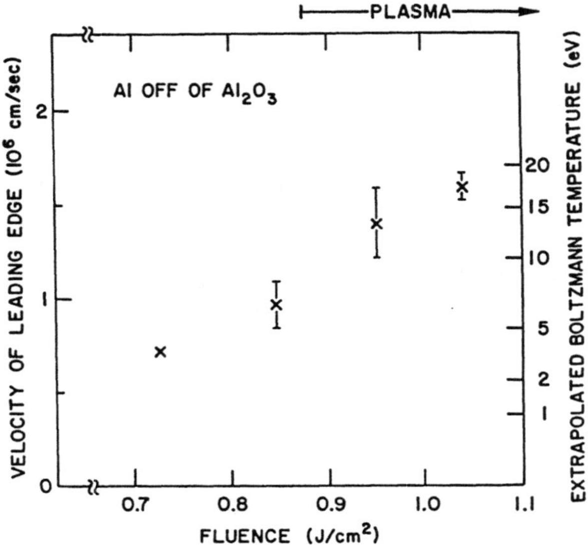


Fig.2. Velocity of Al atoms as a function of excimer fluence above laser etched sapphire surface. Note the rapid acceleration of the Al as a plasma layer forms in the ~1 $J/cm^2$ region.

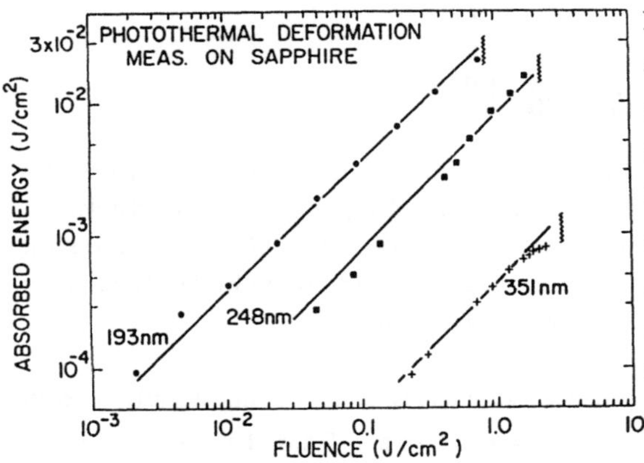


Fig.3. Surface energy absorption as a function of fluence in the below-threshold region for a sapphire surface at three wavelengths. Note the linear, and not higher power-law dependence of energy absorption. Also note that at 193 nm the large absorption is equivalent to an absorption of 3% of the light passing through the surface.

photon absorption process with a large absorption cross-section. This surface absorption is noticeably reduced by annealing and chemical etching as shown in the data of Fig. 5 for two sets of surface conditions and excitation at 248 nm, see also Ref. [3].

Finally, it is shown in Ref. [4] that the 3% absorption observed for the 193 nm etching of sapphire using the PPDT implies the creation of approximately 1 electron-hole pair/oxygen atom on the surface. This very high density of electron-hole pairs is of a mag-

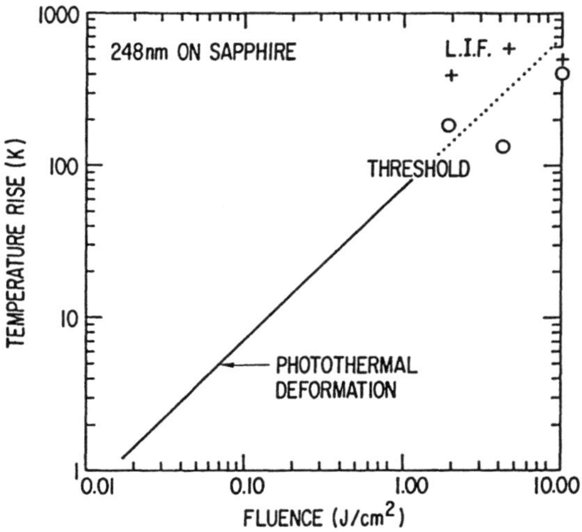


Fig.4. Estimated surface temperature rises as a function of fluence for the two different experiments, e.g. LIF and PPTD. Note that even though the experiments sense temperature in the two different regions, i.e. above and below the threshold for etching, the results are complementary.

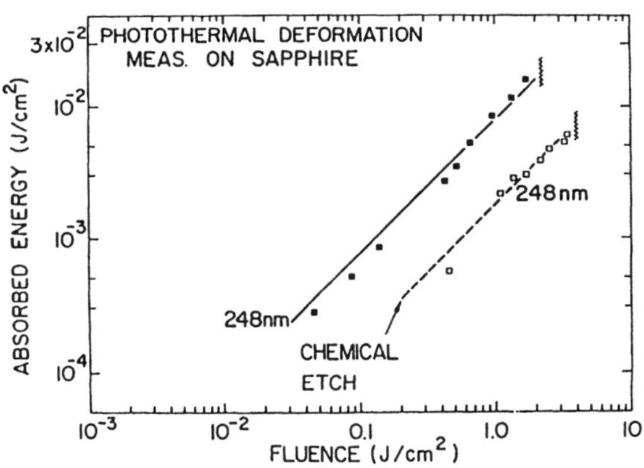


Fig.5. Decrease in the surface energy absorption at 248nm following chemical etching, as sensed by PPTD.

nitude for which one anticipates photochemical etching to occur since the surface bonding is significantly weakened. In summary, these two experiments have provided increased insight and confirmation of the etching pathways accessed by high energy (5.0-6.4 eV) UV photons incident on sapphire surfaces. The results give strong evidence for a photochemical model for the etching of sapphire.


1. R. W. Dreyfus, R. Kelly and R. E. Walkup: Appl. Phys. Lett. **49**, 1478 (1986)
2. R. J. von Gutfeld, F. A. McDonald and R. W. Dreyfus: Appl. Phys. Lett. **49**, 1059 (1986) and **50**, 1491 (1987) and F. A. McDonald, R. J. von Gutfeld and R. W. Dreyfus, Proc. IEEE 1986 Ultrasonics Symposium, p. 403
3. A. S. Barker, Jr.: Phys. Rev. **132**, 1474 (1963)
4. R. W. Dreyfus, F. A. McDonald and R. J. von Gutfeld: in press for J. Vac. Sci. Tech., Sept.-Oct. (1987)

# Multiphoton-Induced Desorption from $BaF_2(111)$

*H.B. Nielsen*[1], *J. Reif*[1], *E. Matthias*[1], *E. Westin*[2], *and A. Rosén*[2]

[1]Fachbereich Physik, Freie Universität Berlin, D-1000 Berlin 33, Germany

[2]Institute of Physics, Chalmers University of Technology and University of Gothenburg, S-41296 Göteborg, Sweden

Wavelength and intensity dependencies of laser-induced electron and positive ion emission from a $BaF_2$ (111) surface were studied in the $MW/cm^2$ intensity range, well below the damage threshold. In the range 500 to 560 nm we find similar shapes for the spectral emission yields of electrons and ions. Electrons are emitted by two- and five-photon absorption below and above 1.5 $MW/cm^2$, respectively. In contrast, ion desorption involves a ten-photon process, suggesting that two holes in the valence band are required. The interpretation of the data is supported by molecular cluster calculations. In the blue spectral range at higher intensities (~50 $MW/cm^2$) we observe rapid bleaching of the irradiated spots and evidence for layer-by-layer ablation.

## 1. Introduction

This contribution addresses the question: what are the underlying processes for the emission of electrons, ions, and neutrals from optically transparent ionic materials at laser intensities well below the damage threshold? A prerequisite for any understanding of particle emission from surfaces is a precise knowledge of the energy absorption mechanism. Here we would like to promote the idea, that electronic surface states due to defects play a dominant role in the initial photon absorption by providing a structured density of states in the band gap which can resonantly enhance multiphoton absorption. The holes created by ionization then lead to bond-breaking. To support this conjecture, experimental evidence is presented in form of a dramatic wavelength dependence of both electron and ion emission as well as their variation with incident laser intensity. The interpretation of the data is qualitatively reinforced by the results of molecular cluster calculations. For our investigation we used $BaF_2$ as a prototype for divalent ionic crystals, going on the assumption that for a non-stoichiometric surface the dangling valence electron of the metal ion favors the formation of both occupied and unoccupied surface states.

Particle emission caused by multiphoton absorption has been reported before. Schmid et al. [1] observed the directional emission of halogen atoms and ions from alkali halides. Their results were thought to originate from multiphoton generation of defects in the crystalline bulk [1,2]. Siekhaus et al. [3] found large emission yields for negative particles from optical materials caused by multiphoton excitation, and mentioned the possibility of enhancement due to defects, impurities, or surface states. Multiphoton absorption in the infrared has been held responsible for the desorption of adsorbates [4]. Also, for modeling optical damage mechanisms, multiphoton induced plasma generation has played an important role [5,6]. With the exception of laser-induced desorption in the infrared [4], much less has been reported about the wavelength dependence of laser-surface interaction. One knows [5] that the damage threshold of optical materials differs with wavelength. Also in laser etching [7] and desorption studies [3] the influence of wavelength has been shown. However, previous studies in the optical range seem to have been limited to fixed-frequency lasers. In our experiments we used a tunable dye laser to study the emission yields with good spectral resolution [8-10].

## 2. Experimental Arrangement

A sketch of the experimental setup is shown in Fig. 1. The $BaF_2$ crystals were mounted and cleaved in ultra-high vacuum (UHV) of $\leq 10^{-9}$ Torr. The (111) surface was irradiated with light from a Nd:YAG pumped pulsed tunable dye laser. The angle of incidence of the laser light was approximately 70°, with the plane of incidence coinciding either with the [$\bar{1}$10] or the [$\bar{1}\bar{1}$2] direction, as indicated in Fig. 2. When measuring wavelength and intensity dependencies, ions as well as electrons were detected with a secondary electron multiplier placed close to the crystal. A metal grid in front of the multiplier was held at ground potential for ion detection, and at +5 V for electron detection. By using a gated integrator, a time correlation to the laser pulses was assured. By independent measurements with a conventional mass spectrometer (emission current off) the positive ions $F^+$, $Ba^+$, and $(BaF)^+$ were identified, as illustrated in Fig. 3 for a laser wavelength of 532 nm. This observation is in agreement with previous results [8] obtained at somewhat higher intensities for the blue spectral

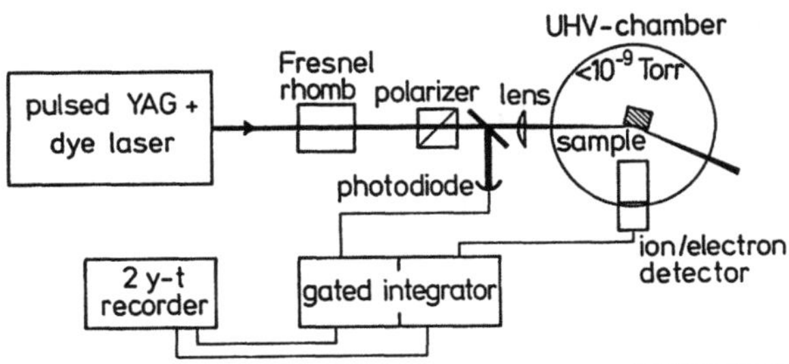


Fig. 1. Sketch of the apparatus used for measuring ion and electron emission following laser irradiation of $BaF_2(111)$.

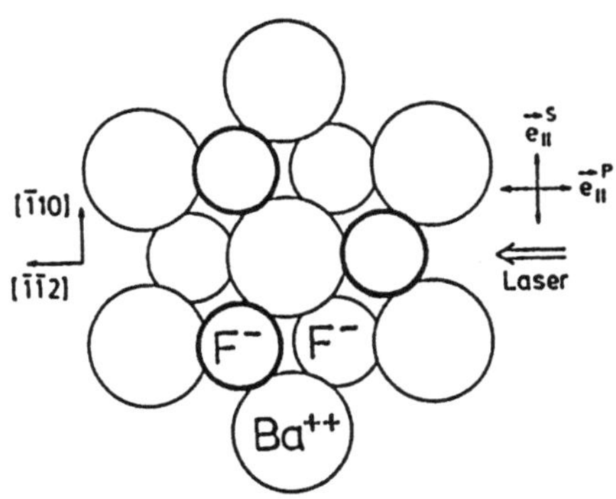


Fig. 2. Structure of the $BaF_2(111)$ plane. Three $F^-$ ions are positioned 0.144a below or above the $Ba^{++}$-plane (a=6.196 Å). The two polarization choices used for our experiments are shown.

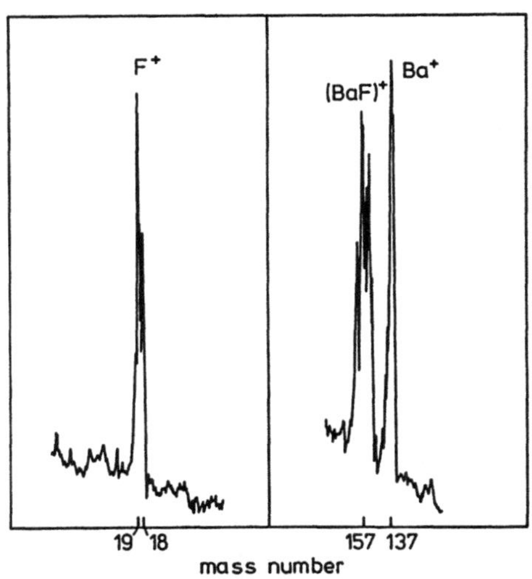


Fig. 3. Typical mass spectra obtained at 532 nm with a power density of about 10 $MW/cm^2$.

range, with the exception of $Ba^{++}$ which showed up weakly in our earlier study and could not be seen at 532 nm, most likely because of the lower intensity. No negative ions were found, neither with the quadrupole mass spectrometer nor with the mass-unselected detection, where negative ions could have been distinguished from the electrons by their longer flight times. Although the flight path from the crystal surface to the detector was only 3 cm we could easily distinguish between the flight times of the light ($F^+$) and heavy ($Ba^+$) ions coming off the surface. Furthermore, the time spectrum changed drastically between desorption and plasma generation at breakdown intensities and thereby provided an easy and sensitive check that the measurements were actually carried out under desorption conditions.

## 3. Experimental Results

As was reported earlier [9] we observed in the blue spectral range a wavelength dependence of the $Ba^+$ desorption yield which changed from scan to scan for the same irradiated spot, as shown in Fig. 4. We attributed this to a change of the chemical environment from which the Ba-ion departs and, according to the layered structure of the crystal, this can be either Ba-rich or F-rich. These measurements were carried out at comparatively high intensities, yet about an order of magnitude below the damage threshold of 500 MW/cm$^2$ for $BaF_2$ in air [11]. It is not clear to what extent this value is valid for cleaved surfaces in UHV. Our experience is, however, that already at intensities of 50 MW/cm$^2$ the spots behave quite erratically, some of them quenching rapidly, others remaining active for longer times. The general scheme at these intensities is nevertheless a

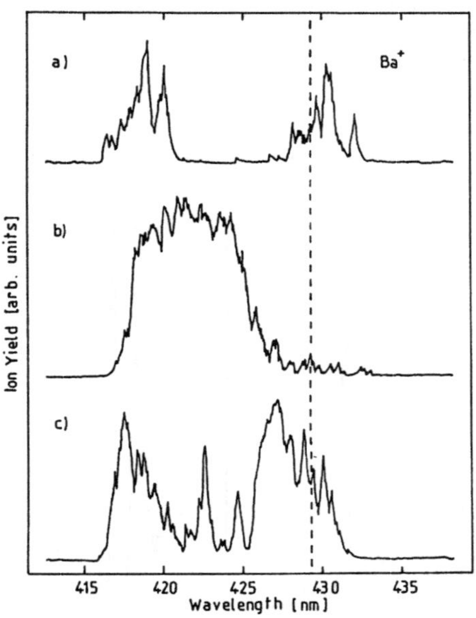


Fig. 4. Spectral dependence of $Ba^+$ desorption from $BaF_2(111)$ in the blue spectral range. The yield spectra a,b,c were taken successively at the same spot with a laser intensity of about 50 MW/cm$^2$.

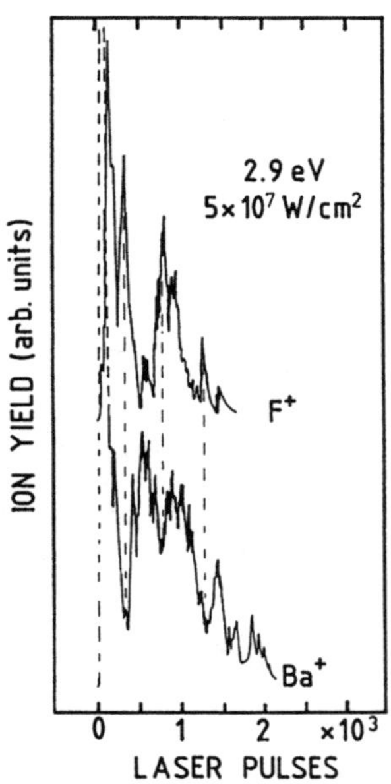


Fig. 5. An indication for layer-by-layer desorption of $Ba^+$ and $F^+$ from $BaF_2(111)$. Pulse frequency 10 Hz, wavelength as indicated by the dashed line in Fig. 4.

decrease of desorption yield with time, which we believe is due to metallization [12] that changes the electronic surface structure and hence the absorption characteristics. A typical quenching of the ion yield with time (equivalent to the number of laser shots) is shown in Fig. 5. In this example the desorption has ceased after about 3 minutes. A closer inspection of the two curves reveals an anticorrelation between the desorption yield of $Ba^+$ and $F^+$ in the sense that one is most intense when the other is at a minimum, as indicated by the dashed-dotted lines in Fig. 5. This is in agreement with what one would expect when peeling off the layer sequence Ba-F-F-Ba along the <111> direction. If this interpretation is correct, we have removed during 2000 laser shots about 5 Ba-layers. The data in Fig. 5 are taken at the wavelength indicated in Fig. 4 by the dashed line and are consistent with the picture of a spectral yield alternating with the chemical composition of successive layers.

The layer-by-layer sputtering might be of interest for applications, but fast decaying spots permit no reliable spectroscopy that can shed light on the basic processes involved in desorption. Hence we decided to use still lower intensities and investigate the electron and ion emission at lower photon energies in the green spectral range. In Fig. 6 an example is shown how a photoemission spectrum can develop with intensity. We believe that the spectrum at 5 MW/cm$^2$ represents a non-stoichiometric surface with some fluorine missing in the upper layer (F centers). During the scanning time at lower intensities neutrals, impurities, and structural defects may be desorbing until a spectroscopically clean situation is reached. However, the processes reflected by the spectra in Fig. 6 are far from being understood and need further investigation.

The observed spectral dependencies of electron and ion yields are shown in Figs. 7a-d. All spectra have clearly peaked structures, evidencing that the emission process is resonantly enhanced. The spectral shapes are highly polarization dependent. When the plane of incidence coincides with the <$\bar{1}\bar{1}$2> direction and the light is p-polarized (cf. Fig. 2), yield spectra as shown in Fig. 7a,b are obtained. With s-polarized light the spectra in Fig. 7c,d occur. When the plane of incidence coincides with the <$\bar{1}$10> direction the polarization dependence is reversed. Hence the spectra in

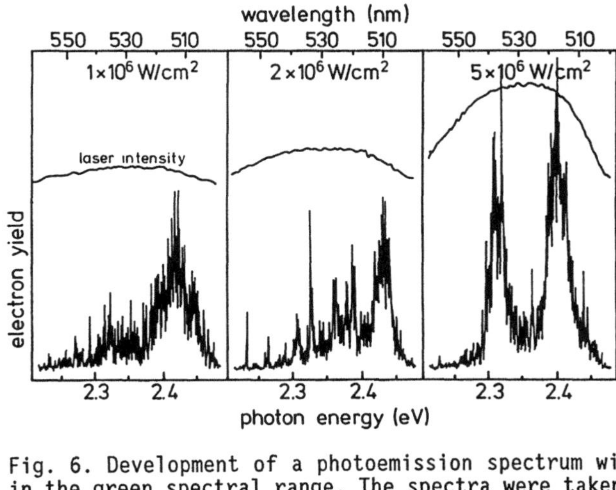


Fig. 6. Development of a photoemission spectrum with increasing intensity in the green spectral range. The spectra were taken with the same spot and successively increasing intensity.

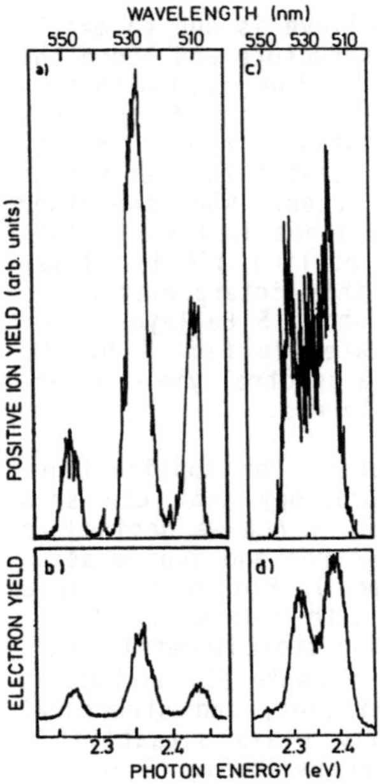


Fig. 7. Spectral and polarization dependence of electron and ion emission from $BaF_2(111)$ at about 5 $MW/cm^2$. Angle of incidence ~$70^o$.

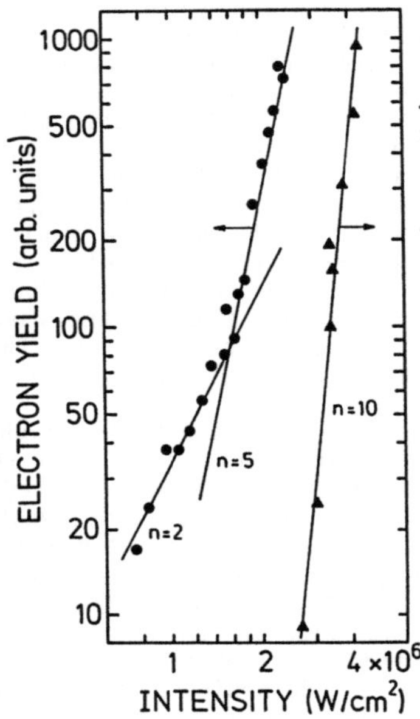


Fig. 8. Intensity dependence of electron and ion yield measured at 518 nm and a polarization corresponding to Fig. 7c,d.

Fig. 7a,b result when the projection $\mathbf{e}_\parallel$ of the polarization vector onto the surface is in the <$\bar{1}\bar{1}2$> direction, and those in Fig. 7c,d when $\mathbf{e}_\parallel$ is in the <$\bar{1}10$> direction. These results show that the wavelength-dependence of the $e^-$ and ion emission is related to the crystallographic structure of the (111) surface. We also note that the spectral yields of ions and electrons have similar shapes and coinciding peak positions. This shows that ions originate from the same excitation process and hence we conclude that ion emission is the result of photoionization of electrons.

The shape of the emission spectra was independent of the laser intensity, i.e. the peaks remained in position and only their amplitudes varied with intensity. At fixed wavelength (518 nm) the intensity dependence of the yields is illustrated in Fig. 8. At low intensities (<1.5 $MW/cm^2$) the electron yield varies as $I^2$, indicating two-photon ionization. At higher intensities repeated measurements at different spots were all consistent with a $I^5$ variation of the photoemission which points toward a five-photon absorption. From the conventional point of view it may come as a surprise that a five-photon process sets in closely above a two-photon process. However, similar effects have been observed in free atoms and the reduced symmetry of the surface may enhance these effects even further. To our knowledge, a comprehensive theory is lacking. On the other hand, five photons bridge the gap of about 10 eV, and this absorption process gains strongly from the high electronic density of the 2p valence band. Also, we suspect that surface imperfections caused by cleaving might generate a certain density of electronic states in the lower half of the band gap

[13] which could lend strength to the 5-photon excitation process. The alternative interpretation that the $I^5$ dependence could be due to avalanche ionization, does not seem realistic. Simulations of Bräunlich and Schmid [6] indicate that avalanche ionization only plays a role at intensities well within an order of magnitude of the damage threshold. Our experiments were carried out at intensities two orders of magnitude below the threshold.

More surprising is the $I^{10}$ dependence of the positive ion emission shown in Fig.8. Since no such power law is observed for electrons and since the emission spectra of ions have the same shape as those for electrons, we interpret this result as evidence that the ion emission is not a genuine ten-photon process but rather a secondary process following the five-photon photoemission of two electrons. Emission of atoms or ions following interaction between two valence band holes has previously been suggested in the literature [14,15].

## 4. Results of Cluster Calculations

To get a qualitative idea about the electronic structure of the surface in different configurations, we have performed cluster calculations for a cluster in bulk symmetry, and for stoichiometric and non-stoichiometric clusters in surface symmetry. The $BaF_2(111)$ surface was modelled by a 13 atom cluster (Fig. 2) which describes a stoichiometric composition and, alternatively, a 10-atom cluster with the upper three fluorine atoms removed, representing a Ba-rich surface. We used the discrete variational embedded cluster approach [16-19] with a properly chosen embedding crystal potential. For the surface clusters the embedding crystal potential was constructed from 21 $Ba^{2+}$ and 41 $F^-$ ions surrounding the cluster in a half-sphere. This number of ions was chosen to assure that the total charge of cluster and crystal is as neutral as possible. To block the transfer of electrons from the cluster into occupied states associated with the exterior potential well, this potential was truncated at an energy slightly above the Fermi level. The molecular as well as the crystal potential were constructed within the local density approximation using the form of the exchange-correlation potential proposed by von Barth and Hedin [20]. Molecular wavefunctions were calculated as linear combinations of atomic orbitals with the addition of some virtual orbitals, i.e. $Ba([Xe]6s^2 6p^0 5d^0)$ and $F(1s^2 2p^5 3s^0)$.

The results of ground state calculations, with no relaxation due to photoemission taken into account, are shown in Fig. 9. Considering a relaxation shift of 1-2 eV, the calculated binding energy of the 2p valence band is in reasonable agreement with the photoemission results of Pong et al. [21]. This gives us confidence that the predicted structure in the band gap is qualitatively correct. The main result is that the calculations for the non-stoichiometric surface cluster predict the existence of occupied states in the upper half of the band gap as well as a rich structure of resonances extending well above the vacuum level. On this basis, the observed wavelength dependence can be attributed to multiphoton ionization enhanced by the electronic resonances in the upper half of the band gap, as schematically indicated by the arrows in Fig. 9. The two-photon process must start from the occupied states near the middle of the band gap. As discussed above, the five-photon process is thought to reflect the ionization of 2p valence electrons. The fact that it shows the same spectral dependence as the two-photon ionization must be caused by an identical resonance enhancement of the last one or two photon steps. It seems not unlikely that the five-photon absorption adjusts itself to res-

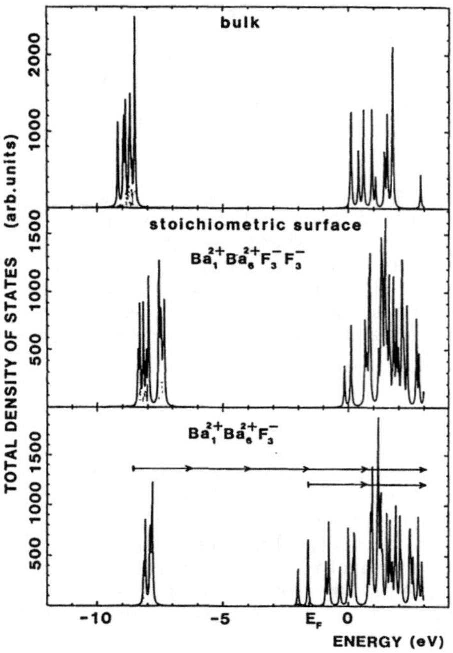


Fig. 9. Calculated density of states for a $F_1Ba_4F_6$ cluster in $T_d$-symmetry describing the bulk (top); for a $Ba_1Ba_6F_3F_3$ cluster (see Fig. 2) describing a stoichiometric surface (middle); and for a $Ba_1Ba_6F_3$ cluster describing a non-stoichiometric surface where the upper F-layer is missing (bottom). All levels have been broadened by a Lorentzian of 20 meV width. Occupied states are shaded.

onant transitions within the cascade since the initial state energy can shift within the width of the 2p band. Precise knowledge of the nature of the states which give rise to the spectra in Fig. 7 is still lacking. Since the features are so sharp, however, we believe that we are dealing with localized surface imperfections. The cluster calculations indicate that these imperfections might be surface F-centers or conglomerates thereof ("non-stoichiometric surface").

## 5. Conclusion

The presented photoemission data in conjunction with the results of cluster calculations indicate that the electronic surface structure plays an important role for the multiphoton absorption on a (111) surface of $BaF_2$. Identical shapes of electron and ion emission yields as a function of wavelength provide evidence that ion desorption is a secondary event following hole creation by multiphoton ionization. The apparent ten-photon process for ion emission is consequently interpreted in terms of two holes in the 2p valence band being required for the ejection of one positive ion. The observation of $F^+$ supports this conjecture. We suspect that, under similar conditions, this pattern applies to other divalent crystals as well. However, to further substantiate our present conclusions more experimental data are required.

## Acknowledgements

This work was supported by the Deutsche Forschungsgemeinschaft, Sfb 6, the Swedish National Science Research Council and the Swedish Board of Technical Development.

References


1. A. Schmid, P. Bräunlich, and P.K. Rol, Phys. Rev. Lett. 35, 1382 (1975)
2. R. Smoluchowski, Phys. Rev. Lett. 35, 1385 (1975)
3. W.J. Siekhaus, J.H. Kinney, D. Milam, and L.L. Chase, Appl. Phys. A39, 163 (1986)
4. T.J. Chuang and I. Hussla, Phys. Rev. Lett. 52, 2045 (1984)
5. T.W. Walker, A.H. Guenther, and P.E. Nielsen, IEEE J. Quant. Electr. QE-17, 2041 and 2053 (1981)
6. P. Bräunlich and A. Schmid, Appl. Phys. Lett. 26, 150 (1975)
7. W. Sesselman, E.E. Marinero, and T.J. Chuang, Appl. Phys. Lett. 49, 770 (1986)
8. J. Reif, H. Fallgren, W.E. Cooke, and E. Matthias, Appl. Phys. Lett. 49, 770 (1986)
9. J. Reif, H. Fallgren, H.B. Nielsen, and E. Matthias, Appl. Phys. Lett. 49, 930 (1986)
10. E. Matthias, H.B. Nielsen, J. Reif, A. Rosen, and E. Westin, J. Vac. Sci. Techn. B, Sept./Oct. 1987, in print
11. F. Rainer, W.H. Lowdermilk, and D. Milam, NBS Special Publication 669, 157 (1984)
12. R. Kelly, Surf. Sci. 90, 280 (1979)
13. L. Ernst, Surf. Sci. 176, L825 (1986)
14. N. Itoh, T. Nakayama, and T.A. Tombrello, Phys. Lett. 108, 480 (1985)
15. D.R. Jennison, and D. Emin, Phys. Rev. Lett. 51, 1390 (1983)
16. A. Rosén, D.E. Ellis, H. Adachi, and F.W. Averill, J. Chem. Phys. 65, 3629 (1976)
17. B. Delley and D.E. Ellis, J. Chem. Phys. 76, 1949 (1982)
18. D.E. Ellis, G.A. Benesh, and E. Byrom, Phys. Rev. B20, 1198 (1979)
19. N.C. Amaral, B. Maffeo, and D. Guenzburger, Phys. Stat. Sol.(b) B117, 141, (1983)
20. U. von Barth and L. Hedin, J. Phys. C5, 1629 (1972)
21. W. Pong, C.S. Inouye, and S.K. Okada, Phys. Rev. B18, 4422 (1978)

# PSD of Excited Hydrogen from KCl

L.T. Hudson[1], A.V. Barnes[1], N.J. Halas[1], R.F. Haglund[1], M.H. Mendenhall[1], P. Nordlander[1], N.H. Tolk[1], Y. Wang[1], and R.A. Rosenberg[2]

[1]Center for Atomic and Molecular Physics at Surfaces, Department of Physics and Astronomy, Vanderbilt University, Nashville, TN 37235, USA

[2]Synchrotron Radiation Center, University of Wisconsin-Madison, 3731 Schneider Drive, Stoughton, WI 53589, USA

## 1. Introduction

The electronic interaction of hydrogen with surfaces is of fundamental interest in the study of DIET processes. The hydrogen-surface model system provides important information on bond-breaking mechanisms at surfaces which lead to excitation and charge transfer processes and desorption. The ejection of protons from surfaces has been studied previously [1-3]. We present first measurements of fluorescence radiation from excited hydrogen neutrals desorbed from KCl under photon bombardment. The observed temperature and time dependences of the Balmer alpha ($H_\alpha$) radiation cast light upon the nature and strength of the hydrogen bond on alkali halide surfaces. In addition, the introduction of hydrogen is also seen to dramatically affect the desorption yield of excited-state substrate potassium atoms. This work is an extension of and complementary to that presented in DIET II [4] on the *electron* stimulated desorption of hydrogen neutrals from alkali halide substrates.

## 2. Balmer Radiation from Photon-Stimulated Desorption of Excited Hydrogen

These experiments were performed at the University of Wisconsin's Synchrotron Radiation Center on a beamline equipped with a grazing incidence Brown-Lien-Pruett grating monochromator (20-1200 eV). The photon beam was incident along the normal of a single crystal KCl sample which had been cleaved in air and baked in vacuum at 400 °C. Fluorescence radiation was collected 90° from the surface normal. The ambient pressure was less than $3 \times 10^{-10}$ Torr throughout these experiments.

Before $H_2$ was introduced, a spectral scan with incident zero order (undispersed) light showed KCl bulk fluorescence [5], but no hydrogen or potassium line radiation. The sample was then dosed continuously at room temperature at $2.4 \times 10^{-7}$ Torr hydrogen partial pressure. To eliminate scattered light in the 2000 Å - 8000 Å region of detection, an aluminum filter was placed in the zero order beam, thereby passing only photons in the range of ≅ 16 - 70 eV. The fluorescence spectrum is shown in Fig. 1. Our ESD spectra taken under the same conditions exhibit similar features: bulk fluorescence, the Balmer lines of hydrogen, and emission from the hydroxyl radical OH. The precursor surface state of desorbed excited state hydrogen continues to be the subject of ongoing investigations. Apparently some incident $H_2$ dissociates on the surface to form OH. Other possibilities include HCl, KH, H and $H_2$ diffusing from the bulk, and hydrogen at defect sites.

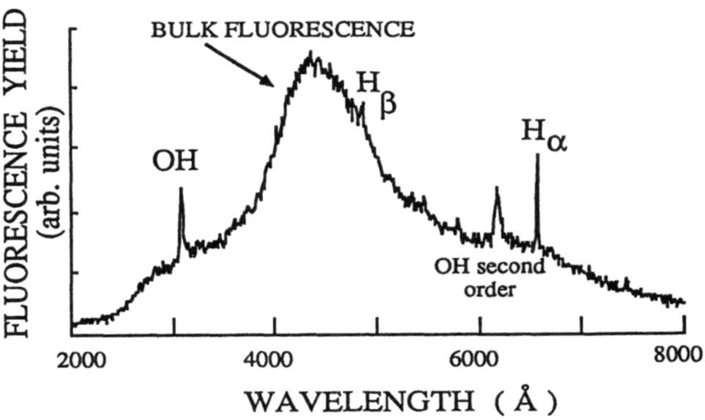


Fig. 1 Fluorescence spectrum upon photon bombardment of KCl dosed with $H_2$

The first order (monochromatic) photon "current" through the monochromator was less than 0.1 μA. Using dispersed wavelengths, we detected no hydrogen Balmer radiation. This is consistent with our ESD studies of hydrogen; the $H_\alpha$ signal decreases with electron current and is barely above background at 0.1 μA.

## 3. Temperature Studies and the Supression of Excited Potassium

Figure 2 compares the temperature dependences of the Balmer alpha and excited potassium yields. In each case off-line background was subtracted and then the signal normalized to synchrotron beam flux. The two signals do not share a common absolute ordinate axis, but each signal was pursued until barely above background. The intensity of the $H_\alpha$ line decreases rapidly with temperature and at temperatures over 200°C it was not discernible under these experimental condition We interpret this behavior as an effect of hydrogen coverage. Hydrogen is expected to bind very weakly to the surface and therefore at elevated temperatures can be entirely removed. The $K^*$ line, however, increases with temperature exhibiting an Arrhenius type behavior. We note that the potassium line only appears after the disappearance of the hydrogen signal. Our ESD studies have further suggested that $K^*$ may be quenched by hydrogen coverage at low temperatures. This is seen in the two spectra of Fig. 3, taken only five minutes apart. The front spectrum resulted from 22 μA of electrons incident on clean KCl at 34° C. The potassium doublet at

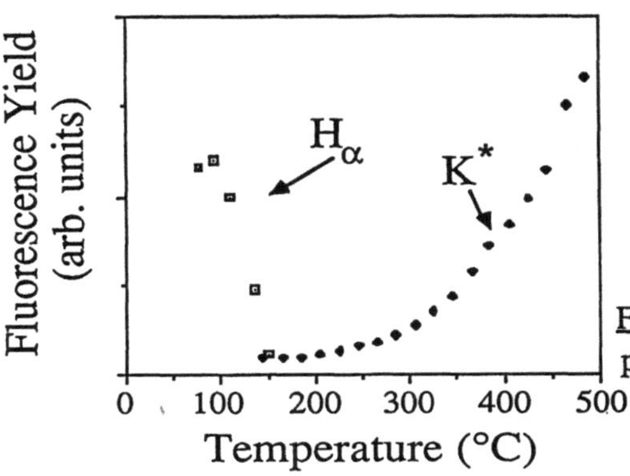


Fig. 2 Comparison of $H_\alpha$ and excited potassium PSD yields vs. temperature.

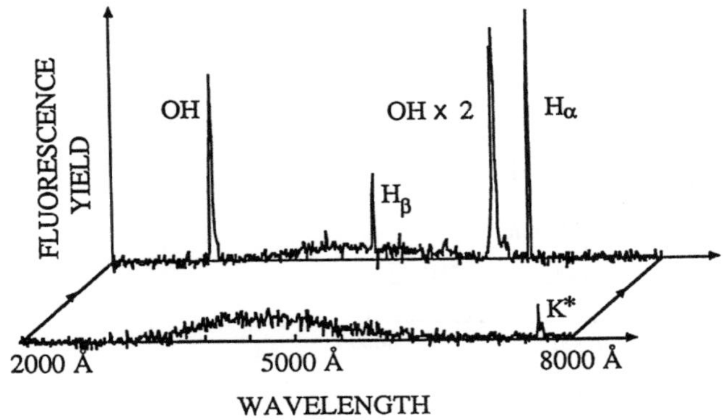


Fig. 3 ESD spectra from KCl before (front) and after (back) dosing with hydrogen. The electron beam was chopped at 100 Hz. On-counts minus off-counts are plotted.

7665/7699 Å is clearly visible. Hydrogen was introduced at 2.0 x $10^{-8}$ Torr, and, as the rear spectrum reflects, $K^*$ was quenched. Both spectra are plotted with the same ordinate scaling.

## 4. Timing Studies

When photons or electrons are first incident on a hydrogen dosed alkali halide sample, we observe a turn-on transient of the Balmer hydrogen fluorescence signal. An example is shown in Fig. 4. It shows the yield of $H_\alpha$ as incident synchrotron radiation is opened to and blocked from the sample. In measurements of this type, the signal generally decays to some nonzero constant value. This corresponds to a steady state between all sources and sinks of hydrogen, the level of which is a sensitive function of $H_2$ dosing pressure and incident photon flux. The absence of a marked turn-on at T=78 s in Fig. 4 is due to the brief time between the second and third exposures of the sample to synchrotron radiation. These measurements indicate that the equilibration of $H_2$ on the surface is a slow process ( ≥ 2s ).

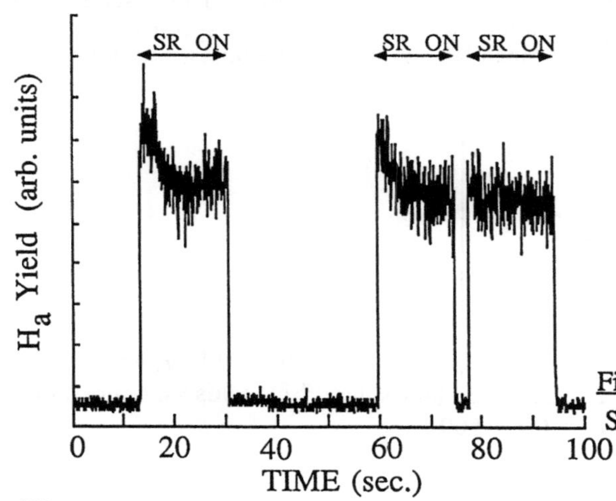


Fig. 4 Turn on transient of $H_a$. SR = synchrotron radiation.

5. Summary and Future Work

We present first measurements of fluorescence spectra, temperature and timing dependences of excited hydrogen desorbed from an alkali halide surface. $H^*$ was not detected with the low fluxes of first order light. We also found hydrogen coverage to supress $K^*$ emission at room temperature. Each of the PSD results agrees with our earlier ESD work. This suggests the involvement of common electronic desorption mechanisms. One primary concern is the precursor state of the desorbing hydrogen. To this end, future studies shall include the dependence of excited and ground state hydrogen yields upon incident photon and electron energies and from substrates with different electronic structure.

References


1. M. L. Knotek, V. O. Jones and V. Rehn, Phys. Rev. Lett. 43 (1979) 300

2. T. R. Pian, M. M. Traum, J. S. Kraus, N. H. Tolk, N. G. Stoffel and G. Margaritondo, Surf. Sci. 128 (1983) 31

3. C. C. Parks, D. A. Shirley and G. Loubriel, Phys. Rev. B 29 (1984) 4709

4. N. H. Tolk, R. F. Haglund, Jr., M. H. Mendenhall, E. Taglauer and N. G. Stoffel: in Desorption Induced by Electronic Transitions DIET II, ed. by W. Brenig and D. Menzel (Springer, Heidelberg, 1985) 152

5. A. I. Bazhin, E. O. Rausch, and E. W. Thomas, Phys. Rev. B 14 (1976) 2583

# Ground-State and Excited Atom Production by Electron and Ion Bombardment of NaCl and $CaF_2$

G. Betz[1], E. Wolfrum[1], P. Wurz[1], K. Mader[1], B. Strehl[1], W. Husinsky[1], R.F. Haglund, Jr.[2], and N.H. Tolk[2]

[1]Institut für Allgemeine Physik, Technische Universität Wien, Karlsplatz 13, A-1040 Wien, Austria

[2]Center for Atomic and Molecular Physics at Surfaces, Vanderbilt University, Nashville, TN 37235, USA

## 1. Introduction

For alkali and alkali earth halides, ion, electron and photon bombardment are efficient processes leading to continuous surface erosion [1,2]. While sputtering of metals under ion bombardment is well understood in terms of a collision cascade emission mechanism based on momentum transfer due to elastic collisions [3], for electron or photon bombardment no such processes can account for any particle emission due to the negligible momentum transfer. Electronic processes have been proposed to explain such electron or photon induced desorption processes [4,5]. Selective emission of the halogen atoms and molecules is ascribed to the H-centre migration model leading to the formation of a metal overlayer on the surface. If the vapour pressure of the alkali/alkali earth metal layer at the target temperature is high enough these excess metal atoms will desorb thermally, thus leading to continuous particle emission. Otherwise the excess metal layer will eventually prevent any further halogen emission and the emission process will be only of transient nature [6]. We have compared the particle emission processes for ion and electron bombardment of NaCl and $CaF_2$ to differentiate between a collision cascade induced contribution and one due to Desorption Induced by Electronic Transitions (DIET) under ion bombardment and, in addition, to compare DIET processes under ion and electron bombardment to obtain a better understanding of the desorption mechanisms involved.

Even though emission of ions, excited and ground-state metal atoms has been observed, evidence exists that the majority of electronically desorbed particles are neutral ground-state atoms [7]. Using high resolution optical spectroscopy [8] and Laser Induced Fluorescence spectroscopy (LIF) [9] we have studied the energy distributions of ground-state as well as of excited Na and Ca atoms at different target temperatures. In addition the yield - temperature dependence for ground-state and excited atoms was studied to gain insight on whether excited atoms are formed by an intrinsic process at the surface or by post-excitation of emitted ground-state atoms in the gas phase due to secondary electrons or the primary electron beam.

## 2. The velocity distribution of ground-state Na and Ca atoms under ion and electron bombardment

LIF, employing the Doppler effect, was used to measure the velocity distribution of emitted Na/Ca atoms from NaCl/$CaF_2$ single crystals under 15 keV Ar ion (current density 5µA/cm$^2$) or 400 eV electron (current density 200 µA/cm$^2$) bombardment in the temperature range from 300 to 700 K. In the case of Na, in essence, the velocity distribution was found to be identical for electron and ion bombardment at all temperatures within the accuracy of our measurements (see Fig. 1) and, in addition, the distribution was purely thermal and could always be fitted with a Maxwell-Boltzmann distribution at target temperature. No cascade contribution was resolved (which should exhibit a maximum at much higher energies around 0.5 eV, which is half the binding energy for Na [10]).

To check for the absence of any cascade contribution under ion bombardment of a clean NaCl target further experiments showed that if the oxygen partial pressure in the vacuum chamber is increased up to $10^{-6}$ mbar the sputtered thermal Na flux decreases by about a factor of 5 and in addition a non-thermal cascade contribution now becomes evident.

On the other hand, in the case of ion bombardment of $CaF_2$ at room temperature and up to 500K the velocity distribution is purely collisional and a fit of the distribution using the Thompson formula [3] for metals gives a surface binding energy of 1.4 eV, in perfect agreement with previous measurements, we have performed for metallic Ca [11]. Starting with temperatures above 500K, an additional thermal component appears in the energy distribution, and for temperatures above 600K this becomes the dominant contribution to the sputtered flux (Fig. 2). The differences between the velocity distributions for Na and Ca under ion bombardment can be easily explained in terms of the much higher vapour pressure of Na. Thus we assume that for NaCl already at room temperature the thermal component is so dominant that the cascade contribution could not be resolved.

For both materials under ion bombardment the thermal component in the emitted flux is ion beam induced and not a pure thermal evaporation, as was confirmed by chopping the ion beam. In addition, for ground-state Ca emission from $CaF_2$ we investigated the time dependence of the LIF signal in more detail. As can be seen from Fig. 3, at least three different contributions to the sputtered flux can be identified for temperatures above 500K:

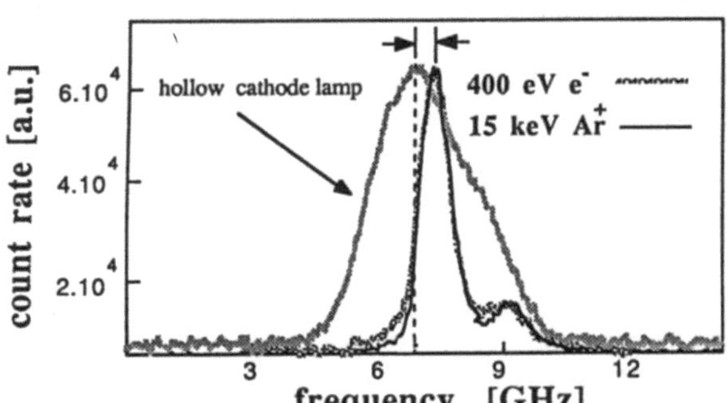


Fig.1: LIF spectra over the velocity distribution of emitted ground-state Na atoms under ion or electron bombardment. The spectra are identical within the accuracy of the measurements. The frequency shift between the maxima of the hollow cathode lamp and the fluorescence yield curves corresponds to 320K, which was the actual target temperature. Any energetic cascade contribution under ion bombardment should be visible as a tail towards higher frequencies.

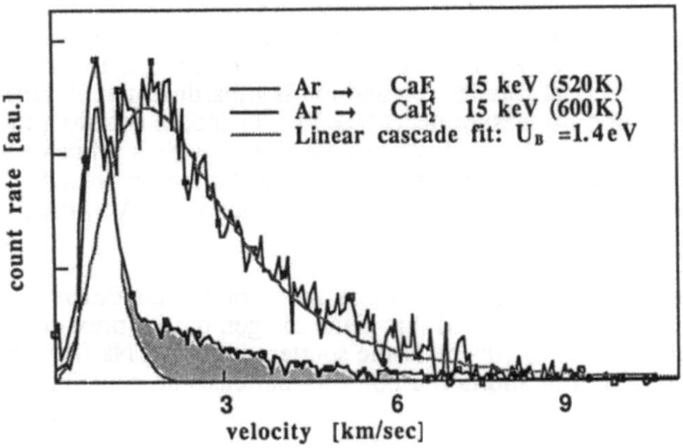


Fig.2: Normalized velocity spectra of ground-state Ca atoms under 15 keV Ar ion bombardment. At 520K the spectrum is mainly collisional and can be fitted by the linear cascade theory assuming a surface binding energy of 1.4 eV. At 600K the spectrum is mainly thermal and the cascade contribution is visible only as a tail in the distribution (shaded area). The thermal part of the spectrum corresponds to a Maxwell-Boltzmann distribution at target temperature.

a) A **prompt decrease** of the signal by almost an order of magnitude was observed immediately after the ion beam was removed from the target (Fig.3, target temperature 600K). Due to the limited time resolution we only know that the decrease is faster than 0.1 s. As at 600K the thermal contribution is dominant (more than 90%), we assume that these promptly emitted atoms consist in reality of two contributions: the **cascade** contribution at non-thermal energies which should not last longer than $10^{-12}$-$10^{-11}$ s [12] after the ion bombardment is stopped and a **thermal** contribution due to $V_k$ centres. These $V_k$ centres cause emission of the halogen atoms and as a consequence evaporation from a Na-enriched surface occurs. As the lifetime of these $V_k$ centres is of the order of milliseconds they will also contribute to the prompt decrease [4].
b) A **slow thermal contribution** exhibiting a life time of the order of 30 s, which we assume to be caused by long-lived trapped exciton states diffusing to the surface and in turn causing Na emission similar to the $V_k$ centres.
c) A purely **thermal evaporation** component, not caused by ion bombardment, which is for the whole temperature range investigated a minor contribution to the sputtered flux.

Under electron bombardment $CaF_2$ showed no emission of Ca ground-state atoms up to 500K, the same temperature for which under ion bombardment a thermal contribution appeared. At temperatures above 500K as outlined above pure thermal evaporation occurs and in addition under electron bombardment Ca desorption (signal size under our experimental conditions about 4 times the thermal evaporation at 550K) occurred. After turning off the electron beam a delayed signal with a decay time constant of the order of 3 min was observed in this case (target temperature 550K),possibly indicating that under electron bombardment either other long-lived excitons are created or that the different decay time as compared to

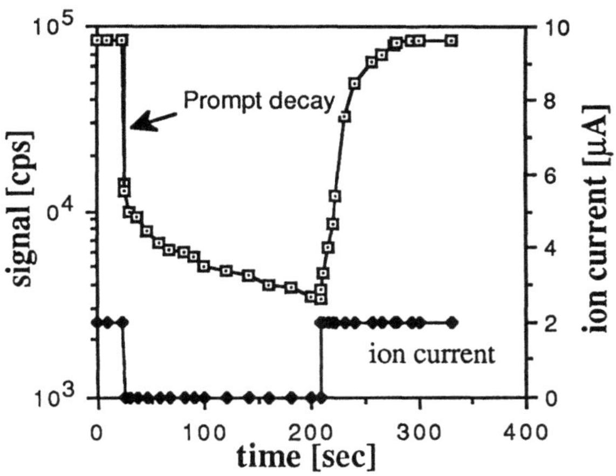


Fig.3: Time dependence of the Ca ground-state signal if ion bombardment is interrupted at a target temperature of 600K. The fully drawn part of the curve labelled "prompt decay" corresponds to a signal decrease within less than 0.1s.

ion bombardment is caused by the greater penetration depth of the primary electrons. In all cases under electron bombardment the velocity distribution was thermal, corresponding to a temperature equal to the target temperature.

## 3. Energy distribution of excited Na and Ca atoms under electron and ion bombardment

To determine which process causes emission of excited Na and Ca atoms, we measured the energy distribution of excited Na (Na I 588.9nm) and Ca (Ca I 422.5nm) atoms using high-resolution optical spectroscopy studying the Doppler line profile broadening [8] of the cited lines. For NaCl under ion as well as electron bombardment strong optical emission is observed and the analysis shows that under ion bombardment excited atoms are non-thermal, exhibiting an energy distribution which peaks at about 3 eV. Contrary to this, under electron bombardment a thermal distribution was found for the Na I atoms [9]. These results are in good agreement with measurements by WALKUP et al. [13], who using a similar technique obtained the same results. The high energies of excited atoms under ion bombardment indicate that these atoms are part of the cascade contribution being excited when leaving the surface during the last collision.

In the case of Ca I from $CaF_2$ under ion but not under electron bombardment emission of excited atoms was observed over the whole temperature range investigated. For ion bombardment the velocity distribution of Ca I was identical to the emission from a pure Ca target, and a mean energy of 6 eV was derived from the analysis [8].

## 4. The dependence of the yield of ground-state and excited atoms on the target temperature

To gain additional information on whether excited atoms under electron bombardment are caused by secondary effects (excitation by secondary or primary electrons) [1,13] or due to an intrinsic process [14], we studied the temperature dependence of the yields by LIF or light emission, respectively. Figure 4 a,b give the results for

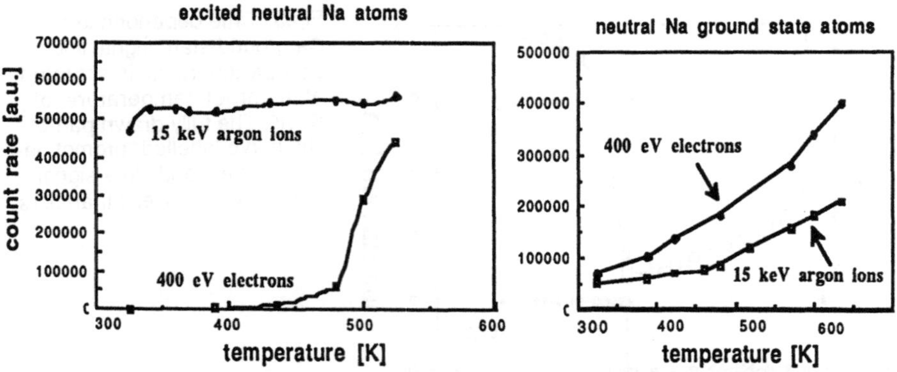


Fig.4: Temperature dependence of the Na ground- and excited-state yields, under ion or electron bombardment.

NaCl, which clearly show that under electron bombardment the excited yield is proportional to the ground-state yield and both increase exponentially with temperature. However, under ion bombardment only the ground-state yield shows a significant dependence on temperature. The independence of the yield of excited atoms with temperature and their non-thermal energies strongly suggest their origin from the collision cascade. The increase in the ground-state yield follows from the observation that the majority of emitted Na atoms are thermal and caused by thermal evaporation following electronic desorption of the halogen atoms. The results for electron bombardment together with the thermal energy distribution of the excited atoms do not contradict a formation mechanism due to gas phase collisions. However, the absence of any excited Ca atoms under electron bombardment of $CaF_2$ contradicts the assumption of gas phase collisions and indicates other processes causing the emission of excited atoms.

Financial support by the Fonds zur Förderung der Wissenschaftlichen Forschung (Project Nos. P5547, P 5577, P5833) and by the National Science Foundation (N00014-86-C-2546) is gratefully acknowledged.

References

1. Z. Postawa, J. Rutkowski, A. Poradzisz, P. Czuba, M. Szymonski: Nucl. Instr. Meth B18, 574 (1987)
2 R.F. Haglund Jr., N.H. Tolk, G.M. Loubriel, R.A. Rosenberg: Nucl. Instr. Meth. B18, 549 (1987)
3. P. Sigmund: Phys. Rev. 184, 383 (1969)
4. P.D. Townsend: Nucl. Instr. Meth. 198 , 9 (1982)
5. N. Itho: Nucl. Instr. Meth. B27, 155 (1987)
6. R. Kelly: Surface Science 90, 280 (1979)
7. N.H. Tolk, P. Bucksbaum, N. Gershenfeld, J.S. Kraus, R.J. Morris, D.E. Murnick, J.C. Tully, R.R. Daniels, G. Margaritondo, N.G. Stoffel: Nucl. Instr. Meth. B2, 457 (1984)
8. G. Betz: Nucl. Instr. Meth. B27, 104 (1987)
9. W. Husinsky, G. Betz: Scanning Microscopy 1, 1603 (1987)
10. W. Husinsky, R. Bruckmüller: Surface Science 80, 637 (1979)

11. W. Husinsky, I. Girgis, G. Betz: J. Vac. Sci. Techn. B3,5, 1543 (1985)
12. R. Kelly: Radiat. Effects 80, 273 (1984)
13. R.E. Walkup, Ph. Avouris, A.P. Gosh: Phys. Rev. Lett. 57, 2227 (1986)
14. R.F. Haglund Jr., R.G. Albridge, D.W. Cherry, R.K. Cole, M.H. Mendenhall, W.C.B. Peatman, N.H. Tolk, D. Niles, G. Margaritondo, N.G. Stoffel, E. Taglauer: Nucl. Instr. Meth. B13, 525 (1986)

# Optical Radiation from Electron, Photon and Heavy Particle Bombardment of Lithium Fluoride and Lithium-Dosed Surfaces

*N.H. Tolk, R.G. Albridge, A.V. Barnes, R.F. Haglund, Jr., L.T. Hudson, M.H. Mendenhall, D.P. Russell, J. Sarnthein, P.M. Savundararaj, and P.W. Wang*

Department of Physics and Astronomy, and
Department of Mechanical and Materials Science Engineering,
Center for Atomic and Molecular Physics at Surfaces,
Vanderbilt University, Nashville, TN 37235, USA

## 1. Introduction

Comparative studies of electron, photon and ion bombardment of the same sample surface probe the various channels through which incident particle energy is dissipated, often leading to desorption. Electron or photon irradiation of alkali halides results in the swift ejection of halide atoms, leaving behind an enriched alkali metal surface from which the alkali atoms thermally desorb. Ion bombardment involves momentum transfer as well as electronic mechanisms, which results in a different surface stoichiometry at the time of desorption. This study explores how the degree of surface metallization influences the choice of the final excitation state of the desorbing particle. An important way to study the desorption products is to monitor the characteristic spectral radiation from the de-excitation of ejected species [1]. In this paper we report recent results of this type. Measurements were made to compare the desorption of excited state neutral lithium from lithium fluoride by electron, photon and ion bombardment and from lithium-dosed tungsten and lithium-dosed glass by electron and photon bombardment.

## 2. Experimental Results

### A. PSD of Li* from Lithium Fluoride

Photon irradiation studies were performed at the University of Wisconsin's Synchrotron Radiation Center at Stoughton, Wisconsin. Zero order visible and ultraviolet photons from
the two meter grazing incidence Brown-Lien-Pruett monochromator (20 - 1200 eV) were incident along the surface normal upon a LiF crystal which was previously cleaved in air and baked under UHV conditions at 600°C.

The photon detection system included a 0.3 m McPherson 218 monochromator which has a resolution of 26.5 Å at 1 mm slit width, a photo-multiplier in a cooled housing, and a multi-channel analyzer or CAMAC crate for data acquisition. We scanned the spectral region of 2000-7000 Å using a 1200 lines/mm grating blazed at 5000 Å. Fluorescence radiation was measured at 90° to the surface normal. The base pressure in the experimental chamber was less than $3 \times 10^{-10}$ Torr throughout these experiments.

The spectrum obtained from irradiation of lithium fluoride (Fig.1) at room temperature shows the first resonance line of lithium at 6708 Å and two broad

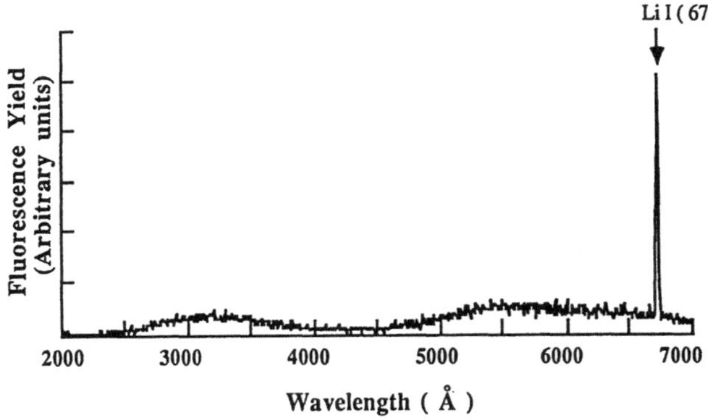


Fig.1 Fluorescence spectrum from photon irradiation of lithium fluoride

continua centered at about 3200 Å and 5600 Å. These continua are due to bulk luminescence and scattered light from the irradiating zero order photon beam. No other line radiation was detected above the background noise. We estimate that other spectral lines have maximum intensities that are at least a factor of 100 less intense than the intensity of the observed resonance line.

B. Ion Bombardment of Lithium Fluoride

In further studies, we irradiated a similarly prepared lithium fluoride sample with a 9 keV $H_2^+$ beam from a Colutron ion source. The de-excitation of the desorbed species from the surface and the bulk luminescence were monitored by the signal detection scheme discussed in part A. The ion bombardment was performed at room temperature. Note that in addition to the first resonance line of lithium 6708 Å (2p-2s) there are lithium lines at 3233 Å (3p-2s), 4603 Å (4d-2p) and 6104 Å (3d-2p). The relative intensities of the excitations originating at $n \geq 3$ after correcting for detection system efficency are greatly enhanced compared to arc discharge data. This indicates a preferential population of these higher excited states via some unknown mechanism. (See Fig. 2).

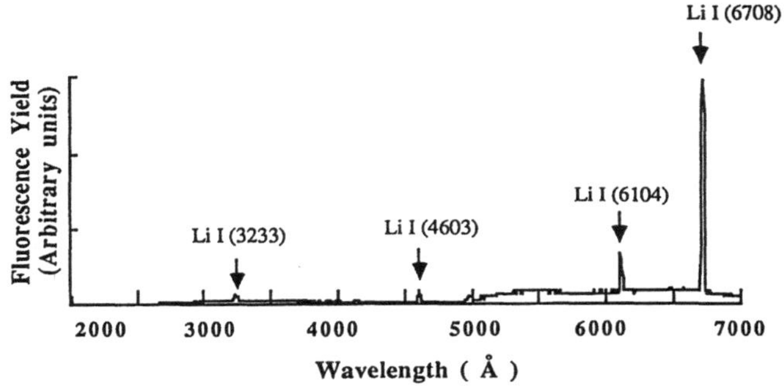


Fig 2. Fluorescence spectrum from 9 keV $H_2^+$ bombardment of lithium fluoride

C. ESD of Li* from Lithium Fluoride

It is well known that electron bombardment of alkali halides can desorb generous amounts of excited alkali atoms [2]. A typical optical spectrum taken at room temperature is shown in Fig. 3. Here, a 300 eV electron beam was incident along the surface normal on a single crystal LiF sample which had been cleaved in air and baked under UHV conditions at 600°C. Fluorescence radiation was collected at 90° from the surface normal; the ambient pressure was less than $10^{-9}$ Torr. The 6708 Å lithium resonance line is clearly visible as well as a trace signal of hydrogen Balmer alpha at 6563 Å. As in the photon irradiation studies, no other excited lithium lines were observed above the background noise; any would have to be at least a factor of 100 weaker than the observed line.

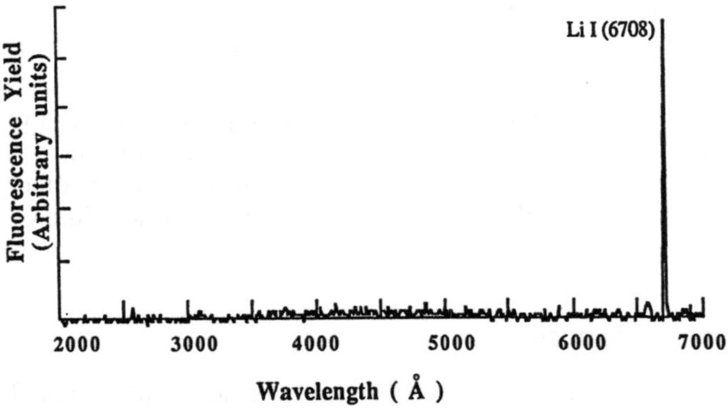


Fig 3. Fluorescence spectrum of LiF bombarded with 300 eV electrons

D. Lithium Metal Studies

We have observed energetic electron and photon induced desorption of excited lithium neutrals from a thickly dosed lithium film, which was possibly contaminated with oxygen. The experimental setup was similar to that described above. In vacuum we dosed a soda- lime glass plate by means of a SAES lithium getter. Since the dosed film was opaque (it appeared shiny gray), we conclude from the attenuation length of light in lithium that it had a thickness of at least 1000 Å. At a pressure of $10^{-9}$ Torr we observed electron stimulated desorption of Li* (6708 Å) within 5 minutes of deposition. The electron beam current was 80 μA and energy 300 eV. Figure 4 shows a fluorescence spectrum taken with the sample at room temperature. The spectrum represents 200 signal integrations over five minutes at 1 mm slit width. Previously observed bulk fluorescence of the glass plate had vanished; the plate is viewed as a mechanical support only. The power density of the electron beam was 10 $mW/mm^2$ which does not cause appreciable heating or thermal evaporation of the lithium layer.

In another study, we used a tungsten substrate. The base pressure was $10^{-11}$ Torr, thus the environment was much "cleaner" than that for the study described above. The lithium resonance line was not seen until several hours after the dosing. Subsequently, zero order synchrotron radiation was used to desorb Li*. The results show exactly the same characteristics as those for electrons.

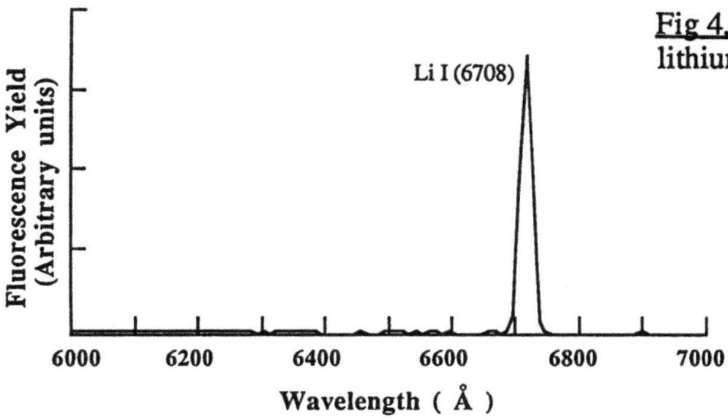


Fig 4. ESD of Li* from a lithium layer dosed on glass

## 3. Discussion

In the DIET experiments ( Figs. 1, 3 and 4 ), we observe only the first lithium resonance line. These results contrast with the ion sputtering results of Fig. 2, where emissions from higher excited states are observed. We therefore propose the following picture.

It has been well established that the surface does not remain stoichiometric during electron or photon induced desorption of alkali halides at room temperature. The halide atoms desorb much faster leaving behind an alkali enriched surface [3]. In our ESD and PSD studies of LiF, we thus expect to have excited lithium atoms desorbing from a surface enriched with lithium metal by a previously discussed mechanism [4]. In the desorption studies of dosed lithium, excited lithium is also desorbing in the presence of a metallic lithium surface. In this case, contamination provides centers of localization for the electronic energy deposited by the incoming electrons or photons leading to desorption of lithium.

We now account for the fact that only the 2p-2s transition is observed in the DIET experiments: as Li* leaves the surface it interacts with the band structure of lithium metal surrounding the desorption site (Fig. 5). Electrons in the 2p level cannot resonantly tunnel to the filled levels of the metal below the Fermi energy. Excitations of Li* to the $n \geq 3$ levels may resonantly ionize ( $10^{-15}$s ) faster than

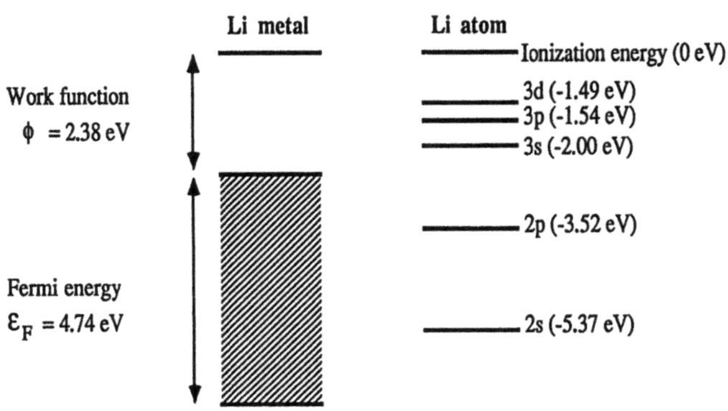


Fig. 5. Schematic energy level diagram of a lithium atom near lithium metal.

de-excitation ( $10^{-9}$s). Hence, higher transitions are suppressed and we observe only the single 2p-2s line in our spectra.

In contrast to this, we do see higher excited states of lithium upon ion bombardment of LiF. Ion sputtering is much more violent than desorption induced by electronic transitions and proceeds through momentum exchange processes. Under ion bombardment, the surface is more rugose and a lithium metal band structure will be much less developed. The desorbing Li* departs with a higher velocity and resonant de-excitation of higher excited states by interaction with the band structure is less probable. Consequently, we are able to observe radiative transitions from the higher excited states of lithium.

## 4. Conclusions

In conclusion, it is clear that desorption induced by electrons and photons proceeds by different mechanisms than does desorption induced by ion bombardment. For the ESD and PSD processes occurring at LiF and lithium-dosed surfaces at room temperature, the presence of a metal-rich surface provides a channel for de-excitation of excited lithium states which are above the lithium metal Fermi energy. This quenching of the higher excited lithium lines is less likely when the sample is excited by ions. Further studies will include careful surface analysis of the stoichiometry of lithium fluoride during ion, electron and photon irradiation. The ultimate aim of these studies is to characterize the final states of all the desorption products and the extent to which they are influenced by the surface, exciting beam and secondary processes.

This work was supported in part by the Office of Naval Research under contract no. N00014-86-K-0735, by the Air Force Office of Scientific Research under contract no. AFOSR-86-0150 and by an AFOSR University Research Initiative contract no. F49620-86-C-0125DEF.

## References

[1] N.H. Tolk, I.S.T. Tsong, and C.W. White, Analytical Chemistry, 49, 16 (1977)

[2] N. H. Tolk, L. C. Feldman, J. S. Kraus, R. J. Morris, T. R. Pian, M. M. Traum and J. C. Tully: in Inelastic Particle-Surface Collisions, ed. by E. Taglauer and W. Heiland (Springer-Verlag, Heidelberg, 1981) p. 112

[3] P. D. Townsend, R. Browning, D. J. Garlant, J. C. Kelly, A. Mahjoobi, A. J. Michael and M. Saidoh, Radiation Effects, 30, 55 (1976)

[4] T. R. Pian, N. H. Tolk, M. M. Traum, J. Kraus and W. E. Collins, Surf. Sci., 129, 573 (1983)

# Long Lifetime of Bulk Luminescence Observed in Spectrosil Glasses Under Electron Bombardment *

*P.W. Wang, R.F. Haglund, Jr., L.T. Hudson, D.L. Kinser, M.H. Mendenhall, N.H. Tolk, and R.A. Weeks*

Department of Physics and Astronomy, and
Department of Mechanical and Materials Science Engineering,
Center for Atomic and Molecular Physics at Surfaces,
Vanderbilt University, Nashville, TN 37235, USA

We have observed four bulk luminescence peaks from Spectrosil glasses stimulated by low energy electron bombardment in ultra-high vacuum. The lifetimes of these bands were measured by chopping the incident electron beam and measuring the fluorescence decay. It was found that one of them has a markedly long lifetime (2.1 ms) compared with the others (<10μs). After annealing at 500° C for one hour and pumping down overnight, characteristic spectral lines from OH*, H*, K*, a molecular band, and an unidentified line appeared at room temperature during electron irradiation. A possible mechanism for this long-lifetime luminescence will be discussed.

## 1. Introduction

Cathodoluminescence from fused silica was generated by the energy loss of energetic electrons *via* electronic excitation processes. One important channel for the dissipation of electron energy is through the formation of excitons in the glass. The exciton can decompose into an electron-hole pair or produce the vacancy and interstitial through the relaxation of the atom which traps the exciton. Hence the luminescence is produced *via* the decay of excitons, the de-excitation process of the excited atoms, and recombination of electron and hole [1].

In the work reported below, we studied the luminescence from silica glasses induced by low energy (320 and 620 eV) electrons in an ultra-high vacuum environment. The cathodoluminescence and lifetime of each luminescence band were also investigated *in situ* during electron bombardment. Four luminescence bands are found centered at approximately 2880 Å, 4500 Å, 5590 Å, and 6370 Å. The luminescence centered at 4500 Å has an unexpected 2.1 ms lifetime. Other three luminescence bands have much shorter lifetime, i.e., less than 10 μs.

## 2. Experimental

The high-purity fused silicas used in this study were Spectrosil and Spectrosil WF, which differ primarily in the OH and Cl impurity content. Spectrosil contains 200 ppm OH and 50 ppm Cl, whereas [OH] is less than 2 ppm and [Cl] is 180 ppm in Spectrosil WF.

Samples were cut to 1 cm x 2 cm x 0.1 cm in size and then cleaned by acetone, methanol and doubly deionized water in an ultrasonic cleaner. Sample were etched by ca. 10% HF and blow dried before inserting into the ultra-high vacuum chamber. The pressure in the UHV chamber was maintained at *ca.* $10^{-8}$ to $10^{-9}$ torr during measurements. Samples were bombarded by 320 or 620 eV electrons to excite the luminescence. The de-excitation of the desorbed species from the surface and the bulk luminescence below the surface were monitored *in situ* by an optical detection system during electron irradiation at room temperature. This detection system consists of a monochromator, a photo-multiplier in a cooled housing, a multi-channel analyzer and a CAMAC for data acquisition [ 2].

The optical axis of the monochromator was at 90° to the electron beam. The slits on the monochromator were set to 1 mm opening, giving 25 Å resolution in a wavelength range 2000 Å - 8000 Å. Since the yield of the photon was obtained by taking the signal difference between electron on and off, tails of long-lifetime signals will be chopped off more than those of shorter lifetime signals when the speed of electron chopper is increased. Therefore the relative lifetimes of luminescence bands can be detected by adjusted the speed of electron chopper. The lifetimes of the luminescence bands were obtained by monitoring the decays of the luminescence peaks after the electron beam was turned off. The flux of the electron beam was maintained constant on Spectrosil and Spectrosil WF during measurements.

## 3. Results

Four luminescence bands located centered at approximately 2880 Å, 4500 Å, 5590 Å and 6370 Å appear during electron bombardment on Spectrosil WF and Spectrosil at room temperature as shown in Fig. 1 and Fig. 2 respectively. Since the bulk luminescence from Spectrosil is weaker, the electron energy was increased to 620 eV and the sample was rotated 46° from the surface normal to obtain the maximum signal. The Spectrosil WF sample was bombarded by 320 eV electrons at normal incidence. It is clear from Figs.1 and 2 that the bulk luminescence from Spectrosil WF is much larger than from Spectrosil.

A coarse search to find approximate decay times was made by measuring the relative intensities of luminescence bands by making the chopper speed 10 times faster. The intensities of the luminescence bands at 2880 Å, 5590 Å and 6370 Å are increased but 4500 Å band is decreased as shown in Fig. 3. This phenomenon indicates that the 4500 Å band has a longer lifetime than the other luminescence

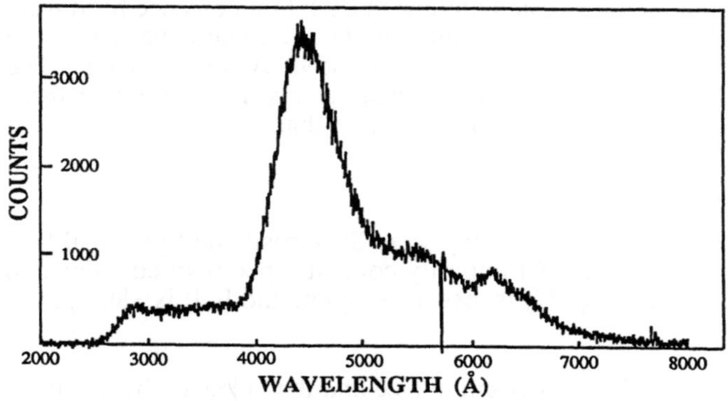


Figure 1. 320 eV $e^-$ on Spectrosil WF. Bulk luminescence of Spectrosil WF due to 320 eV electron beam bombardment at normal incidence. Four peaks located at 2880 Å, 4500 Å, 5590 Å and 6370 Å respectively.

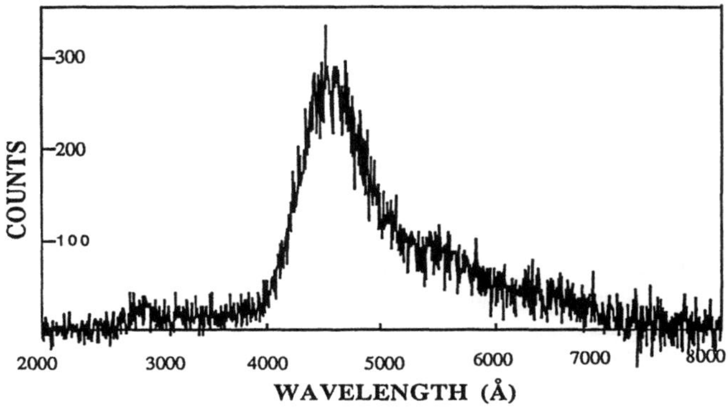


Figure 2. Bulk luminescence of the Spectrosil. The 620 eV electron beam was 46° from the normal of the Spectrosil sample surface.

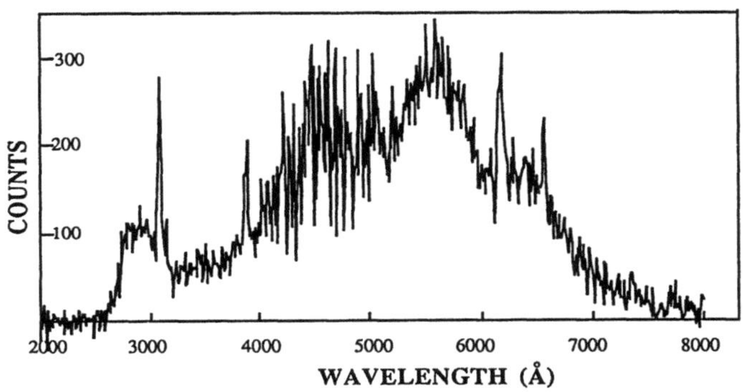


Figure 3. 320 eV e⁻ on Spectrosil WF. The changes of bulk luminescence due to 10 times faster chopper speed. Only bulk luminescence at 4500 Å decreases.

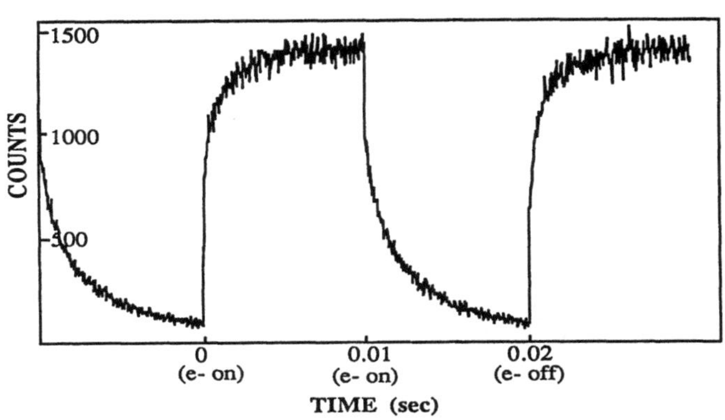


Figure 4. 320 eV e⁻ on Spectrosil WF. The decay time of the bulk luminescence at 4500 Å is 2.1 ms.

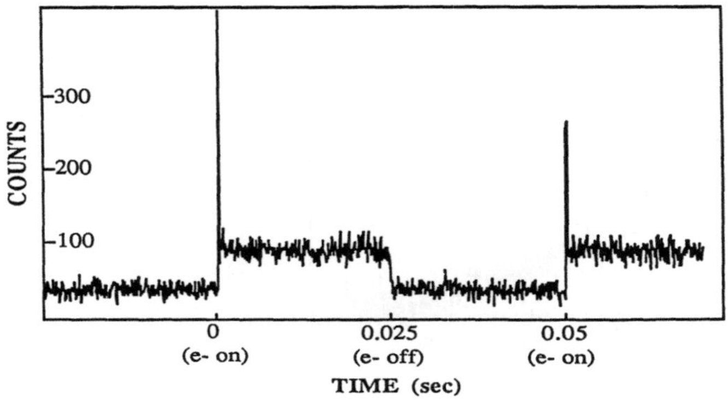


Figure 5. 320 eV $e^-$ on Spectrosil WF. The decay time of bulk luminescence at 2880 Å is less than 10 μs.

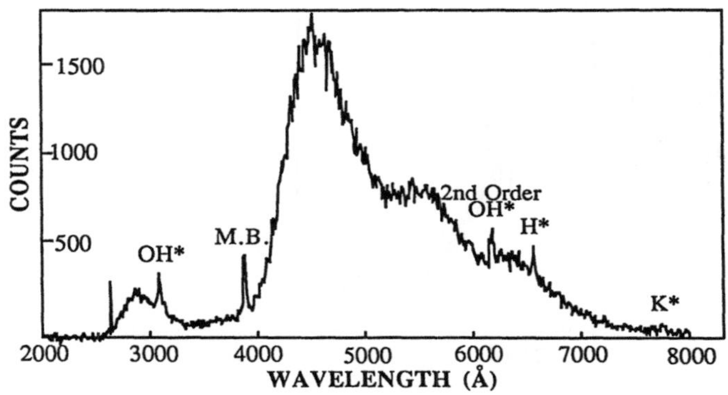


Figure 6. 320 eV $e^-$ on Spectrosil. After annealing at 500°C for one hour and pumping down overnight, characteristic spectral lines from OH*, H* and K*, a molecular band (M.B.), and an unidentified line were found at room temperature during 320 eV $e^-$ irradiation.

bands. Results on the decays of luminescence bands centered at 4500 Å and 2880 Å after the electron beam was turned off are shown in Figs. 4 and 5. Lifetime of the 4500 Å band is 2.1 ms. Lifetime of the 2880 Å band, which is the same as 5590 Å and 6370 Å, is less than 10 μs.

Spectral lines from OH*, H*, K*, a molecular band (M.B.) due to CN° [3] and an unidentified line are found during electron bombardment after 500°C annealing for one hour and pumping down overnight as shown in Fig. 6.

## 4. Discussion and conclusion

The primary result of this work is that the luminescence signal from irradiated Spectrosil WF was greater than that seen in Spectrosil. We believe that this variation in the cathodoluminescence between Spectrosil WF and Spectrosil arises from the differing OH contents of the two glasses. That is to say, the greater the OH content, the less luminescence intensity from silica under electron bombardment. The primary differences of impurities between these two samples are OH and Cl, so either of these impurities might be the cause. However, the presumption of an OH-related variation in luminescence is supported by the work of Mitchell and Denure [4], who found that cathodoluminescence from dry-grown thin $SiO_2$ film is much higher than wet-grown $SiO_2$ film. It has also been shown that hydrogen can passivate defects which may be related to the luminescence centers in the glass [5].

Jones and Embree [6] studied cathodoluminescence by doping imputies, Na, Cl, Ge, and Au into thermally grown $SiO_2$ film and suggested that the band at 2900 Å may be an oxygen vancancy, 4350 Å may be a Ge defect, 5250 Å may be a Cu defect, and 6250 Å band may be a Na defect. Koyama *et al.* [7] reported cathodoluminescence results of $SiO_2$ on Si by the aid of Auger electron spectroscopy. They proposed that bands at 2800 Å and 5300 Å may be due to trivalent Si, while the 4500 Å band may be surface contaminations. We think, in fact, that the 4500 Å band may not be due to surface contaminations because the spectral lines of contaminants are not observed before 500°C annealing. The line radiation may also arise from impurities which diffuse from the near-surface bulk during the annealing process.

The long lifetime of the luminescence band at 4500 Å reveals that a metastable state exists between the excited state and ground state of the luminescence center. This metastable state prevents the direct transition from the excited luminescence center to the ground luminescence center. This slow light-emission process may be related to the diffusion of a defect to some specific position in the glass before relaxation, like the role of F centers in the desorption of $Li^o$ from LiF surface under electron bombardment [8,9].

Characteristic spectral lines OH*, H*, K*, $CN^o$ molecular band and an unidentified line are created after 500° C annealing. This annealing process answers the question itself "Where are those impurities from ?" Due to the extensive studies of ESD on alkali halides previously carried out in this particular UHV chamber, samples are contaminated through the diffusion of the contaminants in the chamber or in the bulk of the sample during annealing.

The conclusions of this work are summarized below.

(i) Greater OH content in Spectrosil glasses is correlated with reduced cathodoluminescence from the glasses.

(ii) The 4500 Å band is caused by radiation defects or intrinsic defects in the glass, but not by contamination.

(iii) Metastable states and diffusion processes are two possible explanations for the long lifetime of the cathodoluminescence at 4500 Å; the present experiments cannot distinguish between these alternatives.

(iv) Contamination of the glass after one hour annealing at 500° C probably results from the in-diffusion of impurities on the walls of the UHV chamber.

DIET experiments show promise for resolving long-standing questions about effects of processes on variation damage susceptibility in glasses.

*This work was supported by Naval Research Laboratory under contract N00014-86-C2546, Air Force Office of Scientific Research (Proj. Nos. AFOSR 86-0150 and F 49620-86-C-0125 DEF) and Martin Marietta Corporation (A 71052).

RERERENCES

[1] See, for example, K. Tanimura, T. Tanaka and N. Itoh, Nucl. Instr. and Meth. B1, 187 (1984).
[2] R.H. Haglund, Jr., and N.H. Tolk in : SPIE Technical Symposium on X-rays in Materials Analysis, San Diego, CA August 21-22, 1986.
[3] M. Mendenhall, A. Barnes, P. Bunton, R. Haglund, L. Hudson, R. Rosenberg, D. Russell, J. Sarnthein, P. Savundararaj, N. Tolk, and J. Tellinghuisen, Chem. Phys. Lett. (in press).
[4] J.P. Mitchell and D.G. Denure, Solid-State Electronics, 16, 825 (1973).
[5] G.N. Greaves, in : The Physics of $SiO_2$ and its Interfaces, ed., S.T. Pentelides, (Pergamon Press, New York, 1978) p.268.
[6] C.E. Jones and D. Embree, in : The Physics of $SiO_2$ and its Interfaces, ed., S.T. Pentelides, (Pergamon Press, New York, 1978) p. 289.
[7] H. Koyama, K. Matsubara and M. Mouri, J. Appl. Phys., 48, 5380 (1978).
[8] G.M. Loubriel, T.A. Green, P.M. Richards, R.G. Albridge, D. W. Cherry, R.K. Cole, R.F. Haglund, Jr., L.T. Hudson, M.H. Mendenhall, D.M. Newns, P.M. Savundararaj, K.J. Snowdon, and N.H. Tolk, Phys. Rev. Lett., 14, 1781 (1986).
[9] T.A. Green, G.M. Loubriel, P.M. Richards, N.H. Tolk, and R.F. Haglund, Jr., Phys. Rev., B35, 781 (1987).

# Index of Contributors


Albridge, R.G. 284
Alvey, M.D. 100,115
Auer, S. 58
Avouris, Ph. 144,258

Bakshi, M.H. 94
Barak, D. 130
Barnes, A.V. 274,284
Betz, G. 278
Block, J.H. 236
Bommannavar, A.S. 94
Burns, A.R. 67,136,167

Ciszewski, A. 236
Cowin, J.P. 190

de Souza, G.G.B. 10
Dösselmann, J. 236
Drachsel, W. 236
Dresser, M.J. 100,115
Dreyfus, R.W. 262

Eberhardt, W. 32

Feulner, P. 58,173
Flodström, S.A. 258

Garrett, R.F. 210
Ghosh, A.P. 258
Gortel, Z.W. 173
Green, T.A. 185
Grizzi, O. 196
Gruen, D.M. 94

Haglund, Jr., R.F. 274,278, 284
Halas, N.J. 274
Håkansson, P. 225
Hanson, D.M. 50
Hedin, A. 225
Hudson, L.T. 274,284
Hulbert, S.L. 73
Hurych, Z. 94
Husinsky, W. 278

Jaenicke, S. 236
Jennison, D.R. 67,136,167, 185
Johnson, A.L. 120,130
Johnson, E. 73
Johnson, E.D. 210
Johnson, P.D. 73
Johnson, R.E. 47
Jonsson, G. 225

Kang, H. 196
Kasi, S.R. 196
Kawai, R. 144
Knotek, M.L. 210
Kreuzer, H.J. 173
Kurtz, R.L. 109,126,160, 182,258

Lablanquie, P. 10
Lang, N.D. 144
Lanzillotto, A.-M. 100,115

Mader, K. 278
Madey, T.E. 109,120,126, 130,182,217
Mancini, D.C. 220
Marsh, E.P. 190
Matthias, E. 266
McDonald, F.A. 262
McGovern, I.T. 205
McGrath, R. 205
Mendenhall, M.H. 274,284
Müller, T. 58
Menzel, D. 2,58,173,236
Morin, P. 10
Murphy, R. 32

Nenner, I. 10
Newns, D.M. 144
Nielsen, H.B. 266
Nordlander, P. 274
Norman, D. 205
Nyholm, R. 258

Pan, X. 73
Pellin, M.J. 94
Polak, M. 120
Puschmann, A. 58

Rabalais, J.W. 196
Ramaker, D.E. 85,182
Reif, J. 266
Rosén, A. 266
Rosenberg, R.A. 220,274
Russell, D.P. 284

Sambe, H. 85,182
Sanche, L. 78
Sarnthein, J. 284
Säve, G. 225
Savundararaj, P.M. 284
Schneider, M.R. 190
Schweitzer, E.L. 94
Senf, F. 258
Sette, F. 210
Shinn, N.D. 217
Simon, M. 10
Soukiassian, P. 94
Stechel, E.B. 67,136,167
Stockbauer, R. 109,126, 130,258
Strehl, B. 278
Stulen, R.H. 39
Sundqvist, B.U.R. 225,230

Tabares, F.L. 190
Thornton, G. 205
Tolk, N.H. 274,278,284
Tranquada, J. 73

Uram, K.J. 100

von Gutfeld, R.J. 262

Walczak, M.M. 120
Walkup, R. 144,160,258
Wang, P.W. 284
Wang, Y. 274
Warburton, D.R. 205
Wen, C.-R. 220
Westin, E. 266
Williams, E.M. 242
Williams, P. 230
Wolfrum, E. 278
Wurz, P. 278

Yates, Jr., J.T. 100,115
Young, C.E. 94

MIX
Papier aus verantwortungsvollen Quellen
Paper from responsible sources
FSC® C105338

If you have any concerns about our products,
you can contact us on
**ProductSafety@springernature.com**

In case Publisher is established outside the EU,
the EU authorized representative is:
**Springer Nature Customer Service Center GmbH**
**Europaplatz 3, 69115 Heidelberg, Germany**

Printed by Libri Plureos GmbH
in Hamburg, Germany